Machineries of Oil

Infrastructures Series

edited by Geoffrey C. Bowker and Paul N. Edwards

Machineries of Oil

An Infrastructural History of BP in Iran

Katayoun Shafiee

The MIT Press
Cambridge, Massachusetts
London, England

This book was set in Sabon LT Std by Toppan Best-set Premedia Limited.

Library of Congress Cataloging-in-Publication Data

Names: Shafiee, Katayoun, author.
Title: Machineries of oil : an infrastructural history of BP in Iran / Katayoun Shafiee.
Other titles: Infrastructures series.
Description: Cambridge, MA : The MIT Press, 2018. | Series: Infrastructures series | Includes bibliographical references and index.
Identifiers: LCCN 2017027986 | ISBN 9780262037044 (hardcover : alk. paper) ISBN 9780262548854 (paperback)
Subjects: LCSH: British Petroleum Company. | Petroleum industry and trade--Iran--History. | Petroleum industry and trade--Political aspects--Iran.
Classification: LCC HD9576.I62 S53 2018 | DDC 338.8/87223380955--dc23 LC record available at https://lccn.loc.gov/2017027986

 Knowledge Unlatched

Contents

Acknowledgments

The research for this project was supported with generous funding from the National Science Foundation and the International Dissertation Research Fellowship of the Social Science Research Council. Archival research in the United States was supported with funding from the Mellon Foundation through the National History Center's International Seminar on Decolonization. Many archivists and librarians helped me navigate the archives in the United Kingdom, Iran, and the United States. I owe a special thanks to Peter Housego, former head archivist of the BP Archive at the Modern Records Center at the University of Warwick. I would also like to thank the archivists and librarians at the British National Archive at Kew and the British Library in London, the US Library of Congress, the US National Archive at College Park, Maryland, and the American Heritage Center in Laramie, Wyoming. In Iran, the archivists and librarians of the Iranian National Archive, the Library of the Parliament, the Foreign Ministry Archive, and the library of the National Iranian Oil Company in Tehran were all generous with their assistance. I am also grateful to employees at the National Iranian Oil Company in Abadan, who graciously shared their memories of oil operations in the 1950s. Most importantly, my research in Iran would not have occurred as smoothly as it did without the support provided by my aunt, Fereshteh Motamedi, and my uncle, Ali Afshar. They sacrificed time from their busy schedules as professors to introduce me to the necessary people and accompany me on various trips to the archives and to Abadan to ensure that my research proceeded without disruption.

The revision of the manuscript was made possible by a postdoctoral research fellowship from the Middle East Institute, National University of Singapore. I am most grateful to Michael Hudson for welcoming me to the institute and to Peter Sluglett for providing guidance and advice along the way. Finally, the Institute of Advanced Studies, University College

London, provided an ideal interdisciplinary setting for finalizing the publication of the manuscript. I am most grateful to Tamar Garb for welcoming me.

The development of the project and writing of the manuscript benefited from the support and criticism of friends and colleagues: Elisabetta Bini, Muriam Davis, Ebru Kayaalp, Fred Meiton, Toja Okoh, and Amal Sachedina. My colleagues at the Middle East Institute also provided a warm atmosphere along with much needed strategic advice: Victor Kattan, Jeremy Kingsley, Linda Matar, Gwen Okruhlik, Zoltan Pall, Madawi al-Rasheed, and Shuang Wen. Petra Shenk provided invaluable editorial advice when I needed it most. I owe a special intellectual debt to Michel Callon for helping me develop the project within an STS framework and to Fred Cooper for his confidence in the project and his encouragement to publish the manuscript. Khaled Fahmy was also consistently supportive of the project. Last but not least, none of the material in this book would have made it to press without the intellectual support and mentorship of Timothy Mitchell over the past fifteen years. I owe him a great intellectual debt.

At the MIT Press, I am most grateful to Geoffrey Bowker and Paul Edwards, editors of the Infrastructures Series, and Katie Helke for ensuring the timely publication of the manuscript. I owe a significant debt to the anonymous reviewers of the manuscript for their extensive and critical comments.

Elements of chapters 3 and 6 appeared in a different version in Katayoun Shafiee, "A Petro-Formula and Its World: Calculating Profits, Labor, and Production in the Assembling of Anglo-Iranian Oil, 1901–54," *Economy and Society* 41, no. 4 (2012): 585–614.

Finally, I could not have completed the book without the love and support of my parents, Haideh and Ali, and my very patient husband, Amer, who encouraged me at every turn. My precious daughter Darya arrived in this world along the way and has made this arduous process most worthwhile. It is to all of them that I dedicate this book.

Note on Translation and Transliteration

All translations from Persian, unless otherwise noted, are mine. In translating titles, key terms, and concepts, I have kept the original Persian words next to their English translation. In transliterating Persian words into the Latin alphabet, I have used common English forms and otherwise adhered to the transliteration standards of the *International Journal of Middle East Studies*. Diacritical marks are provided only to indicate the Arabic letter 'ayn ('). In quoting from old British archival sources, I have reproduced the language and spellings used in the texts.

Introduction

The former headquarters of the international oil corporation that every-one knows today as British Petroleum (BP) were located in the historic Britannic House, Finsbury Circus, in London. The building has a notable history, having been designed between 1921 and 1925 by the celebrated British imperial architect, Sir Edwin Landseer Lutyens, marking his first foray into the design of large corporate buildings (figure 0.1). To the left of the entrance to the building is the stone sculpture, "Persian Scarf Dancer," of a woman performing a traditional Persian dance, by Francis Derwent Wood (figure 0.2). The sculpture, along with the building itself, announces the British imperial origins of the oil company in Iran and its former identity as the Anglo-Persian Oil Company, Ltd. At the time of the building's construction between the two World Wars, transnational oil corporations, including what is now known as BP, emerged as a new kind of political actor in the twentieth century. The building of the world oil industry served as the occasion for one of the largest political proj-ects of technoeconomic development in the latter half of the nineteenth and early twentieth centuries. Along with railroads, dams, electricity and communication networks, and other large-scale technical systems, a vast global network of oil wells, pipelines, refineries, and transoceanic ship-ping resulted from this enterprise. One way social scientists explain the development of this energy system is to think of oil as a natural resource that affects political systems, social and economic orders, and state for-mation from the outside while simultaneously blocking the emergence of democratic forms of politics.[1] Such an account reduces oil to its economic properties as a rent while ignoring the materiality of oil infrastructure in shaping the state and the powers of the transnational oil corpora-tion.[2] *Machineries of Oil* avoids this account of an inside and an out-side to oil operations. It does this by following the transformation of oil through the machinery of oil operations, from the initial development of

Figure 0.1
Facade of the former headquarters of BP at Britannic House, Finsbury Circus, London.
Source: James H. Bamberg, *The History of the British Petroleum Company*, Volume 2:
The Anglo-Iranian Years, 1928–1954 (Cambridge: Cambridge University Press, 1994).
Reproduced with the permission of the BP Archive.

Figure 0.2
Derwent Wood's "Persian Scarf Dancer." Photo by Robert Freidus. *Source: The Victorian Web: Literature, History & Culture in the Age of Victoria,* http://www.victorianweb.org/art/architecture/lutyens/1.html.

the Anglo-Iranian oil industry in the first decade of the twentieth century to the company's dramatic departure and subsequent return as BP during Iran's oil nationalization crisis over fifty years later.

The origins of the first oil industry in the Middle East reside in a little-known part of southwest Iran, now known as Khuzistan Province, bordering the Persian Gulf (figure 0.3). The region forms part of the Zagros basin of southwest Iran, which is characterized by limey-shale formations that stretch across the border into Iraq and other Persian Gulf areas.[3] The carbon-rich source rocks are the product of sea-level rises as well as low-oxygen and nutrient-rich environments stretching back hundreds of millions of years. Tectonic stresses and salt movements in the subsurface produced large "whaleback"-shaped folds, known as anticlines, trapping enormous amounts of oil and gas. But when British investors, led by William Knox D'Arcy, signed the first oil concession agreement with the Iranian government in 1901, they knew little about oil development. That said, British financiers held interests in other oil companies formed earlier—in the 1880s and 1890s—such as the Burmah Oil Company, Royal Dutch, and the Shell groups. American oil companies such as John D. Rockefeller's Standard Oil had been around since the 1870s. However, these companies built their initial successes on the distribution and sale of oil as kerosene for heating and lighting, oil's main marketable use in the early years. The growing popularity of the gasoline-powered combustion engine marked a gradual displacement of coal by oil as *the* power source for the industrial world. Rival imperial navies also transitioned to oil in the early twentieth century.[4] Between 1910 and 1916, oil went from supplying only 5 percent of world commercial energy supplies to over 50 percent.[5] During the course of this global-industrial energy shift from coal to oil, in 1908 D'Arcy's team of drillers first extracted oil from under the ground in southwest Iran. With commercial backing guaranteed by the British Admiralty, the new oil company, the Anglo-Persian Oil Company, produced fuel oil for export by the start of the First World War.

In the following two decades, AIOC (renamed the Anglo-Iranian Oil Company in 1935, abbreviated AIOC) embarked on building research-and-development departments to learn more about the properties of Anglo-Iranian oil, recruiting consulting geologists and engineers with previous oil experience in Southeast Asia, the United States, and Mexico. British business people and drillers designated Khuzistan Province as the site of the region's first foreign-controlled oil industry. In the span of five decades, technologies of oil exploration, production,

Figure 0.3
Map of concessionary activity in Khuzistan Province, 1928. *Source*: R. W. Ferrier, *The History of the British Petroleum Company*, Volume 1: *The Developing Years, 1901–1932* (Cambridge: Cambridge University Press, 1982), xxvii. Reproduced with the permission of the BP Archive.

distribution, refining, and marketing would get worked out with political consequences of sovereignty, government, and revolutionary movements that remain relevant to the Middle East today.

While everyone is aware of the importance of oil to the history of the modern Middle East, surprisingly little is known about how its social and technical properties shaped that history. Histories of oil are dominated by Daniel Yergin's well-known work, *The Prize*, which, while a comprehensive history of the oil industry, typifies the standard history of oil as a story of corporate titans whose actions determine political transformations.[6] No history of oil addresses the ways the technical and social aspects of the industry interacted to shape the state and to open or close political possibilities. *Machineries of Oil* draws on interdisciplinary thinking in science and technology studies (STS) to argue that the technical development of oil in the Middle East did not encounter society and the economy as preexisting spheres.[7] Rather, techniques of extracting and selling oil demanded the development over several decades of new kinds of workers, property rights, engineering skills, geological and economic knowledge, transportation systems, methods of accounting, and political forces. Taking the example of the region's first oil industry, this study reassembles the politics of the modern Middle East through a study of the British-controlled oil industry in Iran between 1901 and 1954. It investigates how the building of alliances and connections that constituted Anglo-Iranian oil's infrastructure reconfigured the local politics of the oil regions and how these arrangements in turn shaped the emergence of both the national state and the transnational oil corporation now known as BP.

In the early years of oil development, the dominant oil companies either purchased oil territories from landowners or signed oil concession contracts with foreign governments in locations extending from Pennsylvania, Texas, and California in the United States to the oilfields of Baku in Russia, Sumatra in Indonesia, Austrian Galicia, and Romania. By 1910, they had expanded their concessionary reach to Mexico, Venezuela, and the oilfields of the Middle East in Iran, Iraq, and Egypt. These companies made their fortunes by monopolizing control over everything from production wells to transport and distribution technologies, such as pipelines, railways, and tankers, marketing arrangements, and filling stations. They were rivals when it came to controlling the flow of oil from the wellhead to the market in the United States, Europe, and Asia, but they also coordinated in managing production and pricing during recurring periods of oil glut by relying on the device of the cartel arrangement (a group of suppliers in oil with the purpose of maintaining high prices and restricting competition). In the first half of the twentieth century, oil managers, government officials, lawyers, accountants, drillers, engineers, geologists, and laborers laid the groundwork that would constitute the powers of the transnational oil corporation and the oil-producing state.

Compared with Standard Oil, the Royal Dutch Shell groups, and the Burmah Oil Company, AIOC was relatively small, making it vulnerable to the growing power of the world oil cartels. It also maintained a peculiar status, having been seminationalized through commercial backing by the British Navy shortly before World War I. As growing world powers, the United States and Russia possessed vast oil reserves within their borders, but the dominant imperial power, Britain, did not. Instead, Britain possessed significant quantities of coal within its borders, making the government's decision to switch from coal to oil highly controversial. Through what techniques and controls would a private British company with special government backing get transformed into a formidable rival of the largest oil corporations already in existence, when it could only rely on the oil regions of one foreign country, Iran? Rather than limiting the discussion to the "hardware" of oil infrastructure alone, *Machineries of Oil* answers this question by tracking the organizational work involved in moving Anglo-Iranian oil from under the ground through a variety of technical, legal, scientific, and administrative networks for its production and sale on multiple temporal-spatial scales.[8] This political process of alliance building and maintenance simultaneously relied on a diverse set of equipment to remain stable.

Sociotechnical Assemblages as Matters of Politics

The reorganization of fossil fuel networks to rely on oil rather than coal in the first half of the twentieth century altered the possibilities for democratic politics in Iran. It is well known that countries that enjoy large oil revenues face unusual challenges in economic development and in creating democratic forms of politics.[9] Yet the reasons for this "oil curse" are poorly understood, because oil is typically studied only in terms of its financial revenues and their impact.[10] An Iranian political scientist, Hossein Mahdavy, first coined the term *rentier state* through an investigation of Iran's economic history in the twentieth century.[11] Iran's historical trajectory was particular, yet generally applicable to other Middle Eastern oil-producing countries relying on oil rents because they faced similar obstacles to democracy. Studies say little about the sociotechnical arrangements through which oil is produced and circulated. For oil to be transformed into profits, it must rely on a set of technical arrangements, human forces, political powers, distribution systems, forms of expertise, and coercive mechanisms.

Through the course of some of the most controversial disputes in Iran's history, AIOC relied on various forms of information management to transform political issues into technoeconomic calculation, thus guaranteeing total control of its profits, labor, and production regimes. AIOC's accountants, managers, and petroleum experts constructed formulas, technical devices, and expert knowledge about oil to build a world, their world.[12] In their different variants, the oil company enacted several political arrangements between which the Iranian government, oil workers, and public opinion were expected to choose. Each chapter investigates the specific ways various technologies were constructed to alter the political possibilities in relation to Anglo-Iranian oil. As rival oil firms scrambled for concessions to protect oil markets and keep profits high, the interwar years marked the eruption of a series of concession disputes in Iran that increasingly involved the question of the national control of oil. AIOC responded by constructing mathematical formulas, legal arguments, and scientific representations about oil that established spaces of negotiation for managing political outcomes in favor of British concessionary control and undermined the sovereignty of the Iranian state. With the exception of Thomas Hughes's investigation of Thomas Edison's use of formulas for the building of an electrification network in the United States and Europe, historians have not accounted for the history of a formula or a concession contract as a technique of political management.[13] The official history of AIOC discusses controversies over profits, labor, and production but makes no mention of the role of formulas or the machinery of concession terms in shaping these disputes.[14] Thus, *Machineries of Oil* goes further by placing the organizational work of formulation and calculation (e.g., technical, administrative, governmental) at the center of the analysis.[15] In doing so, I reveal the pivotal role that the sociotechnical properties of oil played in the co-constitutive process of building the Iranian state and the transnational oil corporation, particularly as they confronted the rising threat of militant nationalism and populist politics.[16]

A central aim of this book is to bring to the surface the differences in calculative equipment of the agencies involved within each battle over oil. STS scholarship has shown the multiple ways technical devices and calculating technologies shape agencies, particularly in markets.[17] The analytical framework of sociotechnical *agencement* takes into account the diversity of calculative equipment of agencies involved in a market, but also in science and politics.[18] For example, in his influential study, "Sociology of Translation," Michel Callon details the advantages of a sociotechnical approach as a mode of inquiry because it enables one to view the simultaneous production of knowledge *and* the construction of a network of relationships in which social and natural entities mutually control who they are and what they want.[19] The approach demands following both human and nonhuman actors through their construction-deconstruction of nature and society, forming relationships and adjusting these connections *only in action*. The outcome reveals how differences in the calculative equipment of the actors is a political issue, shaping relationships of domination, and provides a point of entry into the analysis of power struggles.[20] For instance, as operations expanded, AIOC accountants devised formulas for managing the company's labor recruitment policy to delay the replacement of British workers with Iranian ones at higher skill levels. This tactic, in the shape of a formula, had a legitimating function for a racially organized labor regime with a colonial legacy of hierarchy across the world's mineral frontiers. During the course of concession disputes, this calculating technology strengthened the British side's bargaining position by allowing them to make their case numerically in terms of a lack of Iranian workers with adequate training to support increased economic output. AIOC's labor formula played a pivotal role in this vast political project of sociotechnical engineering by the transnational oil corporations, which aimed to inhibit unions. The ease with which the company managers, government officials, and oil workers passed from the so-called technical world of oil to the world of society, geopolitics, and petroleum economics demonstrates that there is no rupture between the two worlds, but, as Bruno Latour says, it is "a seamless cloth."[21] Technical problems about the oil, its extraction, transport, and refining needed to be solved for decisions about politics, labor, property, and pricing to be made.

The difference that a sociotechnical approach can make in historical and political analysis works in two ways. On the one hand, the book draws on interdisciplinary methods from STS to introduce new puzzles to the fields of Middle East studies and global history concerning oil, state formation, technoscientific expertise, and democracy. Strands of research within STS have examined the ways infrastructures embody controversies, politics, and the constituting and excluding of subjects.[22]

Other studies in the sociology and history of sociotechnical devices have opened up the black box of formulas as market devices. For example, Donald MacKenzie's history of the Black-Scholes option pricing equation, the most fundamental equation in modern financial economics, exposes the considerable creativity and inventiveness involved in the equation's construction. Early disagreements in the design of the pricing formula gave it a kind of flexibility that enabled rival actors to attach to it differently, frame interests in advantageous ways, and construct and maintain the financial worlds they claimed to describe as economic equations.[23] These studies map how the hypothetical world built into the model equation or product formula increasingly becomes more real, but with the understanding that this is contested terrain and at all times uncertain. Yet, in the focus on the performativity of economics and various financial markets, less emphasis has been placed on the role of technical devices in politics.[24] *Machineries of Oil* brings the study of the Middle East to the fields of STS and the history of technology, which show a bias toward the investigation of small-scale economic and scientific experiments, technical systems, and laboratories.[25] Recovering the organizational work of BP in Iran exemplifies the ways sociotechnical devices worked as *political* devices and played a pivotal role in organizing relations between countries of the Global North and Global South in the twentieth century.[26]

Reassembling the politics of the Middle East as sociotechnical, therefore, requires a number of provocations. The first is to think about the oilfields of Khuzistan as a laboratory or site of experimentation for producing knowledge about nature and society.[27] Building AIOC's oil infrastructure on the ground involved the setting up of oil derricks, an 8-inch, 130-mile pipeline, an oil refinery, and housing facilities for employees divided according to race. In addition to organizing social relations, the building of oil infrastructure involved the construction of scientific descriptions for quantifying and qualifying the oil. British geologists arrived in Khuzistan only to discover that the properties of Anglo-Iranian oil, such as its viscosity and sulfur content, were not the same as those of the oil in Burma, the United States, or Mexico. Even worse, the high levels of sulfur made it unsuitable for use as kerosene for lighting. To overcome this costly setback, they needed to work on the environment and coordinate with the international scientific community to generate profits while also overcoming the constant disruptions posed by striking oil workers demanding more control.

The second provocation is to inquire into the operations of the transnational oil corporation, not as an all-encompassing abstract authority based in the metropole, such as BP in London, but sociotechnically, as a long-distance machinery of concession terms, formulas, laws, and disciplinary regimes for managing international oil markets at multiple sites in Iran and abroad.

The third provocation is to open up the black box of not just the financial but the *directly political construction* of this machinery. The political construction of these management technologies was central to the resolution of Iran's oil disputes and often worked in favor of the British company's calculations. When it did not succeed, other devices were invoked, such as US national security, to redirect political outcomes against the threat of militant nationalism allied with nationalization of the oil industry.

The final provocation is to think about these spaces as "technological zones" for the exercise of mechanisms of economic governance and managing political uncertainty by policing boundaries between knowledge (inside oil operations) and ignorance (outside, local society, the state).[28] In this way, it becomes possible to view how local disputes about oil were rapidly amplified through the existence of global zones of law, finance, and security. In many instances, these zones were transformed by their encounter with oil. As discussed in chapters 5 and 6, concession disputes about Anglo-Iranian oil triggered the development of new international legal doctrines concerning the rights of new states in the so-called Third World to confront foreign firms and make claims to nationalization of an industry in terms of permanent sovereignty over their natural resources. On the other hand, local disputes often could not be transposed to a global stage because of the particular interests at stake. Questions about improving the conditions of oil workers, for example, were overlooked using strategies of temporization when addressing the activities of oil operations.

Writing a sociotechnical history that seeks to overcome the widening gap between studies of society across diverse geographies, on the one hand, and studies of technology and the firm on the other, cannot occur without drawing inspiration from certain pathbreaking works. Studies in environmental history, STS, and the history of technology have transformed the writing of history by taking seriously the nature of technological and infrastructural transformations in relation to society and politics.[29] Timothy Mitchell's *Carbon Democracy* is the most prominent example of a work that extends this scholarship on the study of politics in the Middle East.[30] Mitchell argues that rather than viewing mechanisms for democracy as conflicting with mechanisms of carbon energy production, distribution, and usage, or studying how one impacts the other from the outside, investigators should treat the two mechanisms as inextricable. The set of practices involved in transforming Anglo-Iranian oil from a liquid to a profit were simultaneously organizational, economic, and political, contributing to the construction of local, state, and corporate settings.[31]

The recurrence of unexpected objects in the history of BP in Iran, such as formulas, concession terms, and international law, is not a choice I made as a researcher, but one made by the actors themselves. As the chapters reveal, each calculating technology, when successful, can organize heterogeneous behaviors, political decisions, and economic outcomes. But some actors, such as oil workers allied with militant nationalism, are able to deconstruct formulas to open new situations in which alternative formulations are proposed. Working out the mechanisms of this calculative equipment was simultaneously a battlefield of technicality and political contestation, (re)formulating similar but moving problems over the course of fifty years.

Governing Oil Infrastructure in the Middle East and Its Limits

The history of AIOC in the first half of the twentieth century is in part a long history of relations with the Iranian government over the control and distribution of oil. The working of concession contracts, the management of oil workers and production rates, and the controversies they generated, were all events that contributed to constituting the Iranian state and the organization and strategic priorities of the largest transnational oil companies. This history was punctuated by dramatic crises over the terms of these relations organized around oil. In each crisis, the strategy followed by AIOC centered on defining the terms of the concession contract or a formula and policing its limits. AIOC also used other measures to resolve critical moments, such as international law and disciplinary regimes, economic sanctions and boycotts, forms of coercion, and an engineered overthrow of the Iranian government in 1953. In each controversy, the cultivation of strategic uncertainties, forms of ignorance, and temporization played a useful role in the organizational life of the industry because it allowed the oil company to negotiate possible compromises and avoid reaching a more equitable solution.[32] Thus, the forging of the historical identity of the actors involved in the governance of oil infrastructure—the transnational oil corporation and other institutions such as the International Court of Justice and the World Bank—can be thought of as a political process of separating between a well-governed (global/Western) oil economy and forms of (non-Western) state governance, mismanagement, and violence, which are taken to exist outside the space of this economy. The effort to forge a distinction between what the infrastructure of oil governance is and what it is not, as it operates locally in the Middle East, helps constitute its historical identity.

By this account, *Machineries of Oil* challenges the dominant historical narrative about oil governance in Iran.[33] This historical moment is usually framed as one about the role of the British Empire in the Middle East with Iran serving as a buffer region between Russia and India, but British imperial rule can be seen as a byproduct of the global energy shift to fossil fuels.[34] Kenneth Pomeranz explains that this shift, coupled with the ecological relief provided by Europe's ties to the Atlantic slave trade and the acquisition of colonial territories, enabled countries such as Britain to trade a growing volume of manufactured products for a growing volume of land-intensive products such as food, cotton, and other industrial crops.[35] These kinds of developments in land, energy, and labor contributed to the development of fossil fuel–based mass production centered in cities, and thus to the building of world imperial powers such as Britain. But the infrastructure that emerged out of the global energy shift from coal to oil also marked a point of vulnerability for British imperial power. The largest refinery in the world was located in Abadan, Iran, on the Persian Gulf and constituted the British Empire's largest single overseas asset. Britain imported much of its oil from Iran, and with AIOC's exclusive rights to develop and market Anglo-Iranian oil, the country was able to pay for the oil in its own currency, pounds sterling.[36] As chapter 5 explains, the Iranian government's decision to nationalize the oil industry and evict AIOC in 1951 represented more than a challenge to British imperial influence, it risked devastating the pound sterling for the United Kingdom, opening the door to US dollar dominance as the preferred reserve and trading currency in a new post–World War II petroleum order. New strategies such as economic sanctions and an oil boycott were devised, in part, to manage flows of oil and flows of finance for restoring AIOC's former standing in Iran.

Contrary to what the dominant narrative suggests, the energy needs of the British Navy were only a partial connection in the political project to build a British-controlled oil industry in Khuzistan, Iran. During the interwar period, the major oil companies, such as AIOC, were the first and largest of the new transnational corporations of the twentieth century to emerge at this moment of political power jockeying and energy shifting, coinciding with the collapse of an imperialist "old form of empire."[37] But they did not come into being with their agency and powers already formed. The history of AIOC's formation is usually told as if the company operated within its own world, independently of the Iranian state, as a commercial concern, and was then simply boosted by the British government and the navy's decision to switch to fuel oil. Scholarship on oil companies examines the ways corporate managers organize a set of decisions about what to control or not to control. The organizational form of the state and its relation to the transnational corporation are treated as two separate entities, tied to one another by an external relationship.[38] However, keeping these relations undefined allows one to follow which boundaries close and which are left open. The following chapters trace the techniques and controls that organize spaces of immense political and economic significance such as cartel arrangements, international law, oil diplomacy, economic sanctions, and US national security. The outcome is to view the simultaneous construction of the power of the transnational oil corporation as a long-distance mechanism for placing limits on oil supplies, keeping profits high, and undermining Iranian state sovereignty, making it very difficult to view certain actors as states and others as private companies.

Together with the oil companies, oil-producing states such as Iran played a pivotal role in the battle over shaping the oil infrastructure that emerged out of the global energy shift to oil.[39] In the 1930s, these battles intensified as protesting governments and oil workers in Iran, Mexico, Venezuela, and eventually Saudi Arabia and Iraq, demanded better contractual terms regarding profits, production levels, and the treatment and training of domestic labor. The period was marked by the increasing appeal of populist and nationalist forms of politics allied with calls to expropriate and nationalize the foreign-controlled oil industry. The dubious honor of having the first commercially exploitable oilfields in the Middle East affected and perhaps made possible the very formation of the modern state of Iran in this period. Theories of the state, however, tend to focus on questions of sovereignty and executive power.[40] As a result, they overlook the large-scale technical projects of infrastructure and energy networks that have often represented the most widespread, complex, and localized forms of state power and authority.[41] *Machineries of Oil* starts from this premise to examine the building of oil infrastructure in the making of the global oil economy in the Middle East.

However, it uses a specific and local history in Iran to map the anatomy of one company, now known as BP, and the mechanisms through which its political agency was assembled. It is not possible to make sense of the simultaneous rise of mass politics and democracy in the West and of authoritarian regimes in the Middle East without following the material and organizational work involved in building a world to rely on fossil fuels such as oil.[42]

Structure of the Book as the Structure of the Oil Concession

The importance of raw materials, such as oil, to the global economy has always been understood. The problem is that the materials were located principally in the "developing world."[43] When the oil resources of the former Ottoman Empire were distributed among the British and French mandate powers, the resources of these territories were understood as "belonging to humanity as a whole." Equipped with a rhetoric originating in the mandate era, Western imperial powers and financiers competed for contracts in oil to fuel navies, warplanes, tanker fleets, and eventually the cars of ordinary consumers. Thus, the history of different forms of contractual agreement organized around oil, originally known as concessions in the late nineteenth and early twentieth centuries, played a pivotal role in the twentieth-century history of the Middle East. In practice, countries in the region often defined their borders at the point at which "oil companies required maps delineating their concessions."[44] There were maps of nations and maps "showing the region cut up into squares along the coast, marked with initials" of oil companies.[45] Previous foreign concessionary projects in Iran—such as the Reuter Concession, or Russian, French, Belgian, and Austrian propositions for concessions in roads, railways, utilities, and factories, for example—made no significant contribution to the political economy of the country.[46] On the contrary, they often helped trigger revolutionary movements such as the famous Tobacco Revolt, which constituted one of the first organized and popular rejections of the British-controlled tobacco concession in 1890.[47] Thus, the structure of *Machineries of Oil* loosely follows the structure of the Anglo-Iranian oil concession. Each chapter considers a pivotal dispute in the history of Anglo-Iranian oil by examining the terms of the oil concession contract regarding property and mineral rights, petroleum expertise, royalties, labor, international legal arbitration, and the cancellation of the concession in favor of political sovereignty. The goal is not to compare the representation of oil in contractual terms with its reality on the ground, or to view how the two (abstract versus the material) affected each other in terms of an external relationship. Rather, I follow how the terms of the concession embodied a world that became more real, equipping heterogeneous actors with differing degrees of agency but with the understanding that this was highly contested terrain and at all times uncertain.

One of the first sociotechnical problems British investors had to solve was how to attach property rights to a subsoil material owned by a foreign government in a region where local forms of property ownership already existed. The first chapter, "Properties of Petroleum," starts by asking why British investors designated southwest Iran as the site for oil extraction when other regions of the Middle East were known to have more easily accessible reserves. By what political methods did British investors impose the concessionary rule of property in southwest Iran, particularly since this conflicted with older forms of property and local claimants to the land? Through a careful examination of the 1901 D'Arcy Concession's articles on property, I argue that they first appeared as a legal text between two contractually equivalent parties that seemed to guarantee a kind of absolute authority. In practice, however, many rules of concessionary property conflicted with the building of a series of international cartel arrangements among the major oil companies to protect oil markets in Asia by limiting oil production from the Middle East in the early years of the concession's existence. The concession contract was a useful political weapon, but it was highly porous because its articles maintained an ambiguity that could be exploited by both the British and Iranian sides, as well as by local nomadic groups with claims to the land.

Securing access to the subsoil through the format of a concession contract was not enough. AIOC was confronted immediately with another sociotechnical problem in producing and refining the oil that was coming out of the ground. The second chapter, "Petroleum Knowledge," paints a precise picture, at the local and international levels, of the scientific knowledge and technical expertise that allowed for the mastery of the political project to extract and process oil in southwest Iran in the early twentieth century. I track how British oil managers, engineers, and geologists constructed and monopolized knowledge about Anglo-Iranian oil to maintain a distance from political issues such as national control. I argue that placing techniques of standardizing oil at the center of the analysis reveals how power relations were worked out in the oil regions of southwest Iran, Tehran, and London with political consequences for the formation of the Iranian state and its claims to sovereignty over a natural resource. As a business strategy, techniques of standardization and forms of information management equipped AIOC with the tools to build a world in which its oil could thrive.

Through the 1920s and 1930s, the Iranian government continued to push for higher production rates, royalties, and more technical information about its oilfields. Such demands conflicted with secret, international cartel arrangements to place limits on world oil supplies and keep profits high. This controversy culminated in the ruling monarch's abrupt cancellation of the 1901 concession in 1932 and increasing calls for national control. It triggered the largest interwar dispute between a foreign company and a national government. Chapter 3, "Calculating Technologies in Crisis," opens up the royalty clause of the 1901 concession in order to map the multiple sociotechnical arrangements that would help maintain an economy of oil involving an artificial system of scarcity. As a weapon, the company, in collaboration with its largest shareholder, the British government, transported the dispute to the Permanent Court of International Justice to create delays and to ensure the continued blockage of the increased flow of oil from Iran. I elaborate in detail how, with the help of formulas, AIOC managers and accountants worked to manufacture a kind of ignorance about the nature of royalty calculations and the failure to replace foreign workers with Iranian workers. The problem was that such concerns were already entangled with other kinds of calculation, such as the threat of nationalist politics rendering the energy network highly unstable.

The interwar concession crisis between AIOC and the Iranian government helped transform the question of the Persianization of the British-controlled oil industry—the gradual replacement of foreign employees with Iranian workers at higher skill levels—into a major point of dispute that went unresolved. This episode coincided with a series of labor strikes between the 1920s and 1950s, when oil workers transformed the oilfields, pipelines, and refinery of southwest Iran into sites of intense political struggle. These controversies opened up the definition of the kind of worker the oil industry required to survive. In the fourth chapter, "What Kind of Worker Does an Oil Industry Require to Survive?," I track the most pivotal oil worker strikes and the construction of a racial-technical oil labor regime in work and housing on the oilfields. The chapter, on the one hand, emphasizes the importance of social technologies that worked with certain calculating technologies, discussed in the previous chapter, for measuring and controlling the degree of

Persianization of employees at AIOC. The sophistication of these social
and calculating technologies reveals the extreme inventiveness of AIOC
and other oil companies, which transformed political questions of labor
into technical and economic issues and often collaborated in the management of labor in the Middle East and Latin America.

The next sociotechnical problem I investigate has to do with the question of how to manage the economic order of moderate nationalism
in the Middle East while avoiding the more dangerous alliance of militant nationalism, communism, and national control of oil operations.
Chapter 5, "Assembling Intractability: Managing Nationalism, Combat-
ing Nationalization," argues that the resolution of Iran's oil nationalization crisis occurred on terms set by the different international regulatory
regimes and techniques of intervention, such as an oil boycott, involved.
These technologies worked as strategies to block nationalization by rendering the oil dispute intractable and by framing issues in ways that separated political questions from economic and technical ones. Technologies
of temporization and delay constituted the last set of weapons available
to manage oil crises in a way that would strengthen the bargaining power
of Western governments and oil companies to the detriment of the Iranian government and oil workers.

The final sociotechnical problem tracks the machinery involved in
reconfiguring Iran's oil nationalization crisis through the working out of
an international oil consortium between 1952 and 1954. The reconfiguration took place within a post–World War II international petroleum
order that witnessed the building of a peculiar relationship among oil
currency flows, the US arms industry, financial and technical aid, and
the fight against communism. Chapter 6, "Long-Distance Machineries
of Oil," maps in detail the anatomy of the oil corporation as a type of
long-distance machinery for placing limits on new oil discoveries in the
Middle East. A new set of mechanisms and justifications was devised to
place limits on the flow of Iranian oil, namely, the consortium arrangement, US national security, and the calculating technology of the "Aggre-
gated Programmed Quantity," to limit production levels in coordination
with the leading international oil corporations. Anglo-American interests
sought to build a new international order that entertained the wording of
nationalization but only as a means of preserving old and new contractual relations between national governments and foreign firms.

Drawing on primary material from government and business archives
in Iran, the United Kingdom, and the United States, *Machineries of Oil*
comprises six chapters, tracking the oil as it was transformed in southwest

Iran into an elaborate machinery, marking the closure of democratic forms of politics at specific points of vulnerability in the energy system. Each chapter therefore marks the transformation of of oil from a material substance into its representation and abstraction as concession terms, a form of industrial science, mathematical formulas, and a hierarchical labor regime with a specific kind of worker. The excesses of coercion and violence involved in the resolution of the disputes under discussion spilled into national politics. These controversies were explained away in terms of blocking the threat of militant nationalism, the lack of adequately trained domestic labor, and the need to pursue oil operations within global oil infrastructures set up by the seven dominant oil corporations, famously known as the "Seven Sisters."[48]

* * *

What kinds of technopolitical collectives have emerged from this story, enabling the closure of more democratic forms of politics? The oilfields of southwest Iran served as a kind of laboratory in which the political agency of the oil corporation was formatted to fit the hypothetical world claimed to be described in contract terms and inscribed within various zones of measurement, regulation, and qualification. But in practice, oil infrastructure—a set of alliances and interconnections and all the knowledge constructed about them—makes possible the global oil economy. As Andrew Barry has observed, "the question of how and whether particular oil developments are included or excluded from particular zones of measurement and regulatory and quality assessment is critical to the contemporary politics and economy of oil."[49] The process of assembling this machinery, shaping who can act and on what terms, occurs locally starting with a rhetoric that renders the local knowledge of property claimants, oil workers, and a national government as defective, by apprehending them in terms of the economic rationality they lack. Practices of abstraction manifested in concession terms, formulas, and international law are the outcome of the sociotechnical process under examination, allowing relations between politics and international oil markets to remain predictable and stable.

The origins of BP's dramatic success over the course of the twentieth century reside in a little-known region of southwest Iran. By plunging into the technicality of the battle over oil, this study exposes the powers of the oil corporation precisely in the organizational work of each dispute. The power of an account of a boundary between an outside and an inside to oil operations does not arise from the accuracy of its description of the socioeconomic relations of oil. It comes from the tools and arguments made available for handling the organization of oil infrastructure. Decisions about profits, production rates, and labor were directly tied to the ways the control and distribution of oil were organized in the oilfields of the Middle East. These decisions did not determine political outcomes but rendered the authority of oil concessions highly uncertain and the shape of the oil industry open to alternative possibilities. The politics of the Middle East or of any oil-producing region of the world cannot be understood adequately without giving serious consideration to its technical dimensions, and the technical world of oil can be understood properly only in terms of the historical and political forces through which that world has been shaped.

1

Properties of Petroleum

The Government of His Imperial Majesty the Shah grants to the Concessionaire ... exclusive privilege to search for, obtain, exploit, develop, render suitable for trade, carry away and sell natural gas, petroleum, asphalt and ozokerite[1] ... throughout the whole extent of the Persian Empire for a term of 60 years ... .

This privilege shall comprise the exclusive right of laying the pipelines necessary from the deposits where there may be found one or several of the said products up to the Persian Gulf. ... [This privilege also includes] the right of constructing and maintaining all and any wells, reservoirs, stations and pump services, accumulation services and distribution services, factories and other works ... .

The Imperial Persian Government grants gratuitously to the Concessionaire all uncultivated lands belonging to the State which the Concessionaire's engineers may deem necessary for the construction ... of the ... works. As for cultivated lands belonging to the State, the Concessionaire must purchase them at the fair and current price of the Province.

The Government also grants the Concessionaire the right of acquiring all and any other lands ... necessary for the said purpose, with the consent of the proprietors, on such conditions as may be arranged between him and them without their being allowed to make demands of a nature to surcharge the prices ordinarily current for lands situate in their respective localities. Holy places ... are formally excluded. (Articles 1–3, D'Arcy Concession, 1901)

When William Knox D'Arcy signed the first oil concession agreement in the Middle East with the Qajar government in 1901, he immediately encountered numerous technical and financial problems.[2] This was due in part to the costliness of the venture and his failure to strike commercially producible quantities of oil in the first seven years of the concession's existence. The problem was that the 1901 oil concession did not say much about the oil itself or where to find it. Most of the document concerns the question of property or the "Concessionaire's" rights to access and develop the land and petroleum resources underground, as defined within the concession area. The text of the concession was originally written in French and then translated into Persian and English. The English translation of Article 1 of the concession, quoted above, grants D'Arcy automatic and exclusive access to produce, transport, refine, and sell Iran's oil. Article 3 identifies the "Imperial Persian Government" as owner of all lands, "cultivated and uncultivated," "belonging to the State." In addition to granting D'Arcy access to all state-owned lands for the purposes of oil development, the concession also grants him, as "Concessionaire," the legal privilege of purchasing all non-state-owned lands from "proprietors" within the concession area.

But did access to and ownership of the oil and lands above it, as defined in the concession contract, imply the same meanings in English as in Persian? Since the sixteenth century, all natural resources in Iran,[3] including those underground, were owned by the ruling monarch (*shah*).[4] This meant that mining operations such as in oil could only take place with his permission. In Article 1, the ruling government of the Qajar shah grants to D'Arcy complete access and control over the subsoil for the purposes of oil development, but the subsequent article, Article 3, makes a peculiar distinction between lands owned by "the State" versus "other proprietors," implying that property relations in Iran work the same way they do in the West, in terms of a distinction between public (state-owned) and private (local proprietors) ownership. Following this framing of the terms, D'Arcy is granted automatic access to the subsoil to search for, take away, and sell the "natural gas, petroleum, asphalt and ozokerite" contained within the concession area, regardless of whether the surface is owned and controlled by the state or by a local proprietor.

In contrast to those governed by the rule of private landed property practiced in other Western countries—especially the United States, a pioneer in the oil industry and one of its leading producers—D'Arcy and his fellow British investors seemed to be at an immediate advantage.[5] Article 1 guarantees that "for a term of 60 years," D'Arcy and his team of drillers and engineers would have automatic access to the petroleum resources of the "Persian Empire," regardless of the nature of property ownership on the surface. This exclusive access had the potential to change the shape of the oil industry. For approximately the first half of the twentieth century, the United States produced over half the world's output of petroleum, but the US rule of private property posed a problem for the extraction of oil since such a right gave individual owners of the soil automatic authority over the subsoil. Problems related to property laws concerning oil had their origins in US Supreme Court decisions of the late nineteenth century, which drew an analogy between oil and gas and *ferae naturae* of English common law. Thus, "if a wild pheasant ran from one person's property to another's, ownership of the pheasant changed as well."[6] Whoever caught the oil first could keep it. Because a potential oil-field can cover several private lots, private ownership of the subsoil created the problem of multiple producers attempting to extract oil as fast as possible for their own gain, otherwise known as the "rule of capture." For a major oil company such as Standard Oil, "the obstacle represented by small private landed property caused companies such as [John D.] Rockefeller's to monopolize downwards" in transportation, refining, and marketing to avoid the problems associated with production.[7] American oil investors such as Rockefeller also headed abroad because of the high price of US oil. The high price had originally been artificially imposed by the Texas Railroad Commission and later by federal import quotas to confront the high degree of competition resulting from the rule of capture. In order for Rockefeller to maintain control and prevent competitors from undercutting him, Standard Oil ventured into other parts of the world such as Venezuela and eventually Saudi Arabia, seeking to "find new fields and initiate extraction but based on state landed-property which they helped impose."[8] British financiers such as D'Arcy appeared to have avoided the US problem of private control of the subsoil because the concession identifies the ruling government of Iran as maintaining total (state-owned) control over it.

The signatories to the legal text of the 1901 concession recognize two contractually equivalent parties, the British concessionaire (D'Arcy) and the Iranian government. It suggests that along with third-party proprietors, they are the only actors involved in this story. But did the oil company's concessionary terms of representing land in a foreign language and according to the rules of Western property ownership simply get imposed as a legal text between the two parties? This chapter starts from the concession agreement itself, treating the articles on property not just as a text over which the main actors disagreed, but as an apparatus through which actors take on new powers and politics. I investigate the ways the concession contract operates as a kind of device helping to give actors their bargaining power and shaping who can act and on what terms.

Property agreements such as a concession are not simply a question of new rights, a legal document, and a contractual agreement. There remains the problem of what actors will be party to the agreement and what actors will be excluded. One party involved a new kind of actor, the transnational oil corporation. What was the status of this actor? What was its power? The powers of actors are formed partly in the agreements they make. The text of an agreement such as a concession helps define their powers—it is a device that helps create agency.[9] For the concession terms on property to succeed in oil development, the British concessionaire needed to impose and stabilize the proposed property arrangements defined in Articles 1 and 3. Multiple devices such as concession terms but also technologies for measuring and valuating oil needed to be mobilized and attach themselves to diverse actors to achieve the goal of finding, producing, transporting, and selling oil. Over the years, British investors (D'Arcy, AIOC, the British government) increasingly placed the 1901 concession's terms between themselves and all other entities, such as local political groups and the Iranian government. Along with oil workers, landowners and other claimants to the land battled to define their identities otherwise in the pursuit of more equitable forms of oil development, namely, national control. Thus, struggles over a text's details, the properties of petroleum in this instance, matter.

The history of the relations between the oil, the British oil company, and the Iranian government can therefore be understood, in part, as a series of crises over the property rights of an oil corporation in relation to a sovereign government and local claimants to the land. Concessions, formally defined as "grants by a state to citizens, aliens, or other states of rights to carry out specific economic activities and of capitulatory rights on its territory," have existed in Iran since the sixteenth century.[10] However, no existing history of Iran, or of any other country in the Middle East, deals closely with the development of oil in particular by taking seriously its technical and concessionary history. Scholars have not taken into account the peculiar properties of oil embedded in Iran's 1901 concession contract, but rather focus on the date of the concession's signing, the most powerful parties involved, and certain disputes over its financial terms.[11] But most of the document concerns the question of property rights insofar as the British investors can access and develop the land and underground petroleum resources located within the concession area. By placing technologies of accessing, measuring, valuating, and selling oil at the center of this analysis, I retrace the organizational work of British concessionary property within the larger history of modern property in Iran. The approach also makes visible the properties of petroleum, or all the people and things and forms of knowledge attached to the oil but silenced in the oil concession. While not mentioning many of them explicitly, the concession contract enabled the different actors involved—the

Iranian government, the British investors, and claimants to the land—to negotiate possible compromises and open or close alternative political possibilities in the control and distribution of oil. British investors, managers, and engineers arrived in southwest Iran with a peculiar understanding of the properties of oil, and of the terms and politics through which they would access and exploit oil as spelled out in the concession's text. They also confronted what they viewed as a desolate backwater inhabited by a population lacking adequate technical knowledge and skills for modern energy development, yet possessing claims to ownership of the regions targeted for oil extraction. Placed by the company in apparent opposition to the history of democracy and modernity of the West, these local actors created an uncertainty most evident in the problems that emerged around the question of British concessionary control over the oil.

This chapter recounts a different history of oil by exploring disputes over the properties of Anglo-Iranian oil in five parts. The first section tracks how another article, Article 6 of the 1901 concession, concerning the right to construct a pipeline to the Persian Gulf, initially worked in a remarkable way to block rival firms and delay the launching of the first oil industry in the Middle East, in Iran. The analysis exposes how and why British investors and technologists chose to explore for oil in southwest Iran in the first place, knowing that other more easily accessible locations for oil development existed in neighboring Mesopotamia and Egypt. The second section considers in more detail the text of the first three articles on property to expose the problems that emerged around the question of land ownership in southwest Iran. I then consider the specific ways local forms of property challenged the British strategy of accessing oil and its response to the uncertainties built into the concession. The third section looks at how the oil company bypassed its own concessionary authority in favor of alternative arrangements with local actors who worked to redefine the nature of ownership over oil. The fourth section reveals how technical problems of valuating oil shed light on the power of concessionary property in creating or limiting political possibilities. Forms of property ownership and valuation were deemed legal or not depending on the interests involved (local, national, oil company) and the political consequences of more equitable forms of control. This section as well as the fifth one explore the ways AIOC and the Iranian state increasingly cooperated in the deployment of technologies of control, compulsion, and eviction of local land claimants to stabilize concessionary rule. I point to how the powers of the transnational oil corporation were worked out locally, in a southwest corner of Iran, and the kinds of political closure that the machinery of the British-controlled oil concession helped bring about.

Article 6 and the Move to Southwest Iran

The privilege granted by these presents shall not extend to the Provinces of Azerbadjan, Ghilan, Mazendaran, Asdrabad, and Khorassan, but on the express condition that the Persian Imperial Government shall not grant to any other person the right of constructing a pipeline to the southern rivers or to the south coast of Persia. (From Article 6, D'Arcy Concession, 1901)

British commercial interest in mineral development and the Iranian government's interest in revenues from such development combined to establish the first concessionary oil agreement in the Middle East. Antoine Kitabji, Director of Persian Customs, first made the offer to D'Arcy by way of Sir Henry Drummond Wolff,[12] British Minister in Tehran, to acquire the right to explore, extract, and exploit oil in Iran using the legal format of a concession.[13] Having already made a fortune from an Australian gold mining venture, D'Arcy accepted the offer to acquire an oil concession in Iran, not including the five northern provinces (i.e., Azerbaijan, Gilan, Mazandaran, Astarabad, and Khorasan), which were under Russian influence.[14]

Article 6 of the concession agreement stipulated that the privileges granted to the British concessionaire would not extend to the five northern provinces on the condition that Iran's government would likewise prohibit any other company from building a pipeline across the concession territory to the Persian Gulf. The exclusion of the five northern provinces from the British concession was connected to Russia's imperial ambitions in Iran, but perhaps more so to Russia's commercial oil interests in Far Eastern markets. In the pre–World War I oil industry, anyone who could load oil in the Persian Gulf would have a significant competitive advantage in Asian markets over oil exported from the Black Sea through the Suez Canal because it was cheaper.[15] Russia's commercial ambitions centered on a plan for a Transcaucasus–Persian Gulf pipeline route starting in Baku and backed by the major Russian oil producers, namely, the Swedish Nobel family and the French Rothschild banking firm. Such a pipeline would be a serious blow to competitors in Asian markets, particularly Standard Oil, the largest of the world's oil firms and the company to beat from the late nineteenth century to World War

II. There was only one problem: the British concession was blocking the way to Iran.[16]

In the first decade of the twentieth century, Russia exhibited more influence on the Qajar government and in the northern provinces than Britain, which lost popularity and prestige as a consequence of the nationwide tobacco revolt against the British tobacco concession in 1890.[17] The Nobel and Rothschild firms, the leading Russian oil producers, backed Russia's goal of a government-sponsored consortium consisting of the major Russian oil firms, to compete against and collaborate with Standard Oil and consolidate their control of the Russian oil industry. Collaboration often helped oil firms strengthen their monopoly control so oil firms engaged in both competition and collaboration depending on the targeted markets. In 1902, the Russian minister of transportation pursued this goal by making a loan to the Qajar government conditional on a concession for a pipeline to the Persian Gulf to transport Baku oil to Far East markets.[18]

Scholars often rely on an imperialist framing to argue that a strictly imperial pursuit was what drove Great Britain and Russia's quest for oil in Iran and the greater Middle East and that the aim of the investors in pursuing concession agreements was to *discover* oil, not *delay* its development.[19] In its early years, D'Arcy's 1901 oil concession in Iran appeared to work more effectively as a mechanism for blocking rivals and delaying rather than producing any oil. Article 6 of the concession guaranteed that so long as he was exploring for oil, no other investor or government could build a pipeline across the concession area to the Persian Gulf.[20] In the meantime, D'Arcy did not strike much oil and believed that Mesopotamia was a more promising location for oil development.[21] He tried to relieve himself of the financial responsibilities of maintaining the 1901 concession (e.g., drilling for oil and setting up a company) by attempting to organize a syndicate prior to the creation of an oil company, which he was in no position to do without oil. He approached the rival Rothschild firm, with oil interests in Baku, for financial support. But that strategy created a problem for another competitor—the Burmah Oil Company. An alliance between D'Arcy and the Rothschild firm would potentially speed up pipeline construction to transport Russian oil from Baku to the Persian Gulf and thereby disrupt Burmah Oil's monopoly control over the large kerosene market in India. Alternatively, it would have brought Rothschild-owned Iranian production onto the market. D'Arcy's negotiations with the Rothschild firm helped trigger Burmah Oil's decision to "play along with D'Arcy" by providing financial backing to his concession through the formation of the Concessions Syndicate.[22]

In 1905, D'Arcy formally agreed to share his concession rights in Iran with the Burmah Oil Company, the second largest oil company in the British Empire after Shell.[23] However, Burmah's aim, Timothy Mitchell explains, was to rescue the failing British venture in order to gain control of the concession as a mechanism for *delay and blockage* rather than opening up the production of Middle East oil.[24] Burmah Oil was facing its own threat with the creation of another rival company, Asiatic Petroleum, a jointly held subsidiary of the Rothschild firm, Royal Dutch, and Shell, whose goal was to flood the Indian market with oil that could "undersell Burmah's."[25] As a result, Burmah Oil was forced to enter a cartel arrangement (a group of suppliers in oil with the purpose of maintaining high prices and restricting competition) with Asiatic in 1905. For Burmah Oil, keeping D'Arcy's concession intact ensured, at the very least, the blockage of Russian oil from flowing through a Persian Gulf pipeline, and therefore a degree of protection of its interests in the Indian market.

In the first decade of the twentieth century, Standard Oil and the European oil companies "created arrangements to restrict the production of oil, control its worldwide marketing—and simultaneously address the threat of oil from the Middle East."[26] Securing and coordinating the control of concessionary arrangements in each oil region was central to achieving this strategy.[27] A related threat was that the Russian producers at Baku would shorten the route to Asia by building a pipeline to the Persian Gulf. To block these threats, the three largest British and European oil firms purchased concession rights to explore for oil in the Middle East: Deutsche Bank in northern Mesopotamia (the Ottoman provinces of Mosul and Baghdad) in 1904, Burmah Oil in Iran in 1905, and the Shell group in Egypt in 1908–1910.[28]

In the first decade of its existence, the 1901 D'Arcy Concession arrangement lent its British investors and the future British oil company, AIOC, a kind of power to control markets and join international oil cartels to manage oil production (in the Middle East and elsewhere) precisely through its control over concessionary property and the building of a pipeline. But why move to southwest Iran in the first place, especially when more easily accessible fields were already known to exist in Mesopotamia and Egypt? The ruling Qajar government (1794–1925) first commissioned a series of scientific missions in 1890 led by a French geologist, Jacques de Morgan, who published a volume analyzing oil prospects for potential

petroleum development.[29] Boverton Redwood, the leading British expert
in petroleum matters, arranged on D'Arcy's behalf for a subsequent land
survey to be conducted in 1901 by a British geologist, H. T. Burls.[30] Burls
confirmed de Morgan's conclusions on oil prospects in the concession
area: "The territory as a whole is one of rich promise."[31] Based on Burls's
report, Redwood advised D'Arcy to begin drilling not at Masjid Suleiman
where the first commercially producible oil reserves would be found but
further north at Chah Surkh, located near the border with Mesopotamia.
In 1892, de Morgan had first published a report on oil deposits at Chah
Surkh in the journal *Annales des Mines*.[32] However, by drilling at Chah
Surkh, Mitchell explains, D'Arcy was again blocking a potential bypass
route for the pipeline to Baku, obstructing an increase in Caspian exports
to India.[33]

Failing to strike oil at Chah Surkh, drillers eventually moved southeast, in 1906, to Khuzistan Province.[34] Activities progressed slowly, in
part because D'Arcy did not have the funds to drill in more than one
location at a time. He instructed George Reynolds, a British graduate of
the Royal Indian Engineering College with previous oil drilling experience in Sumatra, to head south and explore possible sites for drilling on
the basis of W. H. Dalton's geological report and to make agreements
with local landowners for such activities. Working on behalf of D'Arcy,
Reynolds hoped to drill in 1904, but activities were delayed until a satisfactory arrangement could be made with the Bakhtiyari khans, local
claimants to the land who were demanding a percentage of future oil
profits. Burmah Oil's lack of interest in oil development along with a
series of other delays, such as with the construction of a road to transport
equipment, meant that drilling at Masjid Suleiman did not begin until
January 1908, finally yielding commercial quantities the following May.

The oil at Masjid Suleiman was naturally high in sulfur, which presented an additional set of problems for its use either as gasoline or
as kerosene for illumination—oil's main marketable use in these early
years. Thus, British investors in AIOC and the Burmah Oil Company
"organized the oil's weaknesses and their business needs into an imperial
interest, as fuel oil for steam or diesel engines."[35] The problem was that
most imperial powers were coal based in the early twentieth century. Few
oil-powered engines were used anywhere near Iran and the company faced
bankruptcy in its early years. In a bid to cover its costs, AIOC offered
discounted prices to the British Navy. The British government rescued
AIOC in 1914 by purchasing 51 percent of the company's shares.[36] The
British government's participation as the primary shareholder in AIOC

transformed the company into a leading member of the emerging international oil cartel, the Turkish Petroleum Company (TPC). The threat of potential oil development in Mesopotamia had helped trigger the 1914 agreement among America's Standard Oil Company and Europe's principal oil companies, including AIOC, to control oil supplies and protect international oil markets by agreeing not to undertake oil production anywhere in the Ottoman Empire unless jointly through TPC (discussed in chapter 3).[37] The concluding phase of negotiations to establish TPC coincided with the British government's decision to finance AIOC. The 1901 D'Arcy Concession involved many more actors and cartel agreements than its terms would suggest. It also enabled a foreign national (British) government and oil company (AIOC) to share in the project to hinder the development of both Mesopotamian and Iranian oil, threatening both the European and Asian markets.

Rather than enabling any oil development, at least in the early years, the concession's terms—Article 6 in particular—worked as a kind of management and control tactic through which the powers of the transnational oil corporation were forming precisely by keeping actors such as the British government, Admiralty, and Burmah Oil in, and other rivals, especially Russian oil, out. Article 6 lent D'Arcy the power to ensure that as long as he met the concession's contractual obligations to explore for oil, the Russian pipeline project could not happen. Not mentioned in the concession's terms was the role of a new set of actors such as the Burmah Oil Company and the British government, which rescued D'Arcy's failing oil venture in order to protect Far East oil markets and to purchase fuel oil for the British Navy at discounted prices, respectively. The powers of the future oil company were forming as a long-distance machinery for organizing the energy network. These powers worked initially by delaying the arrival of new oil supplies from the Middle East, precisely through the terms of Article 6 and in coordination with other cartel arrangements, helping to make the international market in oil more predictable. Further, the organizational form of AIOC was emerging as a mixture of so-called public (government) and private (nongovernment) interests from the start. It was not simply boosted by British government funds as the BP company history suggests. The flow of new supplies of oil from the Middle East was a looming threat to world markets.[38] The 1901 concession's terms on property and oil development in Iran would have to be executed in the oil regions to help build a world in which any future uncertainty was eliminated.

Rules of Concessionary Property

1901 was not the first time the central government had granted a concession to a foreign concern. Since the sixteenth century, ruling governments
had granted economic concessions to promote the exploitation of Iran's
natural resources. The granting of the first concession in the history of
Iran related to carbon-based energy was shaped by the British pursuit of
another fossil fuel—coal. In July 1872, the Qajar government granted
Baron Julius de Reuter an extensive concession for exclusive rights in
railroad and streetcar construction, mineral extraction, irrigation works,
a national bank, and various industrial and agricultural projects.[39] Nasir
al-Din Shah, the ruling monarch of Iran, was reported to have granted
concessions such as this with a "lavish hand" during his reign from 1848
to 1896.[40] The British minister in Iran persuaded the shah to issue a royal
proclamation guaranteeing "the sanctity of private property throughout
the realm."[41] In 1889, the minister then set up a new concession on Reuter's behalf (his first concession was canceled in November 1873) with
the right to exploit Iran's minerals for sixty years.[42] Reuter eventually
sold his mineral and oil rights to the Persian Bank Mining Rights Corporation, whose directors included his son, George de Reuter, and George
Curzon, British Member of Parliament and future Viceroy of India. The
bank spent three years looking for oil. Curzon was at the time writing
Persia and the Persian Question, which was to "advert hopefully to Iran's
mineral resources," including coal, but warned about the dangers of
allowing Russia to build a railway to the Persian Gulf.[43] Along with a
group of India-based British imperialists, Curzon helped portray Russia
as a military threat rather than a commercial rival by not mentioning that
the purpose of the railway and pipeline was to export Russian oil from
Baku. Led by Curzon, these British Indian imperialists helped "arrange
for Burmah Oil to invest in D'Arcy's scheme [the 1901 concession] and
keep alive his monopoly rights both to Persian oil and to the building of
a trans-Persian pipeline."[44] A strictly imperial pursuit to discover, produce, and sell oil as quickly as possible did not drive Great Britain and
Russia's quest for mineral concessions in Iran. Rather, Great Britain and
Russia embarked on a kind of battle over the coordination and control
of various concessionary and cartel agreements, managed in part through
D'Arcy's 1901 concession. These arrangements ensured that oil from Iran
and the greater Middle East would not flow immediately or fall into the
wrong hands (Russia), disrupting world oil supplies and the maintenance
of high profits, particularly in Asian markets.

Iranian officials were only able to pursue the 1901 concessionary arrangement after the concessionary privileges of the Persian Bank Mining Rights Corporation lapsed in 1898. D'Arcy requested and received British diplomatic support for his concessionary application, as was normal practice at the time.[45] He secured the concession with the Imperial Government of Persia in May 1901. One of the first problems D'Arcy and his fellow investors and technologists confronted was how to apply the terms of the concession and establish property rights over a subsoil and mobile material such as oil in a foreign country. As discussed at the start of this chapter, the concession's articles on property told D'Arcy nothing about where the oil was located and what his specific rights were in relation to the population inhabiting the lands where he would extract any oil he found. What was the basis or precedent for such a legal right granted to a foreign concessionaire that extended across a vast territory totaling 480,000 square miles (e.g., the equivalent of over 100,000 London Heathrow Airports)? How did this novel form of securing legal ownership over a foreign property—a concession—work in practice?

Prior to the arrival of British investors and technologists in Khuzistan, the particular properties of this region interacted with an entirely different set of political claims to the land, with its organization within the local population, and in relation to the ruling Qajar government. Claims to land were made in terms of the revenue extracted from inhabitants through taxes imposed by the Qajar government on local profits made from the province's exports.[46] The land where the oil was found was not an object to which a single individual claimed an absolute right.

The Qajar government did not directly rule the land that would become the site of oil extraction, transport, and refining. As the largest landowner in Arabistan Province, Shaykh Khaz'al claimed the region bordering the Shatt al-Arab and extending inland to the land dominated by the Bakhtiyari confederation, which extended further inland and found its limits at the Bakhtiyar Mountains. Sixty percent of the Bakhtiyari district was mountainous, and given the relative inaccessibility of the region, was usually beyond the direct control of the government.[47] The region controlled by the shaykh marked the future location of a section of pipelines and the main refinery at Abadan[48] and its port, while that of the Bakhtiyari marked the other half of the pipeline and the main oilfields at Masjid Suleiman.

These nomadic groups, other village populations, and *ra'iyat* ("peasants" or farmers) inhabiting southwest Iran had long experience with *naft* (oil).[49] Material practices of oil development involved collecting and storing the oil from seeps in the ground in round sheepskin containers and distilling it to make *qir* (tar), which was sold as one of the region's exports. Since the early eighteenth century the Persian Empire's monarch, Nadir Shah, was reported to have issued a decree to transport *qir* to Ottoman Iraq for the building of warships.[50] The only substantial control exerted by the central government over the province was through an annual tax on the petroleum seeps in the amount of 2,000 *tuman*.[51]

Local Arabic-speaking and Persian communities and nomadic groups, such as those ruled by Shaykh Khaz'al and the Bakhtiyari khans respectively, exercised their own form of political leadership and system of authority.[52] In the late nineteenth and early twentieth centuries, the relationship between the shaykh and the reigning shah remained nominal: the shaykh paid annual taxes or personal "gifts" to the shah more as a means of maintaining his autonomy than as a show of fundamental political support to the monarchy.[53] The extent to which the Qajar state successfully collected taxes from the Bakhtiyari confederation was also uncertain.[54] Depending on the migration season, these nomadic groups owed taxes either to the provincial governor of Isfahan (spring) or to the governor of Arabistan (winter).[55] In their capacity as large landowners, "tribal leaders or *khans*" collected their "share of the produce of the land or their rents and dues," and as "tribal leaders," collected "levies from their followers."[56] Thus, government control of property simply registered the ruler's right to tax or extract a tribute from the shaykh and the Bakhtiyari khans. Since the rule of the Safavid Empire (1501–1722), there was never a clear distinction in practice between the land held by the ruler as head of state, which Ann Lambton translates as *divani*, and crown land, which she translates as *khasseh* or *khaliseh*.[57]

Afsaneh Najmabadi explains that the Qajar regime (1794–1925) executed the consolidation of landed property; however, a de facto indirect management of agricultural surplus extraction through tax farming (*toyul-dari*) reemerged, and the land tended to remain within families as "essentially private."[58] With an increasing need for income in the first two decades of the twentieth century, the government sold "state land," rather than paying out revenues from the land, and wealthy families and merchants increasingly bought land as "private property."[59] The history of private property was finally consolidated with the building of a modern legal and judiciary regime during the accession of Reza Shah (1925–1941). Najmabadi's history of property for the two periods largely relies on Lambton's account, which translates *milk* as private property.[60] What should be noted here is that modern English concepts of ownership—such as crown and private land in England or state property in the United States, reduced to the singular issue of who owned the land—did not simply emerge in southwest Iran because they did not exist in the Persian language. Lambton does not explain her translation of the terms *khaliseh* and *divani*, but she does note that the distinction was "not so much concerned with the legal status of the land as with control over the administration and expenditure of the revenue."[61] Further, the two categories of land were "not permanent and fluctuated as political circumstances demanded."[62] British officials and surveyors soon learned that the local practices concerning property that persisted in Iran did not correspond to their concept of ownership as an individual's exclusive claim to land, alienable and inheritable by sale, and represented in a contract. How then would *divani* or *khaliseh* come to mean ownership or "lands belonging to the State" as stipulated in the oil concession?

The arrival of British investors and technologists in Iran necessitated a peculiar kind of spatial knowledge about oil and property rights that marked a shift in the nature of political claims to the land. The older forms linked central government claims to Khuzistan's land through the taxation of revenues collected by a network of semiautonomous shaykhdoms, "tribes," and villages. In contrast, the foreign method pursued by a British financier such as D'Arcy focused on the single issue of "who" would control the land by possession of the guarantee or "right" to exploit its resources. The formulation of a written oil concession or contract represented this different form of creating and concentrating political authority. The format embodied a particular relationship to the land and the subsoil designated for exploitation that was not readily familiar to and practiced by local regional groups or governing bodies.

The 1901 oil concession at first appeared as a legal text that seemed to guarantee a kind of absolute authority concerning the rule of concessionary property. In practice, however, concessionary rights were in direct conflict with local practices and forms of property. D'Arcy and his team of technologists soon realized that they could not simply impose the concession's terms by invoking the absolute authority of an abstract concept. Their operation would have to involve the building of connections between oil, the concession terms, and an earlier history of legal devices and renting precedents guaranteed by the Iranian government to local political groups. Managed on the terms of the first three articles of the 1901 concession, diverse property arrangements were suddenly intertwined with larger battles over the control of fossil fuels. The concession stipulations served as a kind of machinery for defining and arranging authority over property and minerals in terms of a single individual owner: state or private control in the British sense. British investors and technologists, who bore no relation to southwest Iran, needed to find a way to reduce land-use practices and production to the single issue of who owned the land and to separate this question of control from local activities.

Translating Mineral Rights

The political method of granting a concession to a foreign concern occurred at a particular historical moment when the corporation's power was guaranteed by the backing of British diplomatic support and exemption from local forms of law and authority.[63] Articles 9 and 10 of the concession specifically granted the formation of a company or several companies in the future to take over D'Arcy's responsibilities. In an attempt to guarantee the financial backing of the concession, D'Arcy set up the Concession's Syndicate in May 1903 with the authority to acquire all rights and privileges originally vested in him as concessionaire.[64]

But the aforementioned local forms of power and authority over land introduced another kind of uncertainty that threatened the concession's stability even in the form of a company. D'Arcy and his fellow investors soon learned that the Iranian government had its own form of imperial edict or royal decree known as a *firman*.[65] The Iranian government had granted D'Arcy the exclusive right to drill for oil in certain parts of Iran in 1901, including the region in which the territories of Shaykh Khaz'al were located. But in January 1903, Mozaffer al-Din Shah granted a *firman* to the shaykh using language in which the Qajar government admitted, "at least by implication," that it had exceeded its powers in conferring certain privileges on D'Arcy.[66] The lands in question had belonged in the past to the shaykh, his tribesmen, and their ancestors. The rule of the *firman* over local property threatened to destabilize the authority of the 1901 concession. To eliminate this threat, British authorities would have to certify all documents.[67] A company official declared that "we shall recognize no one's claim until they have produced the necessary *firman* either in the original or a copy," certified by the British Legation in Tehran.

The British solution was to pursue "an alternative legal route" by paying the shaykh a proportion of profits derived from the oil obtained in his territories and to "deduct an equivalent sum from the amount due to the Persian Government. ..."[68] However, the shaykh really did not have the right to sell or lease the land on Abadan. The 1903 *firman* had granted Abadan and the Karun River banks to "Arab Tribes" and not to the shaykh by name. By law, this forbade individual Arabs to sell the land to anyone. The Iranian government had also undertaken not to sell it unless it was first purchased at a fair valuation from the tribes. The Iranian government would have to buy the land and "sell it again to the Oil Company."[69]

New and old property arrangements mixed and proliferated within the boundaries of the concession area. It appeared advantageous for the company to frame the land as uncultivated in order to acquire it free of charge from the Iranian government, as stipulated in Article 3 of the 1901 concession, while simultaneously compensating the shaykh. Within a year, the company's managing agents argued that the shaykh had no "*locus standi*" on the matter: the 1903 *firman* did not include mineral rights, and the shaykh had not claimed any mineral rights when the company was first established. Thus, it was unlikely he would claim them now with the relevant *firman* to confirm the claim. The company's agents advised that it follow "strictly on the lines of the concession," and nothing more should be paid to the shaykh in excess of the "surface value of the land."[70]

There was an ambiguity in the terms defining the kinds of property that persisted in English versus Persian terms. The concession discussed property as "cultivated" or "uncultivated" lands "belonging to the State." In contrast, the rules of property in Iran at the time did not entail or imply the same concepts in English as in Persian. By the early twentieth century, Lambton says that land controlled or possessed by "the State" was known as *khaliseh*.[71] One of the most significant concentrations of *khaliseh* was found in Khuzistan Province.[72] Under Qajar rule, the lands and waters of virtually the whole of Khuzistan were *khaliseh* and entailed particular conditions and terms of lease. Much of *khaliseh daym* (unirrigated, dry farming land), however, was in the de facto possession of the Arabic-speaking population of the province.

According to Reynolds, the persistence of local forms of property, differing from how property worked in England, was not an adequate explanation for conflicts over property claims with local forces. He explained that "Arab apathy" was really what made the land lay waste, but "he" (the Arab) recognized that "land has value it never had before as he has seen what can be achieved with irrigation."[73] Thus, invoking British concessionary logic, Reynolds concluded that "if then it be Arab apathy which renders the ground waste, it will be Arab avarice which will prevent you getting it at the price you quote."[74] To maintain the stability of the concession's terms, the British side could not factor the legitimate property claims of local groups into their calculations. At any moment, these groups might threaten to deconstruct the concessionary formula for securing access to the subsoil.

British investors, engineers, and drillers acted as spokespersons for the British civilizing mission, arriving in what they viewed as a technologically and socially backward land of disorder. This perspective informed the concessionaire's technical decisions about how to resolve controversies over property, mineral rights, and the pricing of land. The concessionaire, argued Kitabji, was perhaps following too closely "the letter of the contract, forgetting that they were dealing with people who are unacquainted with laws and contracts. … People thought they had a right to ignore their engagements, and resorted practically to blackmail in order to carry out their designs."[75] Establishing the rules of concessionary property started with a rhetoric that rendered the local population defective in terms of a lack of rationality regarding laws and contracts.

Both British and Iranian interests aimed to secure *khaliseh* under the absolute control of the central government. They argued that local political communities with claims to the oil lands lacked an adequate understanding of the rule of (Western) laws and contracts. These communities were said to be characterized by avarice, which obscured their comprehension of the productive and technological use of land as well as its fair pricing. Such technoracialized attitudes toward local political forces that emphasized their flaws, such as a proclivity for blackmail, informed the company's technological choices and were inscribed in its conceptions of legality, valuation, and the hierarchical classification of English law terms as superior to Persian terms.

Local claimants were deficient in their legal understanding of "mineral rights." From the moment the concession was signed, D'Arcy and his fellow investors knew that "land owners or even lease holders" might not permit drilling on their grounds. It may be necessary, Reynolds explained, "to enter into some agreement with them," to offer compensation "over and above that laid down in the Concession."[76] For they must know, he continued, that "the people with whom we will have to deal have but little respect for the orders of the Shah, and have no idea of the mineral rights of the owner of the ground being nil, to him belonging only the surface and its produce."[77] The concessionaire justified the rule of concessionary property (backed by state authority) in part through a lack of understanding, exhibited by the local population, regarding mineral rights.

Article 3 of the concession appeared to guarantee that uncultivated land belonging to the state could be acquired "free of cost." But with regard to sites designated for the oil refinery and pipeline, "nothing could be satisfactorily settled without the Sheikh's good will and cooperation."[78] Thus, the company had every desire to deal with him "on reasonable business principles."[79] For each land acquisition, the company was required to submit an application to the Iranian minister of mines and public works, who would then classify the land and determine its value.[80] This practice, however, would not eliminate the controversy over the hierarchical framing of property ownership in English terms as superior to Persian ones, which was creating uncertainty with regard to the valuation and pricing of the land, and local claims to it.

The concession's articles did not address the relation between state control over cultivated and uncultivated land, which included subsoil minerals, and the older forms of property ownership that persisted among local communities in the oil regions. The origins of modern (concessionary) property in Iran were not separate from older land arrangements because these so-called traditional claims and productive processes were internal to its construction.[81] British investors interested in oil development worked toward tools and arguments with local and government actors that transformed property arrangements and mineral rights over the subsoil, originally secured through *firmans* and *khaliseh*, into the modern English version of state property authorized by the concession.

Porosity of Concessionary Property

In southwest Iran, it appeared to be to D'Arcy's advantage to work with the central government in securing a concessionary relation through which all cultivated and uncultivated land "belonging to the State" was granted to the concessionaire (cultivated land being purchased from the state at the "fair current price of the Province"). In this early period, an emerging modern, large-scale corporation such as AIOC operated by "establishing oligopolies or exclusive territories of operation."[82] AIOC's political method of gaining access to minerals through a concession was tied to an earlier history of the colonizing corporation, which was automatically granted access and rights by the metropole alone to monopolize trade in specific resources. In contrast, the development and implementation of the D'Arcy Concession would not only require financial and technical control, granted through a legal document signed by the ruling state authority or government, but "an ability to understand the Persian environment in which it would be worked."[83] Only in this way would the recreation of the monopoly arrangement be guaranteed.

The Bakhtiyari khans presented themselves as the legitimate political authorities and owners of the lands, including the subsoil, in and around Masjid Suleiman. The D'Arcy Concessions Syndicate "readily accepted them as such,"[84] in direct violation of the terms it had agreed to in the 1901 concession, which recognized the state as sole owner of the subsoil. D'Arcy's British representatives, such as Reynolds, did nothing to contradict these assumptions and undertook negotiations with the khans to secure drilling access. The Bakhtiyari had won a battle over property that was initially challenged by the syndicate. Reynolds had argued previously that with the aid of the British government, they must make the Bakhtiyari understand the authority of concessionary property in alliance with state property, such that "even if the surface of the land is their [the Bakhtiyari khans'] property, that which is under the ground, is the property of the Crown and it is within the Concession that the Shah has granted D'Arcy access to this land for oil exploration."[85]

On November 15, 1905, an agreement was finally signed between the Concessions Syndicate and the Ilkhani and Haji Ilkhani, the ruling khans of the two dominant families within the Bakhtiyari confederation.[86] D'Arcy won the right to acquire land from the Bakhtiyari khans and to exercise drilling rights in their territory. In return the Bakhtiyari acquired a 3 percent interest in any company formed in their territory and agreed to guard the concessionaire's property and personnel.[87]

After oil was discovered in commercial quantities, an alternative arrangement prevailed in which the Bakhtiyari Oil Company (BOC) was incorporated and another company was set up known as the First Exploitation Company in April 1909.[88] BOC was provided with an issued capital of £300,000 and the "specific object of providing the Bakhtiyari Khans with their subsidies."[89] The First Exploitation Company would sell its oil below cost to BOC, whose "sole function was to resell it to AIOC at the normal price."[90] Following the earlier 1905 agreement, the Bakhtiyari families were given 3 percent of all shares in BOC, while AIOC received the remaining 97 percent and all of the First Exploitation Company shares. Building an alliance with the khans was crucial not only for financial and security purposes, but also because the Bakhtiyari provided 4,500 of the 5,000 laborers working in the oilfields.[91]

The creation of multiple companies existing only on paper with local political groups such as the Bakhtiyari khans demonstrated the porosity of the concession, a kind of ambiguity in its articles and clauses that could be exploited by many sides—British oil investors, the Qajar government, and local political actors. Subsidiary companies such as the BOC were part of a mechanism existing only on paper with some people in an office controlling the sites where oil would be explored and extracted. The power of a company was manifested precisely in these local connections, indicating that its power was not as concentrated or absolute as one might assume from the concession contract, and that such ties could easily fall apart. Both the concession and the various subsidiary companies operating under its authority served as a technical machinery, a set of connections, for separating ownership claims that could be made or not.[92]

The Qajar government refused to recognize AIOC's arrangements with the khans. It also insisted on fixing its claim to the ownership of certain lands as *khaliseh* alone.[93] Prior to the signing of the agreement, the Qajar government argued that any contract concluded between D'Arcy and local political groups must be submitted first to the government before being signed.[94] But as Stephanie Cronin observes, the British consul negotiated with the khans and the Concessions Syndicate apparently without any reference to the Iranian government or to rights the government might have secured in the concession.[95]

The company could not ignore the complex ways local communities managed land. It invented alternative forms of property arrangement in its dealings with the Bakhtiyari that overlooked the articles of the concession. In similar fashion, on May 6, 1909, Percy Cox, British political resident in the Persian Gulf, promised Shaykh Khaz'al "a guarantee that Britain would not allow the Iranian government to disturb the *status quo* of himself or his heirs or successors."[96] A conflict between the Bakhtiyari khans and Shaykh Khaz'al would "not lead to very pleasant results" either for Khuzistan or for oil operations.[97] Bypassing the central government and other local claims to land revenues, the shaykh was to receive financial compensation for access to the land, and such financial rights were secured along hereditary lines. On July 16, 1909, Cox signed an agreement with the shaykh granting one square mile of land to AIOC in exchange for an annual rent of £650 for the first ten years, an immediate British government loan of £10,000, and the withdrawal of the Indian Army guard with "the maintenance of security entrusted to local chiefs."[98] As Cox put it, "the Company's best interests would suffer, and our national interests as well, if they are to pursue their enterprise through the Persian government in spite of the Sheikh."[99]

AIOC's actions indicate the porosity of the 1901 concession, a kind of ambiguity and weakness of ties, exploitable by multiple parties pursuing conflicting property arrangements over oil. The company realized that local claims to the land could not be bypassed as easily as anticipated. In spite of the new representation of state claims to the land and its subsoil outlined in the concession, the company understood that *khaliseh* lands were highly unstable. AIOC officials opted to overlook and actively violate the concession interests of the Qajar government by dealing secretly and directly with local ruling elites in alternative arrangements, neither within nor without the jurisdiction of Tehran. This included training and paying tribesmen as guards to patrol the oil extraction sites and protect their boundaries from disruptions posed by the local population.[100] In turn, different property arrangements were invented, extended, and secured between AIOC and local forces with the power to disrupt concessionary rule over the subsoil. Local forces such as the Bakhtiyari khans and Shaykh Khaz'al pursued their own forms of political authority and power, causing them to suddenly appear outside state power and the terms of the concession. However, they were in practice both inside and outside, as their method of controlling the properties of oil and securing alternative contractual arrangements over them was the prerequisite to the production of oil according to the rule of concessionary property in the first place.

Who Is the Real Owner? Pricing the Properties of Oil

"Compensation" for Bakhtiyari land was a controversial issue for AIOC. What was the basis for estimating the value of the land? In pricing the site of the main oilfield at Maydan-i Naftun, Charles Greenway,[101] AIOC's managing director, looked to India, arguing that "the value of good cultivable land in India" averaged about £3–£5 per acre and that "this appears to me about as much as any similar land in Persia can possibly be worth."[102] Reynolds suggested that the company consult a volume on the nature of ownership in Iran by a British adventurer and ethnographer, Austen Henry Layard, titled *Early Adventures in Persia, Susiana, and Babylonia, Including a Residence among the Bakhtiyari and Other Wild*

Tribes.[103] Layard had written about his travels and discoveries in Iran in the late nineteenth century, including a section on the Bakhtiyari, which they later accepted as forming part of their own history.[104] Company representatives viewed the Bakhtiyari with an imperialistic bias, taking on Layard's framing of them as primitive and "wild" and awaiting progress. Certainty resided in the Bakhtiyari khans' promise to ensure security for oil operations by drafting a letter to "Tribesmen, warning them against interfering with the Company's work and of the consequences in so doing."[105] On the question of land and its valuation, however, a series of uncertainties and interested claimants were getting involved.

The complexity of the ecologies of pastoral nomads and their traditional pasture rights did not factor into the British contract agreement with the Bakhtiyari khans concerning the purchase of their lands (the site of the main oilfields and a portion of the main pipeline) for oil development. The power of the khans derived from two sources: on the one hand, from the confederation of which they were the leaders, and on the other, from the land. In their capacity as authorities over the land, the khans collected their share of the produce of the land or their rents and dues, while as "tribal leaders" they collected certain levies from their followers.[106] The central Zagros Mountains were a difficult terrain through which nomads passed from their winter to summer pastures and back again.[107] Groups such as the Bakhtiyari khans claimed arable land in the *qishlaq* (or *garmsir*, winter quarters and site of oil regions) and sometimes in the *yaylaq* (or *sardsir*, spring quarters).[108] There were permanent agricultural villages and settlements throughout the Bakhtiyari territory except at high elevations and near border regions.[109] This resulted in a highly sophisticated ecology of mobility and settlement, which the company chose to overlook in its dealings with the group by simply treating them as settled, private landowners in the Western sense.

AIOC knew that oil operations in Bakhtiyari territory carried on "without any very definite understanding with the owners of the land here."[110] The arrangement automatically assumed that the khans were "the sole owners of the property which we now know not entirely to be the case."[111] The terms of the Bakhtiyari Oil Agreement of 1905 stated that "all arable or irrigated land required was to be handed over by the Khans at 'the fair price of the day.' Non-arable land was to be free." The "ambiguity," Morris Young explained, resided in the definition of *cultivated* and *cultivable* land.[112] But from mapping the regional and local ecologies of southwest Iran, the English concept of cultivated or uncultivated obviously did not fit because it presumed and was designated for land that had been settled, so to speak.

The entire property for the pipeline area, extending from Bakhtiyari territory to the site designated for oil refining in Abadan, totaled 1543.03 acres.[113] In AIOC's view, approximately 1246.03 acres were under "crop and new fallow," and roughly 297 acres of "old fallow" existed that had not been cultivated in three to four years.[114] But a company surveyor from India noted that lands classed as "new fallow" were those the *ilyat* (nomadic groups) were "in the habit of cultivating every alternate year, while even those lands classed by us [AIOC] as old fallow, had been cultivated within quite recent years."[115] In some cases, areas shown as "old fallow" in the previous year's survey were under crop in the current year. On the one hand, the khans appeared to fall into the category of "proprietors" as spelled out in the terms of Article 3 of the 1901 concession, with claims to both cultivated and uncultivated lands that were not owned by the state. On the other hand, Article 3 excluded the khans by making no mention of pasture rights. Through a simple legal-financial transaction, it was thought, the British concessionaire might purchase the khans' land at the "fair and current price of the Province" and build a pipeline to transport the oil. The problem was that local property arrangements were bound up in diverse ecologies, farming practices, and mobilities of settlement, which meant that lands understood in English as cultivated or uncultivated were not fixed categories and took on quite fluid meanings in practice.

The Bakhtiyari khans would not accept AIOC classifications or figures on land measurements. During the 1911 negotiations over property arrangements within the concession area, the khans contested framing "non-arable lands" as "wasteland," citing their value as grazing lands and seeking a fair price for them. As a response, the khans used local definitions and land surveyors to calculate land area and price in terms of *jarib*, not acres.

The *jarib* measurement of land respected the biannual and cyclical process of mobility and cultivation local to the region. The khans had "their own man who walked about with a stick measuring the land, and completed all his work within the last two weeks," making out a "figure far above ours."[116] The khans' surveyor had reached a figure for an area of property fully double that supplied by Scott, the company surveyor.[117]

British agents concluded that a *jarib* was a piece of land measuring 25 by 40 "shah" *zar'*, each *zar'* equaling 42 English inches.[118] A national law was eventually passed in 1926 to equate Persian measurements with the metric system, and it was finally formalized in March 1935.[119] The *jarib* was finally set at 1 hectare (approximately 3 acres); however, this again varied according to local uses and could represent an area ranging from 400 to 1,450 square meters. The *jarib* also varied according to the nature of farming practice. For example, a *jarib* of unirrigated land was larger in certain areas than a *jarib* of irrigated land. Lambton explains that this variation was due to the fact that the *jarib* was not purely a "surface conception."[120] In practice, it was bound up with measures of weight: "Since the amount of seed sown per acre of irrigated land is greater than the amount sown per acre of unirrigated land, the size of the *jarib* varies in proportion."[121] British investors wanted to impose and stabilize their own (Western) hierarchical system of classification of Bakhtiyari land in terms of a fixed and generalizable measurement in English. The problem was that local practices and technologies of designating an owner and measuring and pricing the land did not work in southwest Iran the same way as in the West.

Local forces in Khuzistan and throughout Iran practiced a sophisticated form of measurement that varied with different kinds of land and weights of the seed. It did not follow a grid or frame as British officials assumed. This points to a particular politics of "enframing" or the peculiarity of absoluteness unconnected with the specificity of activities people do.[122] This politics informed the choices of British investors and technologists, who bore no relation to the land, as they reduced practices of land use and production to a single issue of who owned the land and separated this question of control from local activities. However, as evidenced here, local activities defined logics of property measurement in Khuzistan. This was also the case in the West, as with the term *acre*: the amount of land plowed by oxen or cattle in a day.[123] Europeans arrived in Khuzistan, however, with a different notion of measure—a grid system—unrelated to the actual practices and activities of measuring lands that made the process of measurement and therefore valuation possible.

There was even more ambiguity concerning hill tracts. The khans contended that these tracts were of value as grazing ground. There was nothing in the original agreement saying that these lands were given free.[124] James Ranking , the British consul, conceded that the word *free* did not appear in the agreement, and the word in the Persian text translated as *give* in English did not imply "free gift."[125] There was no such thing as *free land* in the sense of "free gift" in Persian, further confirmed by the Bakhtiyari practice of taxing grazing land as a source of income.[126]

This highly complicated and varied system of measurement and valuation points to why the khans would only negotiate by "rate per Jarib" and not English acres. The specificity of the land and crop along with the work activities attached to them shaped the act of measuring and of valuation, rather than starting from an abstract grid or frame, which company officials preferred. Why? These methods of enframing worked as techniques of power to help render the land and its subsoil readable and thus controllable in abstraction precisely through the terms and authority of the oil concession.

AIOC continued to purchase more lands as operations expanded after World War I. Conflicts over land measurements for the purposes of accessing oil involved further controversies over whether a Persian *zar'* was 42 English inches as stated in the 1911 agreement or 40.95 inches as claimed by the Iranian government. In either case, the British conceded that the khans did not "comprehend inch."[127] By 1928, the aggregate land purchased by AIOC from the khans amounted to 6,750 acres.[128]

At the close of property negotiations, the khans accepted Scott's figure of 9.5 square miles (6,131 acres). They sold the fields for a fraction of the original price, giving away over 3,000 acres of pastureland for free.[129] Framing the Bakhtiyari khans as sole owners, says Arash Khazeni, excluded the rights of pastoralists and farmers to the land, particularly the Lurs nomadic group.[130] The 1911 Bakhtiyari Land Agreement forbade the Lurs to access lands on a seasonal basis to graze their flocks and cattle, even though they had inhabited and cultivated these lands for generations. AIOC policy was to keep people of the district "away for at least one season from the boundaries of our property in order to break the continuity of their work" and to make "the common Lur understand that the land which he has tilled for generations is no longer his."[131] Pastoralists resisted by crossing the designated boundaries of the oilfields regularly, setting up tents on company property and continuing to plow at Masjid Suleiman and Maydan-i Naftak.

Tracing the steps involved in transforming the oil regions into concessionary property has made visible a defined chain of translation, although at all points unstable. British investors, engineers, and drillers, the Bakhtiyari khans, and Shaykh Khaz'al worked hard to "modify, displace, and translate their various and contradictory interests."[132] Concession terms worked as a nice device for framing and directing actors' interests toward concessionary control through the stabilization of connections between an invisible, subsoil liquid, conflicting logics of property and ecology on the surface, English and Persian terms for types of land ownership, use, and measurement, and mechanisms of dispossession and violence. AIOC aimed to enforce the authority of the concession while simultaneously enrolling local forces that bypassed the power of the central government. Either way, depending on the strength of the ties, different groups threatened to pursue alternative interests and disrupt the property arrangements enabling the energy network to be built.

The concession's terms were indeed porous and exploitable by multiple sides. Working out controversies over the fair valuation and measurement of lands covering the concession area has revealed the ways concession terms failed to account for semiautonomous political groups who were simultaneously inside and outside of the law and authority of the central state and its contractual relations with British investors. Even living and cultivating conditions moved with the seasons, alerting AIOC to the realization that it was not dealing with single owners fixed to a piece of land embodied in the English terms of the concession. What was certain in AIOC's view, however, was that local claimants to the oil lands lacked an adequate understanding of proper English measurements and pricing. This informed AIOC's technical decisions and techniques for securing access to the subsoil while blocking more equitable political arrangements of the energy system.

Reassembling the Properties of Oil

Reynolds found it "impossible" to obtain an "idea of the market price of the ground" in the lands claimed by the shaykh that marked the future location of a section of pipelines and the main refinery at Abadan.[133] Arabic-speaking nomadic groups residing on both sides of the Shatt al-Arab waterway (marking the border between Iran and Iraq and emptying into the Persian Gulf) owed allegiance to the shaykh. His farmers (*fellaheen*) cultivated his lands, which included Abadan Island, site of the future oil refinery, primarily in dates.[134] As with the older forms of property that persisted in Bakhtiyari territory, the D'Arcy Concessions Syndicate was again confronted with local forms of property that threatened to destabilize the authority of the first three articles of the 1901 concession. Cox recommended that the syndicate make sure the shaykh "has the right to sell land at all on Abadan" and that they should rely on diplomatic assistance to secure terms "as near fair market rate as possible."[135]

The shaykh had agreed to lease, on behalf of his descendants and tribes, land for an annual rent and had agreed that all constructed property would revert to himself and his kin.[136] He regarded the lands as his own by right of inheritance and Arab by right of settlement.[137] Not only did AIOC agree to this, they also assured the shaykh that whatever change might occur in the "form of government in Persia," the British government was prepared to provide the same support against encroachment of his rights as was promised in his *firman*.[138] AIOC acquired approximately 650 *jarib* on Abadan Island, the site of the future oil refinery, in addition to 10 *jarib* at Ahwaz, with riverfronts on the Karun River for erecting storehouses and other buildings.[139]

Oil operations commenced and agreements were signed with the company's full knowledge that the nature of "tenure and rights ... over and in the soil enjoyed by the Sheikh and his tribes" since the granting of the 1903 *firman* required further investigation.[140] Rather than resorting to arguments about the skills and knowledge that local proprietors "lacked," Cox advised the company to recognize in confronting the Iranian government that the shaykh did enjoy "mineral rights" in his territory, so that "any future exploitation of minerals would ... naturally fall to British enterprise alone." This was preferable to conceding that the Iranian government possessed such rights, which could lead to placing a mining concession in another region of Khuzistan "over the heads of the Sheikh and ourselves."[141] Company officials feared that if they failed to make an agreement that acknowledged local groups as legitimate owners of mineral rights in the oil regions, they risked losing their oil concession. It therefore made more sense, in the early years, to bypass a weak central government in favor of strengthening ties with local claimants.

AIOC faced the problem of creating boundaries between advantageous and disadvantageous forms of property rights as well as their measurement and valuation. First, the company advised that the "right to allow boring for oil" be disassociated from the term "mineral rights."[142] Second, it was in the company's interests "not to ... dissect the question of the Sheikh's precise rights" but to have recourse to various alternatives. AIOC's director suggested the shaykh be paid the "surface value of the ground" and allowed some equivalent of the 3 percent on profits derived from the sale of crude oil extracted in his territory, which would be based on the price paid for the oil to BOC. Due to the complex nature of property ownership, AIOC's chairman admitted to the necessity of negotiating with local forces, but proposed that the costs incurred from meeting their demands be deducted from the 16 percent share of profits owed to the Iranian government.[143]

Property arrangements were not fixed. They proliferated and extended across the concession area, threatening to destabilize the energy system. Over the years, British and Russian intervention had weakened the central government in Tehran, reviving the power of local chiefs as well as numerous leftist and Russian-backed movements countering Britain's hold over the country.[144] In the early years, it had been advantageous for AIOC to strengthen semiautonomous political communities in southwest Iran in the event of the government's collapse.[145] The British government dramatically shifted its strategy after 1921, however, engineering the installation of Reza Khan and a strong centralized government to suppress any further threats from Iran's provinces, especially Khuzistan.

Formerly a Cossack Brigade member, Reza Khan changed his name to Reza Pahlavi and formalized the execution of the British-supported coup by ordering the constituent assembly to depose the Qajar government in 1925. The installation of Reza Pahlavi as the shah marked a realignment of power and exposed AIOC's former alliance with local forces and ruling elites as an increasing liability. For example, the government's decision in 1924 to revoke *firmans* issued after March 10, 1903, put AIOC lands bought or leased from Shaykh Khaz'al in jeopardy.[146] Suddenly, these middlemen were an increasing liability in the emerging alliance between the central government and AIOC. The central government eliminated such groups through political violence and repression or by forcing them to sell their land, opening the door for the shah to deal directly with the company and collect royalties as well as land and petrol taxes.

By 1925, the new shah implemented his tribal policy to ethnically homogenize the state through a string of military campaigns against tribes particularly in Khuzistan, ousting Shaykh Khaz'al, confiscating oil shares controlled by the rival Bakhtiyari confederation, and forcing its chiefs to sell all of their land and disarm.[147] The Pahlavi shah sought to build a modern state with exclusive military control of southern Iran down to the Gulf and to maintain "direct responsibility for protecting the oilfields."[148] An Iranian minister ordered the company to apply to local official governors of the provinces for "ascertaining the ownership of the lands and for their own protection."[149] Such agreements must be made with the "true owners of lands and waters with the cognition of the governors and the chiefs of tribes who will no doubt act according to their duties."[150] During the early years of the Pahlavi shah's reign, the state confiscated considerable areas of land as a means of eliminating "tribal power" and that of certain large landowners. Among these were large areas in Khuzistan taken from Shaykh Khaz'al. The confiscated land was transformed into the shah's *khaliseh*. The shaykh fell from power around 1925 and died under suspicious circumstances while under house arrest.[151] The new Pahlavi monarch was reasserting the terms of the 1901 concession by ensuring that property covering the oil regions was state controlled. This would help to eliminate any obstacles hindering his ability to collect oil revenues and build a modern state.

In this shifting of alliances, new groups with claims to concessionary property were plunged into the battle over the control of oil. "Arabs and peasants" rebelled across Khuzistan as the Finance Department ordered tenants to pay a tax on every date palm and owners to pay a tax per tree.[152] Such protests threatened to spread to other provinces. *The Times* of London reported that the cause of the revolt "is attributed to the nature of property holdings in Khuzistan" connected to the shaykh.[153] With the shaykh's deportation to Tehran, his *ra'iyat* (farmers) "seem to have thought that it would be well to claim their holdings as their own, especially as their landlord was no longer in a position to collect his dues."[154] The British Foreign Office confirmed the report: "Arab fellaheen have been quietly disposing of the date produce for their own benefit" and cultivators were refusing to pay the tax on trees in defiance of local finance officials.[155]

The new Pahlavi regime reasserted the rule of state property, above and below the surface, as internal to the authority of the concession through mechanisms of dispossession and eviction, alliance shifting, and debt. For example, the Bakhtiyari khans had started to borrow money against their oil shares, a large portion of which was already mortgaged by 1920 from the Imperial Bank of Persia. With a new British-backed central regime in Tehran, AIOC, the British Legation, and the Imperial Bank collaborated to use the khans' financial difficulties as a means of "keeping their shares under British control."[156] Banking and debt were twin mechanisms of compulsion exercised to fragment the Bakhtiyari confederation by associating one faction more closely with the British. The Bakhtiyari responded in multiple ways: by withdrawing protection and labor from the oilfields and embracing "private property" in land and oil shares to separate the senior khans, the main signatories to the Bakhtiyari oil agreements, from the junior khans, who were demanding their share of the dividends.[157] By 1928, in the midst of the central government's efforts to "clarify property rights" through the enactment of a national law requiring the registration of property and title deeds, AIOC had successfully diverted all dividends and Bakhtiyari oil shares to British control, leaving them with nothing.[158] Whereas the company had previously overlooked the state's concessionary authority over the subsoil in favor of alliances with local middlemen, the 1920s saw a transformative shift in strategy. It suddenly appeared advantageous for AIOC to pursue alternative alliances with a new British-backed monarch who worked to centralize control by eliminating local claimants to the oil regions through mechanisms of debt, dispossession, and eviction. Now working hand in hand with the company, a strong and centralized Iranian government ensured that the 1901 concession terms regarding property and access to oil were reassembled as state control, whether defined in English or Persian.

Early in his rule, Reza Shah introduced new legislative, regulatory, and property arrangements to consolidate his control of the soil and subsoil of Iran.[159] The previous Qajar regime established the first Department of Land Registration in 1922 within the Ministry of Justice, under Arthur Millspaugh, the government's American Financial Adviser, for the registration of property and documents. Land Registration Acts spanning 1921 to 1929 required that all titles be registered with a special office.[160] A uniform land tax was implemented throughout the country in January 1926. A cadastral survey followed by a supplementary law for the registration of property in 1928–1929 fixed the charges incurred on registration.[161] The latter law made registration of real estate compulsory and required the establishment of registration offices across the country. Failure to register within the two-year period after the establishment of the first Department of Registration empowered the government to register the property in the name of the holder at the time. Notices were posted in the oil towns of Mohammerah and Abadan calling on all owners of landed property to register within one month or run the risk of losing their property.[162]

For the whole of Khuzistan Province, which included the oilfields, pipeline, and refinery, the Finance Department intimated that it would oppose registration of all land, "as the whole of Khuzistan is Crown property both in town and country."[163] In 1934, the sale of state agricultural lands was resumed for the specific purpose of commencing modern agricultural development.[164] Much of the land of Khuzistan was taken over by Reza Shah, who, by the end of his reign in 1941, was the largest landlord in Iran with 2,000 villages wholly or partly owned by him. In confrontation with AIOC, the shah eventually shifted his strategy to consolidate his

control of more revenues from the oil-rich subsoil by canceling the 1901
D'Arcy Concession in 1932 and signing a revised agreement in 1933.[165]
On the one hand, the revised terms reduced the concessionary area to
100,000 English square miles, a territory whose boundaries were to be
fixed no later than December 31, 1938. On the other hand, there was
no change as the lands available for acquisition were defined in terms
of utilized and unutilized (rather than cultivated and uncultivated) but
still as "belonging to the Government."[166] Government policy regarding
ownership of mineral rights was indefinite and vague until 1939, when
an act of the *Majlis* (Iranian parliament) formally classified all mineral
resources and granted private ownership of mineral resources *except* oil
and precious metals.[167] Thus, Reza Shah's policies of land confiscation
and registration in Khuzistan introduced a new layer of legal intermediaries connecting the subsoil to the surface, claimants to the law, the
central government, and the revised 1933 concession terms. Introducing
procedures and paperwork for the national-legal classification of the oil
regions helped stabilize the control of the oilfields, pipeline, and refinery
in favor of concessionary control, helped eliminate intermediary claimants, and helped solidify a new alliance between the British oil company
and the Pahlavi government. Reassembling the properties of oil in this
way ensured that any political uncertainty introduced by middlemen or
former claimants to the land was eliminated.

Reassembling the properties of oil from local to concessionary practices of oil development and ownership involved a coordinated movement between the oil company and the Pahlavi-controlled government to
eliminate any instability in the oilfields, in part by introducing a new set
of national laws concerning mineral rights and the registration of private
property. This would protect the terms of the concession contract while
blocking more equitable forms of oil production. AIOC mobilized the
concession contract as a device to implement actions that would translate oil into profits. In a similar fashion, the Pahlavi shah confiscated
land, eliminated nomadic groups, and enacted national laws that could
be placed between the oil, striking oil workers, and claimants to the oil
regions to ensure that they would not define their identities toward more
radical and democratic forms of control over the oil.[168]

As company operations expanded through the 1930s and especially
through World War II, numerous complaints were filed against the government's conversion of land to *khaliseh* as a mechanism for legalizing its
sale to the oil company (after being registered at the Land Registration
Department). Property owners in the oil regions claimed that they had

never accepted the valuation of the lands in the first place.[169] Shaykh Khaz'al's son and descendants of the Bakhtiyari khans also attempted to reclaim land confiscated by the Iranian government and the company.[170] In the aftermath of Iran's oil nationalization crisis (1951–1953) and the creation of an international oil consortium in 1954, the newly established National Iranian Oil Company (NIOC) pursued a similar politics of coercion and resettlement, precluding the resolution of complaints by land claimants.[171]

The government argued that NIOC was legally replacing the former British-controlled oil company in the consortium. Therefore, the lands under dispute were already processed and priced leaving the national company in formal possession of them. Following the procedure built into AIOC's concession terms on property, local claimants might be compensated on the condition that they provided legitimate titles to the land.

Inclusive Exclusion

Anglo-Iranian oil's liquid and mobile properties underground demanded a peculiar set of property rights as well as inevitably leading to political struggles surrounding claims to the land targeted for extraction. The British political project to produce oil in southwest Iran involved the building of new forms of property arrangement, built into the format of the 1901 oil concession, that appeared to overlook the local forms of property and politics persisting on the ground. By scrutinizing the terms of the concession, I have considered its operation as a sociotechnical device that helped create the agency of AIOC, the Iranian state, and others, defining what these actors could do and on what terms including the oil. The concession articles on property maintained an ambiguity that was exploited by multiple sides according to different interests and degrees of strength. The signing of the 1901 agreement, with its procedures for building concessionary property, left the impression that a legal contract had merely been signed between two contractually equivalent parties separate from the world outside. And yet new political actors, organizational forms, and public-private interests were participating in the political project to establish the rule of British concessionary property, while other groups, such as the Burmah Oil Company, were simultaneously working to delay oil development in the first decade of the concession's existence.

The company's technological choices about defining, measuring, and pricing lands covering oil reservoirs were informed and inscribed by technoracialized assumptions about local society and their knowledge of mineral rights, (Western) private ownership, and modern energy development. This indicates that the social and technical worlds of oil were not separate but intertwined from the start. The concession worked by introducing novel technologies of intervention and representation that operated with other very physical tools of control, technical and scientific knowledge, political arrangements, and coercion. These techniques worked to transform the subsoil into a manageable object to which claims of concessionary and state ownership and monopoly control could be attached, proliferate, and develop new objects, interests, alliances, and agents.

Mechanisms of property dispossession and acquisition were organized around the lands targeted for oil extraction and exercised by the Iranian government, in alliance with AIOC and the British government, in the name of crown/state land or concessionary land, respectively. The procedure would continue with the dramatic departure of AIOC in 1954 (renamed British Petroleum) and its replacement by NIOC in an international oil consortium. Following the multiplication of local actors with claims to this land is also important for understanding the different ways older and more recent logics of property rule conflicted and by what political methods the modern concept of state property was constructed in the Middle East to guarantee a foreign concern's rule of concessionary ownership. The nature of the problems that evolved over land and property demands that one take into account the specificity of oil's physical and subsoil properties, the complexity of local ecologies persisting on the surface, and practices of property valuation. These controversies made it necessary for AIOC to request more land to build more facilities, conduct oil surveys, and build boundaries managed by the concession's terms. Establishing boundaries, legal procedures, and technical measurements in English for pricing and quantifying helped legitimize the making of state property and its transformation into concessionary property.

The concession was a technology of political management and control whose machinery did the work of assembling the numerous forces at play, human and nonhuman, in a kind of "inclusive exclusion."[172] Depending on the stakes, various groups such as the transnational oil corporation, the Iranian and British governments, the Bakhtiyari khans, Shaykh Khaz'al, and other political forces had the potential to get interested or abandon the shifting arrangements traced here, enabling the technical and political transformation of an invisible material into concessionary property and a large-scale energy system.

On the one hand, I have suggested that the terms of the British-controlled oil concession described a world very different from local forces at play in southwest Iran. On the other hand, the rule of concessionary property was not separate from the world it claimed to describe in its articles. By recovering the history of the construction of concessionary property over oil, I have exposed the connections that helped displace and transform oil by a new intermediary (e.g., a unit of land, measurement, or law) to render each new displacement easier and more abstract and to establish equivalences.

Infrastructural work, or the building of alliances and connections, is what makes the international market in oil stable and the hypothetical world of the concession a reality. These movements resulted in the universal authority of concessionary property as spokesperson. Thus, the oil concession was black-boxed over time.[173] Its execution was all that mattered, not the organizational techniques, local knowledge, and controversies that went into making it operational in the first place.

The work of the concession also involved the silencing of other claims to the lands and the reconfiguration of local forms of politics. AIOC built boundaries between oil operations inside concession property and the political claims of various groups outside. Their status, however, was neither outside nor inside because the rules of concessionary and *khaliseh* property arrangements were the outcome of a battle fought over control of the subsoil with these so-called outside forces. The various procedures for framing, pricing, and converting property and its subsoil to *khaliseh* were inscribed in the concession articles, national land and mineral laws, and property measurements. The procedures were also connected to local forces, mechanisms of debt, dispossession, eviction, and petroleum expertise. The entanglement of diverse actors in the machinery of the oil concession enabled a kind of centralization of information and domination at a distance from AIOC headquarters in London. The remaining historical connection to help stabilize this arrangement was the energy needs of the British Admiralty and the largest oil corporations, which were seeking to preserve their growing monopoly control over international oil supplies, distribution networks, and markets.

The specific properties of oil played a central role in the creation, establishment, and strengthening of concessionary property in southwest Iran, in turn shaping the larger history of modern property in Iran.

The terms of the concession helped create and equip a new actor in the twentieth century, the transnational oil corporation. Concession terms did the physical work of building the machinery in and through which more democratic forms of oil production and politics were shut down and other arrangements left open. They laid the necessary groundwork, namely, foreign control in coordination with oil corporations and cartels to reorganize British concessionary rule in the novel format of an international oil consortium.

2

Petroleum Knowledge

The Concessionaire shall immediately send out to Persia and at his own cost one or several Experts with a view to their exploring the region in which there exist, as he believes, the said products, and ... the latter shall immediately send to Persia ... all the technical staff necessary ... for boring and sinking wells and ascertaining the value of the property. (Article 8, D'Arcy Concession, 1901)

The Company undertakes to send, at its own expense and within a reasonable time, to the [Iranian] Ministry of Finance whenever the representative of the [Iranian] Government shall request it, accurate copies of all plans, maps, sections and any other data whether topographical, geological or of drilling relating to the territory of the Concession, which are in its possession. (Article 13, 1933 Concession)

The Company shall place at the disposal of the specialist experts nominated to this end by the [Iranian] Government, the whole of its records relative to scientific and technical data, as well as all measuring apparatus and means of measurement, and these specialist experts shall, further, have the right to ask for any information in all the offices of the Company and on all the territories in Persia. (Article 14B, 1933 Concession)

The global energy shift from coal to oil in the early twentieth century helped transform political possibilities and forms of expertise in relation to the control of this new fuel.[1] With the discovery of oil reserves at Masjid Suleiman in 1908, the new oil company, AIOC, was confronted immediately with another sociotechnical problem. Unlike coal, petroleum could not be studied in situ. Geologists could only examine the water, oil, gas, and rock shards that came up the borehole.[2] The high viscosity and sulfur content of the oil that came out of the ground here made refining it not only technically difficult but its yield of liquid hydrocarbons and motor gasoline inferior to many other crude oils, and thus unsuitable for civilian consumers.[3] The heavy oil could be used, however, to fuel ships, making AIOC the British Navy's single largest supplier.[4]

The British-controlled 1901 D'Arcy Concession made little mention of the oil itself, where to find it, and how to extract, produce, and refine it. Article 8 simply stated that the "Concessionaire" would send to Iran all necessary "technical staff" for "boring and sinking wells." The concession terms made no suggestion that the Iranian government possessed automatic rights of access to information concerning the kinds of technical activities that would take place. Articles 13 and 14 of the 1933 oil concession, however, suddenly obliged the "Company" to provide to the Iranian government "accurate copies of all plans, maps, sections and any other data whether topographical, geological or of drilling ... in its possession," whenever requested. The terms also granted to the "specialist experts" of the Iranian government "the whole of its [the company's] records relative to scientific and technical data, as well as all measuring apparatus and means of measurement, and ... the right to ask for any information in all the offices of the Company and on all the territories in Persia." What was novel about this fossil fuel that it necessitated the production of new forms of expertise and concession terms?

Most histories suggest that the difficulties AIOC confronted in extracting, producing, and refining oil were technical in nature. BP's company historian, R. W. Ferrier, frames the history of AIOC in Iran in terms of the success of "the practical application of scientific principles" to local oil operations that in turn created an advantage for the company's reserves and cheaper oil supplies.[5] The implication here is that scientific expertise arrived in southwest Iran from the West and was applied to the oil in a pioneering fashion, with the effect of maximizing both the amount of oil within the concession area and the company's ability to sell it at competitive prices. But the history of petroleum expertise as constructed in southwest Iran involved many more actors and controversies than is suggested by the BP corporate history or by a petroleum engineer's story, framed in the technical terms demanded by his discipline.[6]

Concession terms, corporate histories, and the petroleum sciences exclude the historical and political circumstances in which such knowledge was constructed in the first place. In practice, the construction of expertise around a new source of energy in the early twentieth century required an entire scientific community of petroleum consultants, geologists, engineers, and chemists collecting information about the specificity of the oil's physical and chemical properties and its interactions with the local environment and population with claims to where it was found. These properties were embedded in the emergence and organization of new large-scale oil corporations in the early to mid-twentieth century and the reconstruction of an energy network that would serve as the basis for new kinds of geological, engineering, and economic knowledge.[7] To increase their profit margin, oil companies such as AIOC developed large research-and-development divisions to find applications for their unused petroleum byproducts, distribution and marketing divisions to promote their use, and political and public relations departments to help build societies that would consume them. They also colluded to deny expertise to others, in particular the coal industry.[8]

That Anglo-Iranian oil came to possess its own abstract origin myth had little to do with the organizational work involved in transforming the world into a place where oil and its byproducts would be eagerly demanded. To say that the history of Anglo-Iranian oil can be understood better as a history of the working possibilities of technical and scientific information gathering is impossible in the framework of a conventional political or social history of the period.[9] These accounts do the work of silencing the activities of surveyors, geologists, and engineers and the ways other local and government actors factored into their calculations and decision making about the oil. However, some of the most important battles concerning the Iranian state's demand for sovereignty over its oil resources emerged within specific controversies over the exact nature of Anglo-Iranian oil—that is, its physical, chemical, and geological composition, as well as its lifespan, production rates, and reserve estimates. When, in 1932, Iran decided to cancel its British-controlled oil concession, the above controversies, combined with additional demands for the replacement of foreign workers with Iranian workers and for a greater share of royalties, culminated in a renegotiation of concession terms that might favor a national government over an oil company. Much had changed socially and politically since 1901.

Whereas the 1901 concession remained silent on the question of gathering and sharing scientific and technological information about the oil, Article 14 of the 1933 concession obligated AIOC to make its scientific and technical data transparent and available at the request of the Iranian government. The risk for the British side was that this would set a precedent for other oil-producing countries, such as Mexico, to follow suit and make claims in terms of national sovereignty. AIOC's construction and representation of its science and technology in technical reports, journal articles, and international petroleum conferences greatly underplayed the role of interruptions posed by local political communities such as the Bakhtiyari khans. They also understated all the collaboration required to learn about the behavior of the oil, to establish the concessionary rule of property, and to manage the relentless emergence of oil worker strikes in the 1920s, 1940s, and early 1950s. Such exclusions by petroleum experts of the local knowledge gathering that occurred and its historical inter-connections with social and political groups in the oil regions, Tehran, London, and elsewhere have been reproduced in the scholarship on the topic. The implication in the scholarship is that technical controversies about oil are a separate issue from political and economic concerns about state and society.

In his sociotechnical history of the French oilfield services com-pany, Schlumberger, Geoffrey Bowker tracks how company engineers and geologists transformed oilfields at various sites around the world into their laboratory as a business strategy in the early twentieth cen-tury.[10] This chapter sets out to do the same by exposing the company's business strategy through the transformation of the oil regions into a kind of laboratory for producing scientific knowledge and technologi-cal know-how about nature and society. Rather than viewing technical and scientific information as a benevolent force of Western ingenuity or as an instrument of securing monopoly control, I recover the history of Anglo-Iranian oil's petroleum knowledge as a site of political contesta-tion to demonstrate the interconnectedness between the technical world of oil and the social world that it requires to survive. Each section traces the ways AIOC constructed, gathered, and managed information about Iran's oil as a business strategy to exclude rivals and build boundaries between technical, scientific, and economic concerns and politics. By tracking the oil's transformation from a subsoil material into a repre-sentable object with measurable qualities and quantities, and finally into a publishable and generalizable research finding in London, I reveal the company's truth-making strategy, which sought to exclude the possibil-ity of national control and a more equitable distribution of profits from entering its calculations.

In the first two sections, I map the construction and extension of the energy network for the production, circulation, and standardization of petroleum knowledge about techniques of exploration, extraction, and refining as they accumulated in southwest Iran. Only by recovering this local and organizational work can we make sense of the oil company's strategic success in managing Anglo-Iranian oil production in relation to world oil production, protecting its international monopoly agreements, and silencing the historical and sociotechnical activities that enabled the geological representation of the oil in the first place. AIOC engineers, geologists, and chemists encountered many problems in southwest Iran as they gathered information about the oil, evaluated it with various calculating technologies, and finally manufactured a certain ignorance about its knowability, particularly in confrontation with the Iranian government. In the last section, I examine the relation between the work done in the oil regions to define the properties of oil and the representation of that work in the public arena, in presentations to petroleum experts from other oil companies and at petroleum congresses. By scrutinizing the ways techniques of representing and standardizing oil accumulated, operated, and circulated in southwest Iran, Tehran, and London, I reveal that the powers of AIOC and the Iranian state were worked out within the specific and local history of Khuzistan's oilfields, precisely through the management of technical and scientific information about oil.

Petroleum knowledge equipped AIOC officials, petroleum experts, and Iranian government officials, to varying degrees, with agency in political struggles over profits, labor, and national control.[11] As with AIOC's business history, the company's in-house and official scientific publications about the oil portrayed petroleum knowledge as the product of ingenuity arriving from abroad, only to be confronted with a passive natural environment and technically ignorant society. Forms of information management provided AIOC's managers, engineers, and geologists with the tools to build a world in which the company's oil could thrive in the midst of increasing uncertainty regarding Iran's nationalist demands for control of the oil. This entailed the production of scientific knowledge and technological expertise about Iran's oil, first as industrial science and then as pure science, through a so-called process of invention (the birth of an idea; technological) and innovation (commercialization).[12]

Properties of Oil and Practices of Control

In the first two decades of the concession's existence, AIOC geologists, engineers, and managers suffered numerous unintended consequences as a result of the extractive needs of Iran's oil. The oil at Masjid Suleiman was located in Bakhtiyari[13] territory (figure 2.1) situated within the distinct rock formation of a highly porous limestone. The geologist S. J. Shand first identified the reservoir rock as Asmari limestone.[14] Anglo-Iranian oil was particular in terms of the geophysical environment in which it was found but also in terms of the way it came out of the ground. Oil production in many fields, as in the United States, was sustained through pumping, but the naturally higher gas pressure of Iran's oil eliminated the need for pumps, at least initially.[15] To extract commercial quantities, the engineers needed to drill more wells at Masjid Suleiman but without jeopardizing the high-pressure zones of productive wells. In general, some oil wells will produce for extended amounts of time by flowing due to the pressure from subsurface gas or water that pushes the oil out naturally. Other wells must be pumped to extract the oil. High-pressure zones that characterized the oil regions in southwest Iran ensured that the oil came out on its own, at least in the short term. An equally significant concern was the lack of skilled drillers and engineers to read the gauges that monitored the pressure at the bottom of the wells.[16]

The subsoil, liquid, and mobile properties of oil marked its particularity as a new source of carbon-based energy in the twentieth century. Addressing problems associated with the high sulfur content and viscous properties of Anglo-Iranian oil required the technical expertise of trained engineers and geologists, specialized equipment, and adequate living and working conditions. To overcome the cost of exploration and development activities, AIOC's managers aimed to extract the oil, define its particular qualities, refine it as quickly as possible, and create a market for its products.[17] The first problem emerged from the assumption, among petroleum experts, that Iran's oil would behave like the oil at Rangoon, Burma. AIOC officials faced delays in production due to Anglo-Iranian oil's naturally high pressure and sulfur content.[18] Chemists and engineers with prior experience working in Burma were unprepared, and it took many years to analyze the chemical constituents and physical properties of this new source of oil.[19] The persistent fear was that gas pressures in a particular area of a field would be exhausted too rapidly, causing other production flows to fail prematurely.[20] Data on the physical characteristics of the oil reservoir, as well as its behavior and its control using various methods of production and drilling, were essential if the company was to succeed in predicting where the gas pressure would fall first.

There were many other properties to consider. For example, Anglo-Iranian oil came out of the well not as a simple liquid but as a frothy substance.[21] For every million gallons of oil produced, approximately 20–25 million cubic feet of gas (equivalent to 3.56–4.45 million US barrels of oil) were released mainly as methane and ethane. As the gas bubbled out of the oil, it also released light hydrocarbons such as butane and pentane, which were potentially of valuable use in gasoline. In the early years, the gas was ignored and allowed to escape freely into the air.

Extracting the oil was not the only problem. In the early twentieth century, petroleum and the accompanying natural gas was extracted from the ground and distilled elsewhere, according to boiling-point tempera-ture, into fractions such as gas, kerosene, lubricating oil, and asphalt.[22] Anglo-Iranian oil could not be refined using the same methods applied to Burmese or Pennsylvania oil due to its high viscosity.[23] Through the 1920s, problems created by the viscosity of Anglo-Iranian oil, particu-larly the fuel oil for the British Admiralty's warships, were surpassing the "scientific knowledge of the day."[24]

Company engineers and chemists learned about such properties based on their previous experience with American or Burmese oil, and they assumed that Iran's oil could be extracted, distilled, and used in the same way and to the same extent as other oils. From an economic perspec-tive, AIOC officials felt that to make enough capital to cover the cost of extraction, other marketable products needed to be produced using distillation techniques to replace kerosene such as benzene, motor spirit (gasoline), and fuel oil for river steamers.[25] In the early years of the concession's existence, these and other "unsuspected problems" drove AIOC's profit-seeking oil managers to the verge of bankruptcy. This was also due to inadequate labor and equipment, as well as to the "unknown nature of the Persian crude."[26]

Additional information was therefore required not just to overcome the constant loss of capital but to make sense of the properties of the oil. To achieve this goal, the company appointed John Cadman, future chairman of AIOC (1927–1941), as technical adviser of a newly formed research organization in 1917. Through the 1920s, he implemented an extensive plan to develop the company's activities in research and devel-opment because the whole "geological problem is still obscure."[27] This was due in part, the company believed, to insufficient contact between geologists working on Anglo-Iranian oil and the staff in London. Other problems such as escaping gas from the main oilfields were preventing the extraction of the oil.

For Cadman, the economics of oil and the petroleum sciences were seamlessly intertwined. Finding a solution to the problem of gas seep-ages, he explained, would not only have "very important economic results" but would reveal a "good many scientific problems connected with the economical engineering side of the question."[28] Cadman was a well-informed petroleum expert, having previously served as Professor of Mining and Petroleum Technology at Birmingham University, where he established the first Department of Petroleum Technology in the British Empire.[29] The department would become the destination for the majority of Iranians selected for training under AIOC sponsorship.[30] Cadman also served as consulting petroleum adviser to the Colonial Office. He was a member of the Admiralty Commission to Iran (1913–1914), director of the British government's Petroleum Executive (1917–1921), member of the Inter-Allied Petroleum Committee (1918), and president of the Institution of Mining Engineers.[31] This experience gave Cadman a heightened awareness that a host of conditions, social, economic, political, organizational, and technical, were required for the Anglo-Iranian oil industry to work. The fluidity of the boundary between the technological system of oil and its environment was constantly being raised, as was the question of the extent to which "systems builders," such as Cadman, molded their environments to facilitate the growth of their system.[32]

As AIOC formulated its plans for oil research and development, Iran faced heightened problems of food scarcity, high prices, separatist nationalist movements, and a weak central government.[33] With hopes of increasing its share of oil profits to strengthen its power over the provinces, the government signed the Armitage-Smith Agreement of 1920, clarifying some of the ambiguities of the 1901 concession regarding the calculation of profits.[34] Iranian government officials were demanding more production in order to increase their share of royalties, whereas AIOC officials sought to limit production levels in relation to world oil production and to block politics—that is, national control—from entering calculations that shaped the concessionary arrangement.

Combined with a heightened post–World War I interest in developing rather than limiting Anglo-Iranian oil development (discussed in chapter 1), the appointment of Cadman as AIOC's technical adviser marked the beginning of the construction of a large research-and-development branch within the company and the building of new connections to research centers and science and technology institutes based in London. To increase profit margins and to overcome the obstacles to extracting Iran's oil, AIOC embarked on accumulating more data about the oil through field surveys, laboratory tests, publications, international meetings, and in-house company conferences.[35] The primary impulse behind establishing the Sunbury Research Centre in Middlesex, England, in 1917 was to solve the problem of high viscosity characterizing AIOC's fuel oil.[36]

After World War I, as Iran underwent a series of social, economic, and political upheavals, resulting in the division of the country into British and Russian spheres of influence and the revived power of semiautonomous groups, Great Britain, as AIOC's largest shareholder, moved to consolidate its control over Iran. Defining the properties of Anglo-Iranian oil meant controlling a set of connections that would ensure the oil company's strategic success. First, control meant overcoming the failures and uncertainties that emerged around the high cost of extraction, refining, and marketing. Second, it referred to protecting the secret monopoly arrangements of the dominant international oil companies. Their goal was to control world production rates and prices in the midst of domestic turmoil in Iran and nationalist demands for more oil production. From the perspective of AIOC, maintaining the stability of these connections required accumulating more technical information about Anglo-Iranian oil through the recruitment of experts with experience in Britain, Burma, and Sumatra.

International consulting geologists were important actors in the project to stabilize expertise about Anglo-Iranian oil.[37] After World War I, AIOC commenced the hiring of large numbers of geologists, geophysicists, drillers, and petroleum engineers. They formed associations of specialists and published their findings with the permission of the company's chairman and directors.[38] This was acknowledged in a footnote in every publication. Findings were circulated within the company at annual or biannual production conferences and shareholder meetings, which Cadman initiated.

In the early years, a select few consulting firms in London were the principal advisers of these large-scale business concerns, but their specialists were not employees of the company. Boverton Redwood, for instance, was a consultant for both the Burmah Oil Company and the syndicate of British investors in Iran, the British Admiralty, as well as for other operations in Mexico and Nigeria.[39] At a time when few companies employed full-time geologists, his consulting firm was used by virtually every British oil company.[40] Redwood's "major significance" was his intermediary role between AIOC, the Burmah Oil Company, and the British government (the Admiralty and the India Office), maintaining consulting connections with oil-producing companies and equally "intimate association(s)" with British government agencies.[41] He was also one of the strongest advocates of the use of fuel in naval vessels, which helps explain his connections to the British government.[42] Strongly supporting the development of British colonial oil industries, he had great animosity toward the rival Standard Oil and Shell groups. As a consequence, he helped block their efforts at monopoly control to protect Burmah Oil's monopoly of the Indian market, while also ensuring that Anglo-Iranian oil development did not disrupt this market either.[43] This connection also explains his interest in seeing Burmah Oil get involved as AIOC's parent company, effectively bailing it out in 1905.

The Burmah Oil Company had "pioneered the use of geologists in the search for oil," many of them supplied by Redwood's consulting firm.[44] In 1907 the company began to employ full-time geologists, many of whom, such as E. H. Cunningham Craig, Basil Macrorie, and Lister James, were later to use the experience they gained in Burma in the service of Anglo-Iranian. "There was a steady flow of personnel" between Burmah Oil and AIOC, especially on the technical and geological side of the business.[45] This enabled AIOC to "draw upon the long experience of its adopted parent."[46] Much of the preliminary analysis of Anglo-Iranian oil after 1908 was undertaken in London at the New Oil Refining Process Ltd., a subsidiary of Burmah Oil.[47] The first of what became AIOC's regular geological staff went to Iran in 1913 and were joined by many more geologists in subsequent years.[48]

AIOC's desire for secrecy in all aspects of its oilfield operations was opposed by its desire to share information to build up scientific techniques. Geologists such as Duncan Garrow, a director at AIOC (1914–1924) with previous experience in Rangoon, had observed that Anglo-Iranian oil appeared to share similar conditions to those of Mexican oil.[49] The fear was that the reservoir containing Anglo-Iranian oil might "turn to water" as Mexican oil had done in 1920, or it might go into decline as with Baku's oil in 1901.[50] Garrow consulted with A.F. Corwin of the Standard Oil Company for an alternative opinion. Corwin paid a visit to the oilfields of southwest Iran. He argued that while the general behavior of wells in the main producing zones resembled those of Mexico, there was still no concrete proof of a rising water unit, and yet the uncertainty remained. By the 1920s, importing trained staff and engineers from the oilfields of Rangoon and the Indian Public Works Department was no longer adequate if the company wanted to survive commercially in the world market. A geological survey department existed within AIOC at this time, but to meet the company's expansion, Cadman established a formal Geological Survey Department with permanent staff.[51] Young geologists of European origin (primarily British) were expected to handle the "difficult conditions in oil regions for surveying abroad," and such men must be equipped with good university training in surveying, mapping, and report writing.[52] Some twenty additional geologists arrived in southwest Iran, nearly all with postwar training and many with wartime military service.[53] Petroleum experts also arrived there with previous oil experience in other parts of the world and with university training. The problem was that technical experience and knowledge told them little about the underground properties of Anglo-Iranian oil. Led by Cadman, who hoped to rescue the company from going into bankruptcy, these experts took steps to transform the oilfields of southwest Iran into a laboratory.

Building a Petroleum Laboratory and Constructing Publishable Research Findings

Research institutions, techniques of mapping, and other forms of technoscientific expertise suddenly proliferated to help make Anglo-Iranian oil knowable. In the period between 1920 and 1940, there were many competing ways of discovering and then mapping oilfields.[54] This period of heightened interest and expansion in exploration was triggered by the rapidly increasing demand for petroleum, an acute shortage of oil in the United States after World War I, and international competition for the control of the world's oil reserves. AIOC geologists were stationed in each drilling area of southwest Iran, but remained principally concerned with surface exploration. Drilled wells demonstrated underground folding, a kind of bend or curve in the underground sedimentary rock layers, that coincided with surface folding.[55] Geologists named these folds or arches "domes" and "anticlines," forming the basis of the "anticlinal theory," considered the "backbone of geology" in this period. Domes, anticlines, and faults comprised the most important structures in these oil and gas reservoirs.[56] The principal objective of exploration surveys of the major Fars and Bakhtiyari anticlines, characterizing the oil regions of Khuzistan, was to map the region using a scale of an inch to a mile.[57]

The early 1920s marked a decisive moment for oilfield management in which geologists and engineers could observe the "performance of a large limestone reservoir under controllable conditions," guaranteed by AIOC's concessionary "ownership of the entire field."[58] The extensive mapping and surveying of the oilfields of Khuzistan amounted to approximately 2,500 square miles, helping to "form reliable data" about the oil reservoirs and their geophysical structure.[59] But more data was needed. A company geologist informed Redwood that there was still a "lack of maps ... on the geology and physical landscape of the region."[60] According to Owen, in most oil-producing countries of the time, "little effort was made to acquire accurate subsurface data before 1918."[61] There were older scientific publications from the nineteenth century on oil in Pennsylvania, Ohio, and Indiana in the United States, but most

European and American petroleum geologists did little more than "complain about the unreliability of the drillers' well logs."[62] Subsurface geology was not adapted as general oilfield practice until 1918–1921, after which it "spread rapidly."[63]

The accumulation of more data would help render the oilfields testable, knowable in abstraction, and describable in publications. Bowker has observed that the domain of an oilfield was described using three kinds of maps, each of which helps to produce a new kind of space—a transport map, a map of geological structure, and a map of subsurface structure. The era of formation of the oil, the date of discovery, and the statistical year describe three time periods that, together with the maps, combine into a new kind of time.[64] In southwest Iran, for example, one-inch mapping of the Bakhtiyari oil regions was contoured simultaneously with geological mapping and with the use of American plane-table methods,[65] helping to transform the site into a space of measurement that increasingly operated as a laboratory.

To this end, AIOC's new in-house geologists, H. T. Mayo and H. G. Busk, presented one of the company's first scientific publications on Iran's oilfields at the Institute of Petroleum in London in 1918–1919. AIOC's chairman and honorary president of the institute, Charles Greenway, explained that until now, "little information ha[s] been given to the world on the subject of these remarkable oilfields." But now that their "phenomenal richness and extent" had been demonstrated, the oilfields proved to be "one of the largest, if not the largest source of supply for the world's future requirements of petroleum."[66] It was necessary, therefore, that "a little more light be shed on their geological character," following the research findings of Busk and Mayo.[67]

Busk and Mayo's paper presented the current problems with Iran's oilfields. Their findings suggested that numerous stratigraphic problems persisted and more specifically, in locating the part of the limestone structure in which the oil at Masjid Suleiman accumulated. By providing a "generalized view of the geology, and the building of the oilfield region in historical sequence," Busk and Mayo introduced new temporalities and spatial scales to the oil regions of southwest Iran, framed in the technical terms of their discipline, geology.[68] The main stratigraphic divisions of the oil regions and the epochs in their geological timescales were as follows: A. Asmari Series (Eo-Cretaceous epoch), B. Fars Series (Miocene epoch), and C. Bakhtiari Series (Pliocene epoch).[69] "It is always remembered," they commented, "that it was only through detailed and extensive survey that such a history was finally proved."[70] To help guarantee the truth of this peculiar kind of framing of an abstract history in terms of the geology of Iran's oilfields, Busk and Mayo referenced earlier studies of the Bakhtiyari region first conducted by geologists from the Burmah Oil Company and the Geological Survey of India, such as W. K. Loftus, "Pioneer of Persian Geology," in 1849, by G. E. Pilgrim, who established the "main stratigraphic divisions" of the oil regions, and by Lister James, who with his coauthor, Busk, constructed "the first detailed large-scale map of the Maidan-i-Naftun Field," the largest source of commercially producible reserves in the entire concession area. This local work of experimentation, new representations, and development of a "stratigraphic nomenclature" was helping to organize the oil regions according to a different spatial-temporal scale, a kind of laboratory, while getting to the bottom of the problem of ascertaining exactly where the oil at Masjid Suleiman accumulated.

In practice, Busk and Mayo's research findings were helping to construct a systematic body of knowledge called petroleum geology with its own history. Bruno Latour has observed that in both industrial and academic science, there is a process of destroying one's own past.[71] In a typical scientific paper, two specific historiographic processes occur: the presentation of the paper within the history of the scientist's own discipline, geology in this instance, and the destruction of the historical context. A study of publications, company prospectuses, and other pamphlets challenges hackneyed depictions of oil exploration as lawless and random.[72] However, typical reports of prospecting missions detailed not only geological findings and prospecting methods, but also the ethnic composition of territories covered and relationships between personnel of different ethnic backgrounds.[73] Ethnological conventions about race relations dominated explanations of why missions encountered success or failure. Local political groups, for example, kept slipping into and interfering with Busk and Mayo's construction of a geological history of Anglo-Iranian oil:

The Bakhtiari country, under the jurisdiction of the hereditary Bakhtiari Khans, who lead one of the most powerful political factions in the Shahdom and one friendly to ourselves, is situated on the south-west side of Persia between the provinces of Farsistan and Luristan. It ... merges north-west into the Pusht-i-Kuh country ... and ... the Kuh Gahr country, the inhabitants of both the latter, living as they do upon the plunder of their neighbors, being extremely unfriendly to Europeans.[74]

Busk and Mayo set out to "summarise the chief geological features of the oilfields in the Persian concession" by dividing the subject into three geographic regions, beginning with "The Bakhtiyari Country."[75] Oddly, the geographic description was mixed with information outside their domain of interest, namely, the history of power relations between a "powerful political faction," known as "the Bakhtiari Khans," and "ourselves," the author-geologists, and "Europeans" more generally. Thus, conventions about race relations informed technical explanations of how geological missions in this instance proceeded and encountered success or failure.

The article ended by attributing the success of the geological missions in southwest Iran to the geologist's pioneering spirit in an unfamiliar geographic setting filled with political and technical uncertainties:

Pioneers in Persia are much to be congratulated in their enterprise, in a region fraught with many political and transport difficulties, in their following up the first test well at Maidan-i-Naftun with other wells, which have now proved an area which, on account of its richness and phenomenal productivity, entitles it to be classed amongst the great oilfields of the world.[76]

AIOC's representation of its science and technology overlooked the role of interruptions posed by local political communities such as the Bakhtiyari khans and all the work and collaboration that the company pursued with them in order to protect oil operations. As detailed in the previous chapter, this was most obvious in the company's decision to set up a subsidiary company in 1909, known as the Bakhtiyari Oil Company, to pay the khans for access to their land, to secure the concessionary rule of property over the subsoil, and to recruit their followers as guards to patrol the oil regions.[77] For many years to come, these political forces inhabiting the oil regions threatened to disrupt the security and stability of oil operations. As one company geologist, Lister James, admitted, evidence of oil for the greater part of the concession area was still "lacking," due in part to the "political conditions of the country, and the lack of communication" between geological staff in different areas of the concession.[78]

The political conditions of the country in the early twentieth century— that is, the historical circumstances in which the geological findings such as those of Busk and Mayo were obtained—were characterized by the weak central Qajar government discussed in the previous chapter. The Constitutional Revolution (1905–1911) had meant that alliances with local semiautonomous nomadic groups, inhabiting the site of oil operations, pipeline, and refinery, were essential for AIOC's concessionary control and the geological missions discussed here to succeed. The shift to a new strategy in the 1920s sought to build an alliance with a strong central government that would in turn help suppress any further threats from Iran's provinces, especially Khuzistan.[79] By 1929, Iranian oil workers threatened to mix questions of politics with questions of geology, demanding better wages, treatment, training, and housing.[80] Political conditions were indeed one of the reasons why, on the brink of a concession crisis in 1928, AIOC geologists had traversed only a minute portion of the concession area.

AIOC's information gathering was working to exclude the local historical and sociotechnical activities that enabled the birth of the science of geological representation in the first place. Identifying AIOC's domains of knowledge production and the problems that geologists and engineers confronted in standardizing Anglo-Iranian oil has exposed the technical and scientific work AIOC required to construct its authority and objectivity. In this way, the company could distance itself from political problems with land claimants, oil workers, and national government actors. By 1920, scientific and technical research had become a key feature of modern industry. As I discuss below, it also made sense for the companies to share information in order to build up discovery and production techniques.[81] When compared to the difficult process of trial and error in gathering local knowledge about Iran's oilfields, the presentation of AIOC's expert findings in international journals and institutes abroad produced the effect of technical ingenuity, the birth of an idea, commencing a process of innovation in which economic, social, and political concerns would come into play only at a later stage.[82]

The Secret Life of the Oilfield and the Politics of Unknowability

In the period between the signing of the first oil concession in 1901 and its revision in 1933, AIOC's organizational work involved transforming the oil regions into a laboratory for managing power relations between the British oil corporation and the national Iranian state's claims to sovereignty. These activities were an integral part of, indeed indistinguishable from, its in-house scientific work. During the 1920s and 1930s, the question of the life of the oilfield, or the amount of commercially producible oil available within a defined timeframe, was entangled with the "obscurity of the geological problem as a whole," noted earlier by Cadman. The company's difficulties with understanding the behavior of Anglo-Iranian oil necessitated the accumulation of information in the form of standards, measurements, and parameters about not only its geophysical structure, but also the amount and behavior of oil within the reservoir. Questions about the exact nature of the oil and its lifespan (reserve estimates and depletion rates) generated controversies involving the Iranian government; AIOC geologists, engineers, and managers; the British government; and research centers and technical institutes in London. As already noted, this set of concerns helped trigger the largest interwar dispute between a national government and a foreign oil company, when the Iranian government canceled the British-controlled oil concession in 1932.

Calculating the lifespan of British-controlled oilfields was essential not only to the profit-seeking managers, but to the British Admiralty, namely Winston Churchill, who as First Lord made the decision in 1912–1913 to switch from coal to fuel oil, relying totally on the reserves of southwest Iran. In one of the earliest assessments of the life of the Iranian oilfields, the Admiralty Commission expressed the impossibility, in the absence of more conclusive evidence, of making any definite estimate as to the life of the main oilfield.[83] It was therefore crucial for AIOC to ensure that the organizational work was in place in the event of "an outbreak of lawlessness amongst the tribes," as such a situation would "present certain difficulties which should not be insurmountable."[84] Political communities in the local oil regions created constant uncertainty, threatening to disrupt the technical world of oil that AIOC and now the British government were working hard to keep separate from politics.

New representations and formulas for establishing equivalences between the oil and future calculations proliferated, indicating the usefulness, from the company's point of view, of treating the oil regions as a space of measurement and calculation, a laboratory. AIOC's geologists and engineers employed novel kinds of framings such as lifespan, estimating flow, oil reservoir pressure, and gas pressure to manage and control the behavior of the oil. Cadman circulated a report among AIOC's managers in 1923 in which he discussed estimating the life of the field.[85] This would require calculating when the wells would cease to flow under their own pressure and the amount of gas reserves available for the purpose of prolonging the life of the field. Other calculations included identifying when producing wells would "go to gas," and selecting the wellsites that would continue to produce oil by pumping after gas pressure had dissipated. This organizational work of calculation and representation would help make Anglo-Iranian oil more manageable and controllable, and thus make international markets in oil secure.

Techniques of classification had consequences for the economic decisions of an oil company and the political decisions of the British government and Admiralty. Cadman, as technical adviser, admitted in a confidential report that oil reserve estimates considered in conjunction with present production and estimated future production "will not fail to cause some alarm."[86] He explained that "whilst these figures can only be considered as approximate estimates, it is the best that can be done with data available, and if it does no more than direct attention to the fact that the life of the Maidan Naftun field is calculable and limited."[87] The amount of oil underground and how long it would last under current production conditions were indeed calculable, even though the company worked hard to suggest otherwise, particularly in confrontation with the Iranian government.

While it was difficult to "disentangle what is speculative from what is positive," Cadman was drawing attention to the "supreme importance of supplementing the present reserves by other sources of supply."[88] At the same time, the company required more local data from which deductions could be made with "greater accuracy." Numerous unknown variables were getting in the way of acquiring an adequate knowledge of underground conditions.[89] Even when all the data—the new representations and formulas for establishing equivalences between the oil and future calculations—became available, "there will always be one unknown quantity" and that was the change in structure with depth.[90] The shape of the domes, the thickness of the reservoir rock, and its permeability and porosity, would have to be factored in to infer the value of the always-unknown quantity. This general "lack of data" necessitated more calculative tools such as "formulae from which further conclusions may be arrived at."[91] AIOC experts introduced a set of new representations and calculations to describe the oil. Representations would help render the oil (and limits to reserves) manageable and calculable in abstraction, according to a peculiar kind of spatial-temporal scale. The report proposed multiple new terms and formulas to be worked out.[92]

Iran's oil was simultaneously knowable and unknowable depending on the interests involved and the recipients of the information. The standardization of Iran's oil in terms of lifespan helped AIOC's information flow within controlled conditions of measurement, circulation, and access. AIOC was transforming Iran's oilfields into its laboratory. The oil was undergoing a series of transformations through the work of a set of technical devices and measurements: its origins flowing out of these transformations and not vice versa as claimed in the scientific articles. Scientific articles claimed that science arrived in the oil regions and was applied to a passive natural element, the oil, while simultaneously excluding the circumstances that enabled the effect of this process to occur from the start. Historical circumstances were the devices and measurements mobilized by author-geologists as techniques of representation to extract and purify the oil from its sociotechnical environment and politics, but masked by the rhetoric of scientific discovery. In practice, the scientific origins of the oil were formed precisely in the organizational work of getting it out, namely, in the oil's *movement* from under the ground to a two-dimensional representation as a numerical reserve estimate and outline of a geophysical structure on a map. The stability of this movement and its publication in an article necessitated a collective effort and negotiation involving measurements, geologists, company managers, local political forces, concession terms, and monopoly arrangements.

Additionally, there was never a perfect fit between the figure or reserve estimate and the ground in that results were valid only insofar as they were rooted in a particular site. As Bowker has argued for the early geophysical work that Schlumberger scientists did, "the process of getting enough local measurements to do good science and enough work on the oil fields to be able to take local measurements was essentially a bootstrapping one ... a technological equivalent of the hermeneutic circle."[93] Geologists who arrived in southwest Iran came with a general idea of the subsurface geology based on older geology articles and previous experiences in other parts of the world, such as the United States and Burma. They needed to be aware of what they were looking for and then to find a correlation between what they found and the classic rules of general geology. But the uncertainty always remained. The identity and properties of Anglo-Iranian oil were not given by its intrinsic properties, because these properties depended on the relations with other entities, including information.[94]

Cadman's alarm about the calculability and limits of Iran's oil reserves was confirmed by AIOC's scientific work, particularly that of Hugo de Böckh in 1924.[95] What was the cause of the productivity of the reservoir rock? Was it porosity, permeability, or some other property? In de Böckh's controversial report on the "Principal Results of My Journey to Persia," the maps, analyses, and conclusions were the subject of debate for many years as the Iranian government pressed for more information about production rates in the fields. De Böckh had set out to conduct a general investigation of the geological conditions in Iran and to establish the lines along which the search for oilfields should be carried out.[96] AIOC reports were never published and often circulated among a limited number of scientists, engineers, and higher management such as Cadman. De Böckh's observations built on previous work by his colleagues, including unpublished articles on AIOC's geology of Iran. The controversy arose from one of the main conclusions reached in the report, where de Böckh argued that Maydan-i Naftun possessed great quantities of oil, but "they are in any case limited."[97] He warned that "a great effort must be made to discover reserve fields before the production drops, and the time for this is limited."[98]

AIOC's in-house knowledge production was not simply a mundane technical activity, it was controversial because it had consequences for the monopoly arrangements of the major oil corporations and the durability of its concessionary arrangement with the Iranian government. The accumulation of scientific knowledge and technological expertise was internal to the company's business strategy as it served as the basis for managing what parts were deemed private or public, universal or particular, certain (knowable) or uncertain (unknowable) depending on the political and economic costs.

For example, certain arguments could and could not be made about the oil, particularly in confrontation with the Iranian government but also within the company itself. To manage the ongoing controversy over the amount of oil reserves under the ground, Garrow, an AIOC director, instructed William Fraser, deputy chairman of AIOC, that figures of oil reserves and estimates should "never be circulated to the Board or shareholders."[99] Garrow justified his reasoning based on a claim to unknowability, that AIOC could not actually say what the total proved crude oil reserves of the company were because one of the giant fields, Haft Kel, was "not fully understood yet."[100] Certain calculations about oil reserves, which Cadman and de Böckh acknowledged previously as calculable and limited, could not be made public even within the company. Calculations about oil reserves were dangerous because they factored into the company's economic interests, indicating that the oil would run out along with company profits. Access to this information risked inviting a discussion of how long the company could survive. Thus, the operation of these zones of measurement, and the distinction between what was considered an internal and what was considered an external matter, were always contestable.[101] Garrow did not consider the 22- to 24.5-year lifespan of the largest oilfield, calculated from 1929, long enough because the "expansion of world's markets may necessitate production of crude largely in excess of 231 million gallons monthly after 1937."[102] He advised that all favorable geophysical reports be proved or disproved to get a more accurate understanding of the rate of reserve depletion at Masjid Suleiman, the giant field with total reserves of around 27,805,725,000 gallons, by January 1, 1938. Garrow confessed privately to Cadman that the estimated life of the Masjid Suleiman field was about 20 years, assuming 350 hundred million gallons were there.[103]

On the other hand, the science of calculating reserve estimates was not so certain. Comins, a company engineer, argued that his "minimum" figures were "very conservative," but his "probably figures" extremely speculative, particularly for "Unproved fields," for which "they should only be regarded as basis for discussion which can be prepared from existing data."[104] The "general bases" of reserve estimates were based on assumptions or theoretical framings that excluded certain behavioral factors of oil over other local conditions.[105] These assumptions concerned porosity, the oil-bearing area, and thickness of the limestone, which, Comins admitted, "are very liable to wide error."[106] With the exception of Masjid Suleiman, reserve estimates were based on pressure data and the rate of drop of oil-gas levels—that is, the ratio of the volume of gas that came out of solution, to the volume of oil when brought to the surface. These estimates were "extremely speculative and should only be accepted as a very general guide."[107] In the early stages of the development of a field, companies like AIOC kept their estimates of reserves confidential, and this tended to remain so through the lifetime of the field. As Andrew Barry explains, oil companies maintained multiple reserve estimates "for planning purposes and financial reporting requirements, both of which [were] likely to differ from the uncertain estimates made by company geologists."[108] Thus, the life cycle of an oil reserve went through a number of stages during which the degree of uncertainty and controversy concerning how much oil there was underground vacillated.

Over a decade later, the question of reserves was still masked in secrecy and up for debate. What was apparent, according to the technical adviser to the Petroleum Division of the British government, was that Masjid Suleiman "has been heavily depleted," while the other giant field, Haft Kel, "has had about 50% of its oil extracted."[109] On the other hand, the fields at Gach Saran, Agha Jari, and Lali still possessed large reserves underground. Depending on the interests at stake and the recipients of the information (e.g., company shareholders, the Iranian government), reserve estimates were calculable and limited, but their science and measurement were uncertain. As we see below, this information was not to be revealed to the Iranian government because it would invite further questions concerning production rates, royalties, and the validity of the concession, as well as opening the door to more democratic forms of control.

AIOC certainly knew that if it refused to divulge certain kinds of geological and economic information to the Iranian government, it risked losing its oil concession. Fraser approved a strategy to construct systematic ignorance about the knowability of oil reserves by relying on arguments and the presentation of information that suggested their calculation was complicated and understanding their geology was an obstacle; oil quality was just good to know about for the industry but bad for estimates of reserves that could be made public. Fraser instructed that the full details of de Böckh's analysis and conclusions not be revealed to the Iranian government and approved the alteration and omission of certain geological data from the de Böckh report to the government. As the company saw it, the revised version ("Persia. Geological Report by Dr. H. de Bockh and Others. 1925") had a "less contentious title" than the original title ("Preliminary Report on the Principal Results of My Journey to Persia").[110] The report also omitted the introduction, the travelogue format, any geological hypotheses, and the route taken by the author-geologist. Along with a handful of other less contentious geological studies, the report was presented to the Iranian government as "the most comprehensive review in existence of the geology of the Company's concession area."[111]

As Iran's Majlis ratified the revision of its 1901 oil concession in 1933 (discussed in chapter 3), the new director of the petroleum department, Nasrullah Jehangir, and the geologist for the Iranian government, Aghababoff, requested geological data and maps.[112] These were presented during their first visit to the oilfields, but E.H. Elkington revealed privately that the company had failed to provide maps showing the underground contours of the main oil areas. Elkington reported to AIOC's board in London that "you will be glad to learn that in spite of one inquiry respecting our oil reserves we were able to avoid a reply and any further reference to such a subject."[113]

In the meantime, AIOC also made sure that no copies of the original de Böckh report left the company.[114] AIOC traced all seventy-two copies of the original report, including to whom they were distributed. The plan was to revise the originals found and to keep a few originals in "strict custody in London."[115] But Jehangir remained unconvinced, insisting that the original report be published separately from the one AIOC provided him. He demanded more maps, plans, and a copy of de Böckh's report on "Persian Geology."[116] In response, the company claimed that they had "no copy of such report." AIOC officials instructed that all maps as well as engineering and geology reports submitted to the Iranian government should be devoid of analysis, controversial commentary, and underground contour maps.[117] The details of whether oil wells were in use, mudded off, or had gone to water were also misrepresented to the Iranian government.[118]

AIOC's knowledge-brokering strategy was manifested in the circulation of papers, measurements, and scientific work. The strategy involved gathering maximum knowledge about reserves to make economic calculations, while simultaneously constructing systematic ignorance about it to the Iranian government and perhaps even to other oil companies. Cadman was fully aware that oil had a limit and that it was calculable. These calculations would factor into the company's economic interests. By the late 1940s, a policy of scaling down production at the main oilfields was put forward to extend what remained of their lifespan.[119] AIOC's production department admitted that there had been "no improvement in our methods of producing limestone reservoirs," and "very little of a fundamental nature" had been learned about mechanisms of recovery from limestone in the past fifteen years.[120] AIOC's organizational work did not exemplify the "practical application of scientific principles to local oil operations." Rather, organizational work transformed the oilfields of Iran into a laboratory of calculation and measurement, equipping managers with the tools to build a world in which the oil would thrive inside oil operations, and the politics of national control would be kept out.

Political decisions were internal to AIOC's business strategy, understood here in terms of its organizational work. They involved the element of secrecy and with it the introduction of black boxes, not only to prevent geologists and engineers from taking away the information they produced, but to block the Iranian government's access to oil information and thus the political possibility of national control and higher profits. This work of secrecy would help separate the act of measurement from the act of interpretation.[121] Technical information—reserve estimate numbers and new representations of the oil—could be made available to the Iranian government, but the details of their construction and the considerations that went into them could not be made public because they would invite the government to interpret problems unfavorably. Thus, the oil company did not want its opponent, the Iranian government, to know how it obtained a particular piece of information or made a particular decision because this would put it in a better position to ask questions.[122] AIOC built black boxes "as exclusion zones" into which they could "stuff information handling and organizational techniques, local knowledge, and technical innovations."[123] Local knowledge and organizational techniques in this box were masked by the rhetoric of scientific discovery that the company deployed in its research findings. Only by recovering the local and organizational work—the politics of unknowability and the manufacturing of technical ignorance that went into estimating oil reserves and publishing the geological properties of the oil—can we make sense of the oil corporation's strategic success.

AIOC in the Public Arena

AIOC coordinated local forms of petroleum knowledge, gathered in the "private arena" of southwest Iran, with formal and public associations of petroleum knowledge abroad. The first World Petroleum Congress was held in London in 1933 a few months after the signing of the revised 1901 oil concession in Tehran. The Council of the Institution of Petroleum Technologists initiated the meeting to provide "a platform for the discussion of standardization" among the leading geologists, engineers, and petroleum chemists, "particularly those of European countries."[124] Representatives of "all countries specially interested in the subject of petroleum" were invited, with funding support coming from the major oil companies based in the United Kingdom.

Oil companies acted individually and in secret, but also coordinated through private and public associations. The congress appointed Cadman, chairman of AIOC, and J. B. August Kessler, manager of the Royal Dutch Shell Petroleum Company, as honorary vice presidents of the meeting. Along with British government representatives, five international bodies and the delegations of twenty-eight countries were represented at the congress.

The start of the 1933 congress marked the move from standardization within the oil company to the control of that standardization through the industrial research laboratory and an International Standards Association. The congress passed two resolutions on standardization. First, a committee of twenty-eight from the International Standards Association would serve as the coordinating body with respect to all activities connected with the standardization of tests of petroleum products. Second, the Congress would be held triennially with the "hope of ultimately developing a World Empire of Petroleum Technology."[125] As Bowker has observed, this movement marked "the recognition of the independent value of industrial research as standardizing the natural world in the image of the new social world."[126]

The point here is that organizational work preceded the construction of industrial science and operated locally, indicating that the kind of science produced in the industrial setting of oil operations was inherently political and helped constitute the political agency of the oil corporation. AIOC's first move toward standardization involved the partial transformation of the oil regions into a laboratory for enacting measurements and representations of the geophysical structure of the subsoil, calculating reserve estimates, and describing the qualities of the oil. This work facilitated the control of the information flow in the oilfields of southwest Iran, or what could be made public and private depending on the economic and political costs. The second move toward standardization involved the setting up of industrial research laboratories in the United Kingdom, and eventually in Abadan, Iran, in 1935, to test and assess the qualities of Anglo-Iranian oil. The last move was the recognition of the independent value of petroleum science research presented by Cadman at the first World Petroleum Congress. As Bowker says, "rather than look for ways in which science is grafted onto industry," we must "look for ways in which science is a natural extension of industrial processes."[127] Thus, achieving the dual goal of enrolling the International Standards Association as spokesperson for coordinating standardized tests for petroleum products along with the pursuit of a "World Empire of Petroleum Technology" was a simple extension of the organizational process that first occurred in the oilfields of southwest Iran.

Cadman presented his article "Science in the Petroleum Industry" at the congress and located the origins of petroleum expertise in the history of the rise of industrial science. He highlighted the petroleum industry's debts to its "founding fathers" in science, Michael Faraday and James Dewar.[128] He then noted two landmarks in the history of petroleum: the last plenary meeting of the International Petroleum Commission held in Bucharest in 1912, followed by the founding of the Institute of Petroleum Technologies in 1913. Cadman's chronological history of the oil industry continued by noting the technical achievements that came with the invention of a higher-efficiency internal combustion engine, the diesel engine, and the multitude of petroleum products leading to the growth of new industries such as in petrochemicals.

Through a series of narrative movements, Cadman transformed the history of petroleum expertise as an industrial science into the rise of pure science and scientific investigation applied to modern industry. This transformation was afforded by the development of geophysical methods to determine the nature of underground structures. Cadman recalled early difficulties in determining the exact form of underground oil structures in abstraction. But he neglected to recall the local conditions and problems with measuring and controlling Anglo-Iranian oil and all the political forces involved, which had shaped his understanding of the technical difficulties. Two historiographic processes occurred at once: the presentation of the paper within the new discipline of petroleum science, and the destruction of the historical context.

In his historical account, Cadman neglected to include the local historical and sociotechnical activities that had enabled the birth of petroleum science from the start. Such activities included managing oil worker strikes (discussed in chapter 4), protecting the terms of the oil concession, negotiating with political groups for access to the oil, and protecting the company's involvement in international cartel arrangements to limit Middle Eastern oil production. Working through and eliminating any uncertainty attached to these controversies was indeed necessary for AIOC's representational and organizational work to occur in the first place.

The secrecy of the company's technical and scientific work in southwest Iran was made more evident as Cadman proceeded to demonstrate a "working sectional model of an oil field" for his audience (figures 2.1 and 2.2). The model demonstrated wells producing from the oil reservoir and ultimately going to gas and water, according to their position on the structure. Standing at the podium at the front of a lecture theater at the Royal Institution,[129] Cadman made the independent authority of industrial science as pure science visible through the representational work of a model and the various ways it could be applied to resolve technical obstacles. And again, this performance simultaneously excluded the local activities in which such "technical achievements" had been worked out.

Cadman then shifted from a discussion of inventions or the birth of industrial-scientific ideas to one of innovation. "Unit operation"[130] exemplified "applied science" at its best, enabling one organization to control the oilfield as a single unit with advantages "not confined to the purely commercial" but derived from the "avoidance of competition," also known as the monopoly arrangement.[131] Such an operation, Cadman explained, enabled the collection and correlation of data to a "high degree of accuracy" on "uniform and consistent lines," regarding the analysis of reservoir conditions and the calculation of reserves to ensure economical methods of maximum recovery. To ensure the authority of

Figure 2.1

Figure 2.2
Working sectional model of an oilfield, presented by John Cadman, chairman of AIOC.
Source: John Cadman, "Science in the Petroleum Industry," *World Petroleum Congress Proceedings* 1 (1933): 563–570, esp. 564–565.

industrial science, Cadman's lecture excluded the identification of all the political activities of managing information and concessionary disputes, repressing oil workers, revising geological reports, and the uncertainties attached to the calculation and measurement of reserves. In practice, these entities, tools, and arguments were necessary for making the presentation of the working model of oilfield operations possible.

In his conclusion, Cadman noted that it would be "very difficult for the petroleum industry as a whole" to follow his company's practices.[132] Nevertheless this is precisely what his participation in the public arena, as AIOC's spokesperson, demonstrated. Petroleum science was an extension of the sociotechnical processes and organizational work that occurred in the oil regions of southwest Iran, enabling the global oil industry in markets and production to be predictable and stable. Most of the presentations at the meeting consisted of articles published by AIOC's geologists, engineers, and chemists. The various byproducts of oil—natural gas, gasoline, kerosene—each possessed its own abstract origin myth that had nothing to do with the organizational work involved in transforming the world into a place in which such products could thrive and be consumed.

The World Petroleum Congress in London served as the occasion for the extension of the laboratory first built in the oilfields of southwest Iran to the oilfield model presented to an audience of political and international petroleum experts. Cadman concluded that the congress above all demonstrated the need for "encouraging the application of scientific principles and methods to every branch of industrial activity."[133] The "old empirical methods" were giving way to "ordered thought and investigation," and "there is today a clear[er] appreciation of benefits to be derived from the union of theory and practice than ever before."[134] On the ground, however, organizational work operated by controlling social and natural time and space and made industrial science possible.[135]

Distinctions between AIOC and other oil companies in public/private arenas were blurry. After all, AIOC was a cartel or joint organization with interests in the Iraq Petroleum Company (IPC), Royal Dutch Shell, and all the other major Anglo-American oil companies controlling global oil marketing and production. "Petroleum technologists" served either as experts from private companies or arrived from universities and state institutions. These experts emerged in professional associations at institutes, World Petroleum Congresses, and academic and in-house production conferences. This practice and expertise was internal to the system of exchanges of information of oil companies that acted individually but coordinated with each other through associations on pricing, production, and labor.[136] AIOC's domains of knowledge production constituted the machinery of the transnational oil corporation and its historical identity. Cadman's universalist claims[137] to petroleum knowledge as industrial science in the public arena (to the exclusion of the local organizational work of producing that knowledge) enabled the company to step outside of local constraints in southwest Iran, and to ensure that relations between politics and international markets in oil remained predictable.

Extrascientific Origins of Petroleum Knowledge

One of the most important differences between the 1901 and 1933 concession contracts concerned the rights of access and terms of exchange of technoscientific information between the oil company and the national government. In this period, concession terms and the construction of petroleum knowledge actively concealed their extrascientific origins in the need to maintain the effectiveness of British imperial and corporate power over a new source of energy. Through an investigation of the company's local activities of assembling petroleum knowledge, this chapter has illuminated the working out of a transnational oil corporation's power in southwest Iran. New technologies were responses to problems caused by earlier projects, explained in geology articles and conference proceedings as unexpected complications, failed theories and uncertain assumptions, and the need for more data. But following the activities of petroleum experts has revealed their attempts to learn from the failures of oil production and refining, and to reformulate goals from the start.[138] What this means is that technical expertise did not work, as the BP company history argues, by bringing science and technology from abroad to develop a natural resource in a faraway land of inhabitants deficient in technological knowledge. Rather, petroleum knowledge was formed in the organizational work of the battle that preceded and made possible the effect of Anglo-Iranian oil as standardized and separated technically from political questions of national control. Technological zones of measurement and qualification, and the staging and circulation of this knowledge in global terms, generated spaces of immense political and economic importance.

Following the activities of AIOC's managers, geologists, engineers, and chemists exposes the various domains of petroleum expertise through the multiplication of journals, congresses, and in-house company meetings about Anglo-Iranian oil. Each site served as an occasion for the production of technical expertise about oil as well as the erasure, silencing, and destruction of the historical circumstances in which the finding was made possible. The final result of this political project was to present industrial science as "pure science" through a so-called process of invention and innovation.

AIOC's experts devised management methods that were an integral part of the history and politics of Iran in the early twentieth century. The production of scientific knowledge and technological expertise and the ways this knowledge circulated and was monopolized are key elements for understanding the battles over which the Iranian state's demand for sovereignty over its resources emerged. Tracking the construction of technoscience around oil is also necessary for understanding the relations between the British ruling elite of AIOC and its employees, as well as the history of Khuzistan, particularly the oil regions inhabited by political groups threatening the powers of the ruling monarch and that of the company.

Standardizing and managing technical information in public-private arenas was a vital aspect of AIOC's business strategy in the midst of a global energy shift from coal to oil. It involved dealing with the tensions between local and total knowledge, between scientific and business knowledge, and between necessary secrecy and publication. Contrary to what scientific articles and technical and business histories of oil lead us to think, the universal history of oil was not superior to and separate from the local, rather it was embedded in local activities and the particularities of producing oil. AIOC needed to take the properties of Anglo-Iranian oil seriously to make decisions about further exploration and oil policy, how long to keep its concession, how much labor it needed, how much this would all cost, and how to manage the information to control the nationalist demands of the Iranian government. Rather than seeing nature and knowledge on one side and interests, stakes, politics, and power forces on the other, divides between social and technical were the outcome of the political process examined here and not the start.

3

Calculating Technologies in Crisis

[The Company shall pay] ... the said Government [Imperial Government of Persia] the sum of £20,000 sterling in cash and an additional sum of £20,000 sterling in paid-up shares of the first company founded by virtue of the forego-ing Article. It shall also pay the said Government annually a sum equal to 16 per cent of the annual net profits of any company or companies that may be formed in accordance with the said Article. (Article 10, D'Arcy Concession, 1901)

In the event of there arising between the parties to the present Concession any dispute or difference in respect of its interpretation or the rights or respon-sibilities of one or the other of the parties ... such dispute or difference shall be submitted to two arbitrators at Teheran ... and to an Umpire who shall be appointed by the arbitrators before they proceed to arbitrate. The decision of the arbitrators or ... that of the umpire, shall be final. (Article 17, D'Arcy Concession, 1901)

On November 28, 1932, the Iranian government canceled its 1901 D'Arcy Concession with the backing of the ruling monarch, Reza Shah.[1] The Iranian parliament (Majlis) immediately endorsed the decree four days later. AIOC opposed the cancellation as illegal, according to the terms of the 1901 concession, referring to Article 17, concerning the resolution of disputes using arbitration. However, canceling the oil concession was a decisive step taken by the Iranian government because normal forms of control for profits, stipulated in the royalty clause of Article 10, seemed to have broken down and were not working in its favor. AIOC would have to rebuild and restabilize those apparatuses controlling profits, pro-duction, and labor, to ensure its continued control of oil operations and to eliminate any threat to the terms of its 1901 oil concession.

Through the 1920s and 1930s, the Iranian government pushed for higher production rates, royalties, better treatment of its workers, and more technical information about its oilfields, framing these demands in increasingly nationalist terms about sovereignty. The Iranian prime minister, Abbas Mehdi Hedayat, first announced to the Majlis in 1928 that the British concessionaires might consider some revision of their 1901 concession. In the following years, an outpouring of concerns emerged in national newspapers that portrayed the 1901 concession as perpetrated by ignorant government officials of the former Qajar regime, who had been bribed to do so by foreign financiers.[2] While John Cadman, chairman of AIOC at the time, was displeased with the "orchestrated press campaign,"[3] ordinary Iranians expressed support in local newspapers for their government's decision to negotiate a revision of the 1901 oil concession.[4] By ignoring the concession terms that stipulated workers must be "Persian," AIOC was depriving Iranians of employment and thousands of foreign workers were collecting their pay to the detriment of the lawful rights of Iranian nationals.[5] Finally, the company was carrying away the oil, refining it, and selling it back to Iranians at exorbitant prices. In this period, AIOC had already built large refinery facilities in the United Kingdom and one in France. After World War I, it completed the acquisition of both shipping and tanker fleets, enabling it to transport oil and its byproducts "at its own cost and in its own time."[6] Thus, Iranian public opinion urged that in the absence of a resolution, the government cancel the 1901 D'Arcy Concession, in particular because the company was actively taking steps to set up operations outside Iran.

The increasingly nationalist demands of the Iranians conflicted with the secret monopoly arrangements set up by the international oil corporations, including AIOC, to place limits on world oil supplies to keep profits high. During the interwar period, nationalist demands in Iran, Mexico, and Venezuela clashed with the national interests of foreign (Anglo-American) governments, which either held a direct stake in or were home to the largest international oil companies. For example, as the largest shareholder in AIOC, Great Britain's investment in the oilfields was much greater than all Iranian investment in trade and industry.[7] Such connections extended further with regard to the employment of domestic labor in oil operations. As a private British company, AIOC employed more workers in the oilfields than all other Iranian-controlled industries combined.

The shah's cancellation triggered the most controversial confrontation between AIOC and the Iranian government in the interwar period. Its resolution would set a precedent for managing future conflicts, particularly Iran's oil nationalization crisis in 1951 (discussed in chapter 5). As with the question of concessionary property discussed in the first chapter, this chapter follows the oil by starting from the stipulations of the concession concerning profits and labor, then examining the ensuing disputes. I investigate its activities as a sociotechnical device and map the political agency and machinery it offers. Each section probes the technical properties of oil—that is, the organizational work and calculating technologies that went into stabilizing the controversy advantageously, or canceling and replacing the 1901 concession with a new one in 1933.

Canceling the concession did not mark the first time the Iranian government had complained to AIOC about royalty payments and other financial issues. The origins of the interwar concessionary crisis reside in earlier financial disputes that emerged after World War I regarding the definition and calculation of the phrase "16 per cent of the annual net profits" inscribed in Article 10 of the royalty provision clause. Article 3 of the 1901 concession stipulated that the Iranian government would acquire profits from "all companies" formed or working for the concession.[8] In this period, AIOC formed numerous subsidiary companies such as the British Tanker Company, national British oil refineries, and the British Petroleum Company.[9] However, the Iranian government claimed it was not receiving any profits from these companies.

In 1928, the Iranian government was demanding a percentage of the profits from these subsidiary companies and an increase in the proportion of higher-level positions occupied by Iranian workers.[10] An attempt had previously been made to work out the dispute in the Armitage-Smith Agreement of 1920, meant to establish a way to calculate 16 percent of annual net profits. But the Iranian government overlooked the agreement in 1928, and the Majlis never ratified it. Thus, the interpretation of ambiguous terms, "16 per cent of the annual net profits," in the royalty clause of the 1901 concession was at the heart of the controversy over the years. Were revenues to be based on royalty per ton, a percentage of profits, dividends on shares, or oil in kind, and would there be shareholding? How should "profits" be defined?[11]

The interwar concession crisis was the first of its kind between a foreign oil corporation and a national government. Its resolution would set a precedent in international law as two national governments, British and Iranian, and a private British company would present very different views of international law and sovereignty in arguing their respective positions. The dispute also involved the sudden breakdown of operations that were necessary for AIOC to maintain a particular oil economy in Iran and a social world constructed through the framework of the oil concession. Opening up the 1901 concession's articles concerning the formulation of profits, labor, and the legal terms for arbitrating disputes over economic resources exposes the concession's role as not just a text but a political weapon, enrolling multiple actors, procedures, and machineries. By enrolling such remarkable actors, AIOC hoped to eliminate the instability generated by this crisis and restore its concessionary authority, ensuring that relations between politics and oil markets remained intact.

Scramble for Oil: Entangling Anglo-Iranian Oil in the Anatomy of International Oil Markets

In the years preceding the concession crisis, AIOC and the Iranian government wanted to enter into (re)negotiations for a revised concession but with different interests at stake. At a meeting in Lausanne, Switzerland, in 1928, Teymourtash, the Iranian minister of court, based his government's demands for more royalties on the £1.4 million royalty payment received in 1927 and with prospects for further increases from their shareholding.[12] But Cadman argued that the royalty for 1927 was exceptional, "which they may not see again for a long time in view of over-production today and more to come."[13] He was right. The royalty plummeted to £502,080 in 1928 based on 5,357,800 tons of oil produced compared to the 4,831,800 tons in 1927.[14]

AIOC claimed its negotiating position was not only based on their interpretation of the concession but on "hard facts"—that is, their "perfect liberty to divorce from [AIOC] all outside assets, and to all oil not on f.o.b. Persia-basis" to companies that were "complete outsiders."[15] Subsidiary companies of AIOC that collected profits from operations in other parts of the world were not considered internal to AIOC operations in Iran. "F.o.b. Persia-basis" referred to oil that was purchased and shipped from the port in Abadan. Oil that left other ports and was sold to "outsider" companies, even if they were subsidiaries of AIOC or partners, such as the Burmah Oil Company, did not constitute legitimate profits from operations in Iran and thus would not factor into additional royalties paid to the Iranian government. The company claimed that it was already suffering under the disadvantages of the British government participating in distribution and shipping. Accepting the Iranian government as a shareholder would result in a series of drawbacks—that is, a precedent would be set in which the company would have to accept Argentina, Colombia, Venezuela, Albania, and other countries as shareholders. Participation granted to the Iranian government in the event of an increase in the net share of profits must not undermine the British government's preponderance (a 51 percent stake in AIOC), "a thing they would never allow." "Hard facts" referred to a set of distinctions the company reserved the right to invoke in determining which of its activities to include in oil operations and which to exclude.

The Iranian government took issue with the company's position. Royalty figures had dropped even further to £300,000 in 1932, and the government had refused to accept the royalty payment for 1931. The reasons it stated for its decision to cancel the concession were the following:[16] (1) the government had not been allowed to check company accounts; (2) no royalty had been paid by the company during World War I and the interpretation of "16 per cent of the annual net profits" was unfairly arrived at; (3) the company had refused to pay income tax to the government; (4) the cost of oil in Iran was excessive; (5) the company had engaged in reckless expenditures in other parts of the world to the disadvantage of the Iranian oil industry; and (6) the 1901 concession had been obtained through coercion.

AIOC had its own set of concerns. First, the period remaining in the sixty-year concession contract, thirty-two years, was inadequate for sound economic development and for the quantity of oil reserves available for extraction. Second, political concerns and criticism from the Iranian press and public were growing: "There is a steadily growing political feeling with respect to the national significance of its (Iran's) oil production." Such political agitation had "developed to unwise and sometimes expropriatory oil legislation (e.g. Romania, Argentina, Mexico, Colombia)." Third, the company was increasingly perceived as exhibiting monopolistic tendencies, controlling over 500,000 square miles of territory. Finally, the company's marketing policy needed to be adjusted "to a view of the future in which production from Iraq, and possibly from other centres of production, may play an increasingly important role." It was also essential to "realize the menace in terms of Persian resentment should such developments take place to the detriment of Persian production and consequently (as concession terms now stand) of Persian oil revenue."[17]

Iranian government officials were demanding more production in order to increase their share of the royalties, whereas AIOC officials sought to limit production levels in relation to world oil production to protect their own profits. These connections entangled Anglo-Iranian oil with international oil markets and the need to adjust the company's "marketing policy" in relation to future production from Iraq. The structure of foreign ownership over Middle East oil was getting worked out in these connections, which were simultaneously entangled in the

Anglo-Iranian royalty controversy. But over the course of preliminary concession revision negotiations with the Iranian government in 1928, AIOC managers did not mention these concerns upfront. In reality, while there was a major peak in oil discoveries in the 1920s relative to earlier and subsequent discovery rates, it was not in Iran. Oil supplies from the United States, the Soviet Union, Venezuela, Mexico, and Romania were flooding the world market, threatening competition.[18] Prices were collapsing worldwide on the brink of the Great Depression, and this induced a price war among the major oil companies.[19] Like the other leading oil companies, AIOC had two options: compete to win new markets for investment or set up joint ventures with other companies and divide markets among them. In response, AIOC actively pursued a parallel policy on world oil production, prices, and synthetic fuel technology. In August 1928, during the same period as his meeting with Iranian government officials in Switzerland, Cadman traveled to Achnacarry, Scotland, and agreed with the heads of the major international oil companies such as Standard Oil of New Jersey, Royal Dutch / Shell, Gulf Oil, and Standard Oil of Indiana to enter into a "Pool Association" or "As-Is" agreement.[20] This monopoly arrangement was designed to manage the glut of oil supplies by establishing a uniform selling price so that participants would not have to worry about price competition. The group agreed to control world oil production as well, enabling the companies to increase their output above volumes indicated by their market quotas, but only so long as the extra production was sold to the other pool members.

The "As-Is" agreement additionally formed part of a much larger "hydrocarbon cartel" concerning not just oil but the chemical and coal industries.[21] The goal was to control the chemical industry and block the coal industry from accessing the patented use of a hydrogenation technology known as the Bergius process that could be used to convert coal into oil and develop synthetic fuels. The agreement ensured that chemical firms were blocked from using the new technologies to make chemicals, synthetic rubbers, and fuels from the conversion of coal into synthetic oil. Such arrangements would help maintain a particular economy of oil through the construction of an artificial system of scarcity.[22] This system had been in place for a while, but with recent discoveries in Iraq, Saudi Arabia, and east Texas, companies like AIOC would have to work much harder to limit oil production. As the Iranian government began to push for higher production rates and profits during concession revision negotiations, the major oil companies were relying on a system based on the exclusive control of oil production and limits to the quantity of oil produced—only an antimarket arrangement could guarantee their profits.[23]

Tracing the wider set of connections between Iran's dispute over royalties, stipulated in Article 10, and the coordinated activities of international oil companies reveals that AIOC made significant gains from the gradual elimination of foreign competition to its own interests and British imperial interests in the Middle East, public and private. As discussed in chapter 1, this interconnected history between a public and private British entity extended back to 1914, when the British government rescued AIOC by purchasing 51 percent of the shares in the company. The British government rescue enabled AIOC to be a leading member of an emerging oil cartel, Turkish Petroleum Company (TPC), and to share in the project to hinder the development of both Mesopotamian and Iranian oil, threatening both the European and Asian markets.[24] The registered office of TPC was officially transferred to AIOC's premises in 1921. H.E. Nichols, an official in AIOC, was made managing director of the consortium until his replacement in 1929 by Cadman, now chair of AIOC, as head of a newly renamed TPC consortium, the Iraqi Petroleum Company (IPC).[25]

In a period of world oil glut, Cadman's activities at multiple negotiating tables helped build a set of connections between the royalty clause of the 1901 D'Arcy Oil Concession and the "As-Is" agreement to control global oil markets and keep profits high. But this was not the end of his journey because he also needed to control global oil markets in relation to Middle East oil production as a whole, and in light of recent discoveries in Iraq, for example.[26] With British investors having delayed oil development for nearly a decade to protect oil markets in Asia, oil was "discovered" in large quantities in the Kirkuk area in October 1927.[27] With these details in mind and within the same timeframe as his other negotiations, Cadman participated in the secret signing of the "Red Line Agreement" on July 1, 1928, in Ostend, Belgium. AIOC, Royal Dutch / Shell, and the major French and American oil companies each took a 23.75 percent stake in TPC or IPC, as it was known in the subsequent year, and participants agreed not to engage in any oil operations in the major oil-producing fields of the Middle East that were contained within the Red Line. This excluded Iran and Kuwait, which were already under British control.[28] No Middle East oil would be produced except in cooperation with the members of TPC/IPC. At the insistence of American oil interests, AIOC had relinquished its position as the dominant shareholder in TPC/IPC for 23.75 percent in the Red Line Agreement.[29] Final arrangements were concluded between the oil companies, but as with AIOC operating in Iran, no company wanted to produce oil in a period of world oil glut.[30]

AIOC's calculations during concession revision negotiations were complicated because they were bound up with demands by the Iranians for more control over profits and oil production rates that might favor Iraqi production over their own, as well as the company's obligations to control Middle East oil production in alliance with the international oil companies. As chairman of AIOC with a share in IPC, Cadman needed to coordinate his view of both AIOC and IPC's control over production as a single entity unrelated to human-made political borders. But it was precisely these borders, or the threat of national control, that might disrupt the stability of the company in limiting Middle East oil production at a time of world oil glut. Questions of national control, a world oil glut, and protecting the terms of contractual agreements concerning oil production and pricing were some of the intertwined concerns informing AIOC's negotiations for two months in Switzerland, Scotland, and Belgium in 1928.

Opening up Iran's royalty clause dispute has therefore revealed a powerful connection with many new actors and organizational forms involved such as Iraqi oil, the largest transnational oil corporations, and AIOC's entanglement with the British government in "antimarket" consortium arrangements and subsidiary companies. Anglo-Iranian oil was entangled in multiple contractual arrangements, putting AIOC in a powerful bargaining position.

For several reasons, the Iranian government's cancellation of the 1901 concession was not just about the interpretation of "16% of the annual net profits" stipulated in Article 10. First, the royalty clause worked in practice as a kind of apparatus that, when initially opened up, was entangled with numerous subsidiary companies, founded by AIOC to impose limits on the calculation of net profits, the basis for calculating the Iranian government's annual royalty. Second, as we see in more detail below, this apparatus also helped impose limits on Iran's oil production even though the government was demanding an increase. Third, placing limits on an increase in royalties was made possible through a series of larger arrangements that connected AIOC to both the British government and the largest foreign oil companies in a kind of monopoly arrangement over the production and marketing of oil from the Middle East.

In the years preceding the crisis, the articles of the concession were doing much more than a paper contract between two parties would have done. AIOC needed to operate the concession as a kind of technology of control, not just a text, by defining and associating entities to forge alliances that would remain stable through this and future controversies.[31] During concession revision negotiations, company managers neglected to mention that AIOC was working very hard to preserve the rule of the royalty clause in order to limit Iran's oil production and preserve monopoly arrangements over Middle East oil production as a whole. The concession thus served as a kind of managing technology that could be placed between AIOC's allies and all other entities that sought to define their identities otherwise, especially, as we witness below, the Iranian oil workers and a national government demanding fair treatment, more production, and a higher percentage of royalties.

Sovereign Entanglements in International Law

With early concession revision negotiations having failed, Anglo-Iranian oil mixed with nationalist politics, new organizational forms, and novel actors, creating a situation of immense uncertainty. AIOC protested the Iranian government's cancellation of the concession in 1932, claiming that it was illegal and that the Iranian government should withdraw the decree immediately. However, the Majlis ratified the cancellation decree on December 20, 1932.[32] Company officials framed the reports of damage to AIOC property in the oilfields as an "Anglophobic campaign" inspired by the Iranian government. The British Foreign Office finally responded by informing AIOC that the dispute had been transformed into an "intergovernmental question."[33] The British government advised AIOC against pursuing further negotiations until the concession was reinstated.[34]

On behalf of the British government, the British Foreign Office reached an agreement with AIOC that the company should seek a solution initially through its own efforts but simultaneously object to the restoration of the concession "in order to permit resumption of negotiations upon a satisfactory basis."[35] Negotiations had to proceed on the assumption that the 1901 concession would remain in place until change was agreed on. This would block the emergence of an "unfortunate precedent" in which other countries holding British concessions would be encouraged to "act in a similar fashion" if international law was breached.[36] In the meantime, the Foreign Office commenced examination of the juridical aspects of the dispute, concluding that the cancellation was an unlawful act, because Article 17 of the 1901 concession provided for arbitration. The dispute potentially involved a "confiscatory act of sovereignty committed against a foreign company." The latter point constituted a breach of international law and enabled the government of the injured party to make the matter the subject of a diplomatic claim. Framing the dispute in terms of international law and the cancellation as a hostile "act of sovereignty" gave the British government the right to get involved in Iran's concession dispute, on behalf of its "injured party," AIOC. Iran had played no role in the formulation of international law, but its participation in the international system translated into acceptance of the existing rules of international law, including the law of state responsibility with regard to foreign investment.

Through the 1920s, two key articles of the Covenant of the League of Nations had been formulated to enable the Council of the League of Nations to do whatever appeared possible in circumstances where disputes arose between governments (Article 11). When a dispute threatened to lead to a "rupture," the dispute could be submitted to the Council for settlement (Article 15).[37] According to the report by legal advisors to the Foreign Office, Britain would derive certain advantages from an appeal to Geneva: the action would placate League opinion and strengthen Britain's position should the Council rule in its favor. The Council's "moral pressure" would persuade Iran to accept its proposals for settlement, although force might become necessary "should such pressure founder upon the rocks of Persian nationalism." In the event of the use of force, the League decision would justify any British action against Iran and enable Britain to solicit help from other members. Britain preferred a pacific solution, however, as many in the Foreign Office feared that proceedings at the League would help Iran gain the sympathy of several South American countries that "resented concessions held by great powers in their territories." In fact, much of the legal doctrine concerning state responsibility under international law with regard to foreign investment was generated by disputes between American and European investors, on one side, and Latin American states, on the other, insisting that investment be ruled entirely by local, national law.[38]

With the dispute now framed as "intergovernmental" and concerning the question of sovereignty in relation to a foreign entity, an oil company, the British government sent a note of protest to the Iranian government declaring that it would not hesitate to adopt necessary measures to protect British interests and that no damage should be inflicted on the company's property.[39] The British government, in its view, had no choice but to report the Iranian government's cancellation to the Permanent Court of International Justice (PCIJ) at The Hague.[40] The PCIJ was created in 1920–1922 as one of the four principal organs of the League of Nations.[41] In its replies, the Iranian government denied responsibility for any damage to AIOC interests.

The Iranian government sought to frame the relation between international law and state sovereignty advantageously by keeping the British government out. It claimed that the intervention of the British government was hindering the settlement of the concession dispute and that the Permanent Court possessed no jurisdiction in this question, referring to Article 36.[42] The Iranian government also threatened to notify the Council of unnecessary pressure exerted by the British government. At this moment, the British government made the decision to prevent Iran from "forestalling a British appeal to Geneva," because it was "vital" for Britain to appear before the League as the "plaintiff rather than as the defendant" to allow for a "full consideration of the case."[43]

On December 19, 1932, the dispute was formally transformed and framed in terms of international law when the British appeal to the Council was made under Article 15, concerning a "rupture" in relations between two governments, defined as a situation that caused Britain to contemplate movement of either ships or troops.[44] The British government was presenting its case before the Council, not in its capacity as "shareholders in the company," but as the "Government of a State that has thought it necessary to take up the case of one of its nationals whose interests have been injured by acts contrary to international law committed by another State."[45] The British government urged the Council to take the appropriate steps to "ensure the maintenance of the *status quo* and to prevent the interests of the company from being prejudiced" while proceedings were pending before it. The British government made its legal case in the "British Memorandum." The document reviewed past conflicts over the interpretation of "16% of the annual net profits," which the British government argued were understandable given the "ambiguity in the text" of the 1901 concession and the "expanding complexity of financial and accounting arrangements of the company."[46] The Iranian government's refusal to retract its cancellation decree justified the British government's right to intervene on an international stage to "protect the rights of a British national when injured by acts contrary to international law, committed by another State, and ensure … respect for rules of international law."[47] AIOC was not simply a private British oil company operating abroad, it was a foreign national whose interests had been violated by an act of sovereignty.

In a September 19, 1932, resolution to the League, however, the Iranian government reserved the right to require that proceedings at the Permanent Court be suspended in respect of any dispute submitted to the Council of the League of Nations.[48] Article 36 of the Statute of the Court obligated members of the League of Nations to settle disputes concerning treaties, conventions, and questions of international law within the jurisdiction of the Court. As a signatory member to the League, however, Iran had passed a resolution in 1932 reserving the right to require that proceedings in the Court be suspended "in respect of any dispute" submitted to the League.[49] This was also the case in 1951, when the British government submitted the oil nationalization dispute to the International Court of Justice, which replaced the Permanent Court after 1946 (see chapters 5 and 6).

In its memorandum to the Council, the Iranian government reviewed past disputes over the calculation and payment of royalties. It argued that the British concessionaire was continually extending its activities outside Iran while confining itself to "restricted exploitation" within the country. This was "a line of action unacceptable" because the conceding party received a sum varying according to the extent of exploitation.[50] Further, AIOC was infringing on the terms of Article 12 indirectly by constructing refineries and other works outside Iran and infringing directly by employing Indian workers in Iran, despite government protests.[51]

Transforming the dispute into national legal terms, the Iranian government argued that AIOC had consistently refused an arbitrator in concession disputes. Therefore, British intervention removed from the jurisdiction of municipal courts a dispute that "naturally belonged to them" and constituted an infringement of "Persia's jurisdictional independence."[52] Respect for jurisdictional independence was the reasoning behind the Iranian government's appeal to the League (Appendix VI). Article 15 of the Covenant required a "dispute likely to lead to rupture," but the Iranian government argued that such a dispute could only exist when a government has "by means of diplomatic protection taken up the cause of its nationals." This presupposed "a violation of general or conventional international law and the previous exhaustion of municipal remedies." Hence the procedure of cancellation based on nonfulfillment of a contract was not a violation of international law, and if it was unfounded, "diplomatic protection could only come after municipal courts had been given an opportunity of dealing with the matter." The remedies of Iranian municipal law had not been exhausted by the British government as the prerequisite to diplomatic intervention. For this reason, the British government did not have the right to make a diplomatic claim in this case.[53]

An entire legal order was being invoked here, for the first time, with the resort to the League. This would lay the groundwork for the nationalization crisis case in 1951—a landmark in international law. Transforming the dispute in terms of international law by bringing it to the League of Nations and the Permanent Court was, for the British government and AIOC, about connecting the petroleum order to the machinery of international law.[54] The world of Anglo-Iranian oil was constituted "inside" the infrastructure of oil operations, its measurements, regulations, contracts, and now international law, or so company lawyers and British government officials claimed. Legal knowledge about Anglo-Iranian oil derived from the transformation, translation, and movement of a material object, Anglo-Iranian oil, from the oilfields of Iran, into multiple forms of representation that included the text of Articles 10 and 17 on royalties and arbitration, respectively, to court proceedings at The Hague. This chain of meetings, papers, legal arguments, and rulings constituted the machinery of the oil industry. The enterprise maintained the traceability of legal knowledge about Anglo-Iranian oil "inside" oil operations, moving from the center at the League and the Permanent Court to the periphery, or its original context in the oilfields where AIOC had first set up its laboratory (discussed in chapter 2) to exclude the local.

This analysis has not sought to provide a wider "legal context" to the Anglo-Iranian concession dispute. Rather, international law occurred as a set of connections that, at first glance, provided a guarantee to AIOC and the British government of a certain traceability and certainty of control over the oil, from global arenas to the local oil regions and back. The site of the League, an international institution in Geneva, and the various proceedings, statements, and decisions about the terms of the oil concession and its relation to international law and the question of sovereignty enabled AIOC to temporarily step outside local constraints of the oilfields and national institutions in Tehran. The preliminary groundwork was now laid for the return of Anglo-Iranian oil to international law in a more dramatic confrontation marked by the Iranian government's decision to nationalize its oil industry. The Anglo-Iranian oil dispute served as one of the critical occasions on which new legal doctrines concerning domestic jurisdiction, state responsibility regarding the protection of foreign investment, and international arbitration were devised for structuring relations between the contracts of an expanded community of newly sovereign states and foreign corporations in the economic exploitation of their natural resources.[55]

Following the construction of international law around Anglo-Iranian oil highlights how the British government and company officials, as well as Iranian delegates and their representatives at the Council, helped build an extension to the energy network, in the form of legal texts within which the so-called facts about oil would survive. Thus, the British government was not just using law as an instrument of policy, as Beck has argued.[56] Rather, the law worked by arranging the social world— that is, attaching to people, events, oil operations, and papers as they shaped decisions about building the oil industry and the rights of a state to claim political sovereignty over a natural resource.[57] International law made domination and action from a distance possible. It did so by reinforcing the faithfulness of the actors shaping the crisis to achieve a resolution that granted AIOC and the British government the authority to speak on behalf of the many silent actors (e.g., oil workers) in the oilfields.[58]

Different actors were attaching themselves to the law in advantageous ways to eliminate rivals and secure control of Iran's oil. At the Third Meeting in January 1933, Sir John Simon, British Foreign Secretary, made a statement in response to the "Persia Memorandum" reiterating the British government's position on the suitability of the case for jurisdiction of international law.[59] He again argued that the cancellation had adversely affected the company and could develop into a "still more serious situation leading to a rupture between the two countries," both of which desired to maintain friendly relations.[60] The "real reason" for the cancellation decree, Simon argued, was that the Iranian government hoped by this means to "dictate to the company a new concession while the company is in the adverse and unfair situation of having its concession cancelled."[61] This was an "indefensible use by the Persian government of its sovereign power.[62] International law would have to reconfigure itself according to a new set of standards for managing the uncertainty attached to a new social reality in which (postcolonial) states claimed domestic jurisdiction and sovereign rights to take over foreign investment entities, such as an oil industry, operating within their borders.

The concession dispute was not just an econotechnical issue. Rather, it was bound up with political questions of sovereignty and social questions of labor that transformed the terms of international law. Having challenged the economic arguments about royalty payments and other dues stated in the Persian Memorandum, Simon shifted to the question of labor.[63] In response to Iran's allegation that AIOC had failed to observe a stipulation in the concession concerning "the employment of work-men who are Persian subjects," Simon responded with a statement about "the facts":

I think the company has some right to feel a little aggrieved that such a sugges-tion should be made against it. … It is not a ground on which anybody has ever sought to cancel a concession, and has nothing to do with the case. The actual facts are these. The skilled labour employed by the company in Persia includes 118 categories of employees, managers, engineers, etc. When the company began operations, few artisans in any of these categories were available among the Persian population, and during the Great War rapid development led to the importing from India of skilled labour not available in Persia. Since then the number of non-Persian employees has been consistently diminished, and 90% of the company's non-European employees in Persia are now Persian subjects. In order to fit Persian subjects for employment, the company has spent over £100,000 on education in recent years. Apart from artisan-training centers, it has built schools in the Persian province of Khuzistan, where none existed, and for six years it has provided free university education in England for two Per-sian students annually. Further, the Persian Government has benefited directly and indirectly by the expenditure … by the company. On the medical services alone in South Persia, the company has spent over £550,000 since 1924, tens of thousands of non-employees receive free medical treatment every year … but I am entitled to say on behalf of my fellow countrymen, that the reproach that they have not in this matter shown a sufficient regard for the very proper needs of the population of Persia in the neighbourhood concerned has not the smallest scrap of foundation.[64]

The British government was arguing, on behalf of AIOC, that the com-pany had been operating as a benevolent, civilizing social force in the oil regions of Khuzistan Province and was making full use of its Iranian labor. As Simon remarked, this was the first time anybody had made use of a social argument concerning the adequate employment and treat-ment of labor, as opposed to economic arguments, as a basis on which to cancel a concession. But he quickly brushed the labor point aside as ultimately having "nothing to do with the case." The Anglo-Iranian oil dispute introduced social arguments about the treatment of labor that embodied a new social reality of postcolonial nation-states in which claims to bringing Western civilization to a technically (and racially) defi-cient population, as a justification for the operations of foreign firms, were no longer acceptable. International law would have to be reformed and redeemed from its colonial past by developing a novel language of paternalistic development for managing disputes between foreign corpo-rations and newly sovereign, non-Western governments.

The colonial legacy of international law provided AIOC with the tools with which to discredit the agency of oil workers, who were at the same time credited with legitimacy to act as arguments presented by the Iranian government. International legal arguments put forward by both sides provided the tools with which to determine who could act and on what terms, such as Iranian oil workers, whose presence threatened to destabilize and redefine the structure of relations between foreign corporations and non-Western governments concerning the exploitation of natural resources.

The Arabian-American Oil Company, Aramco, as Robert Vitalis has shown, developed a similar set of moral and civilizing arguments in subsequent years to justify its operations in Saudi Arabia. Likewise, large firms in fields such as mining and oil deployed "paternalism" in their efforts to defeat union building in the American Southwest and the oilfields.[65] Even "while racism's ethics governed the hierarchical distribution of benefits," Vitalis explains, "a firm's beneficence ostensibly demonstrated how so-called outsiders and third parties, including the local state," in this instance Iran, "were second best (or worse) options for securing a decent life." As with AIOC, firms brought this model with them when they began producing oil beyond US borders. Thus, Simon's rebuttal was intended to emphasize not only AIOC's beneficence toward the people of "South Persia," building schools and training centers "where none existed," but also how the Iranian government represented a worse option for securing a decent life for its population.

In response to Simon's rebuttal, Iran's minister of finance, Davar, argued that the contract had been made between the Iranian government and a private company, not between states.[66] Furthermore, despite the Iranian government's protests, the company still employed thousands of foreigners who were not skilled workmen but "mere laborers."[67] AIOC could go ahead and "withdraw its benefactions" in supporting two Iranians annually for study in England because the government was itself supporting the study of hundreds of Iranians abroad, particularly in France.[68] Such symbols of beneficence were mere tokens and not practices intended to meet the criteria of the concession.

The Council never expressed an opinion on the legality of the cancellation, the role of jurisdiction for the case itself (whether constituting diplomatic protection or a question for Iranian municipal law), or the legality of claims made by either side concerning the role of the company in its treatment and employment of Iranian labor and its social impact on Khuzistan as a whole. At the Sixth Meeting of the League in February 1933, M. Benes, Foreign Secretary of State for Czechoslovakia, appointed by the Council as rapporteur to work with both parties to reach a "friendly and equitable solution,"[69] reported that the court proceeding involved "important questions of law." He also stated that a provisional agreement had been reached in which both parties agreed to the suspension of further proceedings before the Council until May 1933. The two parties agreed that the legal standpoint of each as stated before the Council remained "entirely reserved."[70] In private negotiations for a final settlement to the conflict, AIOC concluded that operations must continue as they did before the cancellation decree of November 27, 1932.[71] The British government announced its satisfaction with the Council's provision and that a settlement could now be negotiated on "equal terms."[72] On April 29, 1933, the Iranian delegate to the League, Anoushiravan Sepahbody, announced that a new concession had been signed and the Majlis announced its ratification to the Council in June.[73]

In international legal terms, the 1932–1933 concession dispute was the first case of its kind, laying the groundwork for a future milestone case in international law, Iran's oil nationalization crisis in 1951–1953. The relationship between international law and sovereignty was transformed by this first encounter with Anglo-Iranian oil as it triggered the reformulation of a number of doctrines for organizing relations between the transnational corporation and host governments, including "the doctrine of diplomatic protection and state responsibility for injury to aliens."[74] Whereas the British government invoked the law to secure its authority and legitimacy as spokesperson for AIOC's operations, the Iranian government hoped to transform it. In practice, international law worked as a series of memorandums, meetings, decrees, arbiters, and legal texts that extended the energy network from southwest Iran to Geneva, The Hague, and London, helping to elaborate positions of strength and weakness. The British government, on behalf of AIOC, wanted to utilize the law to help bolster its position of strength, claiming that Iran had committed a "confiscatory act of sovereignty," whereas the Iranian government rejected this association in favor of national and municipal laws to bolster its own position with regard to sovereignty over a natural resource. Many of the controversies regarding the impact of "new states" (formed out of the Mandate System of the League of Nations) on the rules of international law emerged in disputes generated by the "doctrine of state responsibility as related to the protection of foreign investment," as in the case of AIOC in Iran.[75] Thus, the law was not simply a way of extending existing power relations and of maintaining the Iranian government in a position of weakness. It was transformed by its encounter with oil, generating new doctrines and linkages that attached international law to questions of national sovereignty over a natural resource and the right of a government to intervene on behalf of a private, international oil corporation framed as its national.

The proceedings at the League occupied an important place in British thinking in that the League's sanctioning of an agreement according to universal standards of international law would make it more difficult for Iran to rescind it.[76] Private discussions to resolve the dispute were not only about the economic question of royalties but also included technical questions of oil production and social questions about the replacement of British labor with Iranian labor at higher skill levels. The Council made no formal decision except to encourage that a resolution be reached strictly between the company and the Iranian government. Transporting the dispute to the League was not just an instrument of British policy. It was a necessary step for the (re)stabilization of the British-controlled energy network in the midst of uncertainty. As such, the groundwork was now laid for managing future controversies involving the nationalization of the British-controlled oil industry.[77] The uncertainty resided in the dispute's susceptibility to a technolegal rearrangement of the so-called real facts about oil by diverse actors, including Iranian government officials, oil workers, legal doctrines, and the oil, connecting into other logics and conflicts such as national sovereignty and revolutionary movements.

Calculating Iran's Oil Royalties

AIOC needed to stabilize the authority of its 1901 concession in Iran by devising various legal formulations and calculating technologies for resolving the concession dispute advantageously. The concession crisis unfolded during the course of two years, commencing officially with the Iranian government's cancellation decree of November 27, 1932, followed by the British government's appeal to the Council of the League of Nations on December 19, 1932. With the suspension of international legal proceedings and the resumption of negotiations in February 1933, company management were called on to prepare the "necessary data" for the review of the London board of directors, which would be involved in negotiating the terms of the revised concession.[78] Company opinion noted the instability of any scenario. For example, all sorts of objections might be raised as to a definition of the "reasonable cost of the production, refining, transporting, and marketing" and might "involve us in admitting a right for the Government to control such costs and may lead to all kinds of difficulties for the future."[79]

The company's first step in avoiding the uncertainty attached to any future compromise with the Iranian government was to define a new set of parameters for the calculation of oil production rates and profits. AIOC officials proposed a revised basis for calculating royalty payments that would satisfy a series of conditions: (1) a substantial minimum annual payment to the Iranian government; (2) an annual payment that under normal working conditions would prevent violent conditions; (3) a scale of payment in a reasonable ratio to the company's total profits; (4) a scale of payment to remove the incentive for the Iranian government to press for increasing tonnage; and (5) a method of payment that carried no shareholding, but gave the Iranian government some share in the profits of the company without depriving it of a minimum in lean years or in the event of a declining Iranian tonnage.

AIOC's "accountancy experts" formulated a set of alternatives for calculating royalties with the goal that "the field of suspicion or dispute would be confined to narrow limits and, consequently, the cause for any desire to examine accounts would disappear."[80] The Iranian government, for its part, was demanding minimum annual production levels of six million tons of oil.[81] From AIOC's standpoint, such production levels were not possible in a period of global economic depression, but also in relation to the new monopoly arrangements set up between the international oil companies. To make matters worse, 90 percent of Iranian oil production was refined within Iran.

AIOC formulated that all royalty schemes should be based on a set of variables, namely, a combination of a tonnage royalty and a royalty tied to the profits of the company, derived from all sources (e.g., company and subsidiary company operations in Iran and abroad).[82] The Iranian government's demand for higher production levels led company accountants to reason that a tonnage royalty would "necessarily increase with increasing production and provide the urge to the Iranian government to press for increasing tonnage." A "means" therefore had been found to apply the "Profit Royalty" in such a way that it "exerts a counteracting influence on the increasing tonnage Royalty." The profit royalty on any "fixed profit decreases as the tonnage increases." Further, the profit royalty, though dependent on profits from all sources, was purposely linked to Iran's oil production on the assumption that it would be "sound in principle" to express the total royalty payable to the Iranian government in terms of its production.

Working out the question of royalties as stipulated in the concession was about securing a particular arrangement of information and politics. The Iranian government had been demanding access to accounts for many years. AIOC officials feared that the introduction of "profits" in any shape might risk an examination of the accounts and thus jeopardize the British concessionary control of oil. A peculiar ordering of information was constructed to block the threat of instability seen in the Iranian government's demand to view accounts and receive a larger share of royalties, whether calculated on the basis of increased tonnage, revenues, or a combination of the two.

AIOC accountants built a series of proportions and limiting variables into the royalty formula to manage the Iranian government's demands for more oil production and eliminate attempts by the Iranian government to deconstruct the formula in order to strengthen its bargaining power and its labor according to an alternative formulation of the energy system. The company calculated that by setting an annual payment (which within the most likely combinations of tonnage and profit varied slightly) and by guaranteeing a minimum royalty that was less than 25 percent of company profits, the incentive for the Iranian government to push for increasing tonnage would be "very largely removed."[83]

Company experts elaborated on their formulation: "It will be seen for instance, it is as much to the advantage of the Persian Government to see the Company remain on a steady tonnage and strive for increased profits as it is for them to keep profits steady and increase tonnage." For example, the first "scheme," as the company called it, proposed an arrangement in which the royalty payable with profits of £5 million would be: £1.15 million on three million tons production; £1.1 million on four million tons production; £1.15 million on five million tons production; and £1.25 on six million tons production.[84] Thus, according to the scheme, increased production would not necessarily lead to an increase in royalties payable to the Iranian government. AIOC's draft clause for calculating royalties proposed the development of a formula for each of two possible "schemes":[85]

$$K = RP + S\left(20A\!\!\big/_{P} - L\right) \text{ in Scheme 1}$$

$$K = RP + {}^{(20AS)}\!\!\big/_{P} \text{ in Scheme 2}$$

where R = tonnage royalty in pounds sterling ($£$) per ton, P = production in tons, A = gross profit ($£$) of the company, S = amount ($£$) paid as "Profit Royalty" for every Sh.1/- total gross profit (all sources) per ton of production from Iran, L = limit in shillings per ton of profit below which no "Profit Royalty" is payable, and K = total payment (tonnage royalty plus profit royalty).

In both schemes, whereas tonnage royalty (RP) increased with production, the profit royalty $\frac{(20AS)}{P-SL}$ or $\frac{(20AS)}{P}$ decreased with production. The purpose of the "limiting factor" L in scheme 1 was to "cause the profit payment to cut out entirely if production increases out of proportion to profit." Most importantly, in each case, if total payment was to be uninfluenced by production, K must remain constant for all production amounts, meaning that the rate of change of K with production P must be zero. Thus, $\frac{dk}{dp} = 0 = \frac{R-(20AS)}{P^2}$ or $R = \frac{(20/AS)}{P^2}$—the condition to be satisfied if total payment was to remain constant irrespective of production. R and S would vary with production and profit according to the two formulas. These values were tabulated over a wide range of production and profit levels. For any condition of production and profit selected, the values of R and S tabulated against this produced a total royalty payment equal to the desired 20 percent of the profits (see figure 3.1).

Each proposed scheme corresponded to a particular arrangement, with the ultimate aim of narrowing the field of dispute by removing the possibility that the government would demand access to company accounts and an increase in production and profits. By excluding what AIOC considered disproportionately high profits (deriving from minimum and maximum levels of oil production) as well as coordinating the profit and production variables in a dependent relationship, AIOC officials designed formulas to help stabilize their hypothetical world. It needed to be stressed, however, that the scale of royalty had been "arbitrarily fixed" in the illustrations, which could be "adjusted to suit [the] monetary consideration it is intended to concede," and "production is, in any case, within the control of the Company."[86]

The calculations built into formulas implied an arrangement whereby numbers played only a secondary role. The variables mobilized by each formula were different but similar in that certain critical variables, particularly the volume of production, were consistently the same. The volume-of-production variable was present on multiple occasions

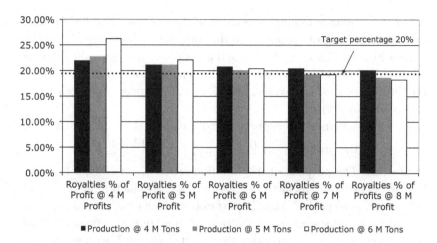

Figure 3.1
Source: Graph of Proposed AIOC Schemes for the Calculation of Royalty Rates to the
Iranian Government at Varying Levels of Profit. Based on figures from "Scheme 1. Draft
Clause," "Scheme 2. Draft Clause," and "Investigation of the Problem in General Terms,"
n.d., 70223, BP Archive, Coventry, UK.

regarding profits, the company's recruitment policy, and production rates
(discussed in chapter 6), and could be seen as determined by the actors
themselves among a hierarchy of variables. Variables built into formulas were also strategic—volume of production became a variable that
allowed actors on the British side to control outcomes and behaviors.
Political possibilities were opened up by formulas in order to organize
certain predictable relations between many critical variables and one
strategic variable, volume of production. This enabled AIOC to coordinate simultaneously with other firms in monopoly arrangements and
to organize relations. Thus, the importance of the strategic character of
formulas turned around one strategic variable and the subordination of
other parameters to this one, making the market in oil manageable and
predictable.

With the help of formulas, AIOC managers and accountants worked
to manufacture a kind of ignorance about the nature of royalty calculations and to deny access to their accounts. Such concerns were already
entangled with other kinds of calculation that rendered the various sites
and actors in the energy network highly unstable. For both sides, the
success or failure of any future compromise would depend on securing
control of Iran's oil production. Cadman, chairman of AIOC at the time,
refused to participate in any profit arrangement defined strictly in terms

of the production variable because it would translate into a significant loss in company profits that also depended on coordinating with other oil companies in limiting the production of Middle East oil as a whole. He instead proposed a minimum annual payment of £750,000.[87] In the end, the ruling Iranian monarch, Reza Pahlavi, accepted most of AIOC's proposals and signed the revised oil concession in 1933. Why? Brought to power with the funds and backing of the British government, first as minister of war under the Qajar regime in 1921 and then as shah (king) of Iran in 1925, the self-proclaimed Pahlavi monarch was in a position of weakness. Without the necessary funds to build a modern state, the shah risked public criticism and larger political threats coming from the Soviet Union.

Article 10 of the 1933 concession agreement stated that royalties were to be calculated on the basis of physical volumes of oil and the financial distribution that the company made to its shareholders, not on profits alone.[88] Starting in January 1933, the royalty was set at 4 shillings per ton of oil consumed in Iran or exported, plus a sum equal to 20 percent of the dividends paid to the company's ordinary shareholders in excess of £671,250. These terms guaranteed an annual royalty to the Iranian government of at least £750,000, just as Cadman had proposed to the shah. This reformulation was applied retroactively to recalculate royalties for the years 1931 and 1932.

Central to their strategy of manipulating scales and formulas for establishing the basis of royalty payments was AIOC's control of the oil production variable. The strategy also involved putting limits on the working possibilities in the calculation of royalties (e.g., controlling certain strategic variables such as minimum and maximum tonnage and royalty payments). This, company officials hoped, would build boundaries, blocking access to accounting information and producing the effect of technoeconomic concerns as separate from political questions of sovereignty and national control.

Reza Shah accepted the terms of the revised concession's royalty clause into which was built the peculiar formulation outlined above. On numerous occasions, Iranian government officials had attempted to deconstruct the formula by demanding more production, but they were bargaining from a position of weakness and uncertainty. Semiautonomous groups throughout the country threatened the central government authority, as did striking oil workers and national public opinion angered by the British exploitative presence.[89] The temporary resolution of the conflict in the format of a revised concession formula was the product of this battle among actors in differing positions of strength and access to calculative equipment. The differences among the strategies of the different actors had nothing to do with their nature and everything to do with the skills, tools, and means available within the network to mobilize other actors and allies and exert power over them.[90]

Thus, the oil disputes traced here do not just concern the ties between the oil company and the government. Rather, they occurred among a series of nonhuman and human actors in the energy network. The Iranian government accepted the formula's terms because its powers were not stable. The state lacked the necessary tools to control the pipelines, the wells, the refinery, oil workers, and semiautonomous groups, and to determine the political outcome at least for a while. Later in 1951, the government's dramatic move to pass the nationalization law would again reopen the formula to scrutiny but within the larger context of reassembling the state in the framework of reformist nationalism.

In practice, the company was controlling elements of the royalty formulas and concessionary articles to establish political control over various forces and to eliminate controversy. All these different actors—Iranian government officials, AIOC managers, local political groups, the oil—were connected to the formula and needed to be enrolled, interdefined, and stabilized in a particular way to make the operation of the formula possible.[91] As an element internal to the concession, the scheme of formulas for calculating royalties was operating as a kind of device with defining terms and variables that needed to be controlled, but that would always have multiple logics conflicting and tying into other logics.

The royalty formula played a central role in shaping the outcome of the most important interwar dispute between an oil company and a sovereign government. To say that the history of Iran's concession dispute might be understood better as a history of the working possibilities of the royalty formula is impossible in the framework of a conventional political or social history of the crisis. These conventional analyses overlook the technicality of the royalty dispute and situate it within the larger context of the Iranian monarch's campaign against foreign influence.[92] As Thomas Hughes has shown for the building of the electrical industry, Edison and his colleagues invoked two formulas, Ohm's and Joule's laws, to make electricity consumption competitive with gas. The technical specificity of two formulas with their variables and interdefined relations helped Edison work the problem out by identifying the need for a high-resistance and durable filament.[93] Likewise, the construction of the royalty formula was shaping the building of a transnational oil company by equipping it with a peculiar kind of agency that could determine who could act and on what terms. This would have political consequences for the powers of the Iranian state and its oil workers.

(In)Calculability of the Oil Worker

(II) ... *the Company* shall recruit its artisans as well as its technical and commercial staff from among Persian nationals to the extent that it shall find in Persia persons who possess the requisite competence and experience. It is likewise understood that unskilled staff shall be composed exclusively of Persian nationals. (III) The parties declare themselves in agreement to study and prepare a general plan of yearly and progressive reduction of the non-Persian employees with a view to replacing them in the shortest possible time and progressively by Persian nationals.[94]

Another formula appeared during the course of the 1932–1933 concessionary dispute, one even more interesting because it was unexpected and unusual. It was a proposal for measuring and controlling the degree of Persianization of AIOC's employees as stipulated in Article 16 of the 1933 oil concession quoted above. Besides the question of royalties and rates of oil production, Persianization or the gradual replacement of British workers (particularly managers and technicians) with Iranian ones at higher skill levels was the major point of dispute during the concession crisis of the 1930s and continued to be contentious through the 1940s.[95] AIOC's attempts at stabilizing its labor regime with the use of a formula for managing the rate of Persianization were entangled with the reassembly of the Iranian state. Connected to this issue was increasing pressure from oil workers and public opinion for the government to nationalize its British-controlled oil industry. The company's Iranian oil workers had first disrupted oil operations by going on strike in 1929 to demand better treatment, including higher wages and living standards. In 1931, they expressed their support for the government's cancellation and revision of the concession, particularly as it addressed the question of Persianization.[96]

Article 16 of the 1933 revised concession concerning "Personnel in Persia" redefined the working possibilities and limits of the oil worker in technical terms that appeared to overlook any racial differences. In practice, technical terms worked by delineating the unsuitability of local labor at higher skill levels and managerial positions precisely in terms of race. Prior to the signing of the final agreement in 1933, a series of drafts of

Article 16 were exchanged between the company and the Iranian government in which the latter demanded the following formulation: Manual workers, laborers, artisans, foremen, overseers, mechanics, typewriting clerks, accountants, and all other junior employees "must be of Persian nationality." AIOC must undertake to "replace all non-Persian" employees in these categories before January 1, 1934.[97] Regarding categories of employees that included engineers and other members of "higher technical personnel," the company must implement a plan for the "progressive annual reduction of non-Persian employees in order to substitute … Persians within a period of 10 years." After this period and according to the proposed formula, the nationality of only one-fifth of the employees in the said categories would be "non-Persian."

AIOC dealt with the controversy by transforming a labor issue into an econotechnical problem. The company proposed that it would recruit artisans and its technical and commercial staff in Iran from among Iranian subjects "to the extent that it shall find Persian subjects who possess the requisite competence and experience."[98] All unskilled staff must be of Iranian nationality. Yet, both parties also agreed that it was in their "mutual interest" to maintain the highest possible "measure of technical efficiency and of economic conduct" in oil operations. Therefore, the calculability of the Iranian oil worker was possible where economic or technical arguments about efficiency needed to be made.

No deadline for Persianization or execution of the General Plan was actually stipulated in the revised concession, but Iranian negotiators had proposed a deadline for 1934. In a private discussion on the company's General Plan, "not for communication to Government," the company hoped to "build up a better and more intelligent type of skilled [Iranian] workman and one who can be trusted with more responsibility, particularly in plant operations."[99] Such a plan, it was argued, would enable the company to "reduce ultimately the British and Indian supervisory personnel." The scheme was "not therefore primarily one of Persianization but rather a scheme for the more economical operation of plant and processes by the provision of a better type of workman with the essential initial training," and for the "provision of a more economical shift labor in the higher grades."

The direct participation of Iranian elites in plant operations and managerial positions might be possible, but this would have to be managed. At all levels, except the lowest grades, AIOC's training would require instruction in the English language.[100] Requiring spoken English in all skilled positions in the worker hierarchy would help secure an island of economic enclaves separate from so-called society on the outside. The strategy worked by excluding and creating proximities and alliances, not just among the elite but between workers and with their managers. In response, Iranian apprentices, trained to fill graded posts, went on strike on multiple occasions, demanding that courses be taught in Farsi instead of English.[101] Thus, AIOC used language proficiency as a tool for creating categories of difference among workers. It was a strategy for managing and integrating the Iranian workforce, in particular by relying on technologies of difference built into the use of English and Farsi at particular places and levels in the worker hierarchy.

The technical capacity of the oil worker in terms of achieving the "requisite competence and experience" was simultaneously calculable and incalculable. In another private report, the company revealed its plan to control the replacement and reduction process by proposing to offer positions as artisans, technical specialists, and commercial staff to Iranian nationals "possessing the requisite competence and experience—this will, of course, never happen."[102] The General Plan would be "reviewed" again in 1943. The company was making its arguments in technical terms to guarantee that Persianization would never happen, claiming that an Iranian oil worker capable of replacing a British or Indian worker of a higher grade was impossible because of the economic costs to the company. To build an oil labor regime in terms of racial-technical difference, Elkington, an AIOC operations manager, advised that the company make use of its calculative equipment, a formula

implying [that] a definite accomplishment [should] be accepted, indicating lines upon which the Company will do its best to fulfill its obligations. ... In other words, we give practical evidence of our good faith ... and at the same time arrange the affair so that we can turn round at any time and in justice say that the results have not been up to expectations for such and such reasons—reserving to ourselves the right to retain this prerogative always. The formula therefore to which I refer should take the shape of this implication. ... Production plan alone is far too ephemeral, but a production plan plus a formula has the semblance of some substance. Our reluctance to imply that we can make a definite numerical reduction in our foreign employees each year ... is due to uncertainties brought about by fluctuations in the programme of work and throughput.[103]

Just as the mechanism of the royalty formula provided a means of managing Iran's interwar concession crisis, the company's strategy for dealing with the labor issue and avoiding the increasing threat of national control of oil was to build another formula. Certain variables would be connected to imply a reduction in the number of foreigners employed in relation to annual expenditure and production. The "great advantage" was that in reality, "any reduction at all can as a rule only take place" if the competent Iranians were available.[104]

Elkington explained his ideas regarding the mechanism of the formula to create the effect of a reduction in foreign employees:

Is it not possible then to relate the number of foreigners whom we are required to employ to the annual expenditure, i.e. the budget, and to the annual production. Thus we can say that a total Capital and Revenue expenditure of £1 million requires the employment of 400 foreigners today, and using this as a yardstick create a formula which would imply a reduction in the number of foreigners employed per million pounds—also we can relate the numbers employed to production and work out a formula which would be in conformity with past results, that is to say the formula applied to the past would give us the position we are in today. In other words, can Mylles [the accountant] work out a relationship between the number of foreign employees, annual expenditure, and production over the past number of years. If so, instead of implying a reduction in accordance with figures above, which are quite arbitrary and make no allowance for flux in program, throughput, or expenditure, we might be able to imply a basic reduction on an agreed datum level which would be influenced arithmetically by the variations in the expenditure and production.[105]

Elkington continued,

For example, let the expenditure be £5 million per annum and the number of foreign employees 2000. Then under present circumstances an expenditure of £1 million implies the employment of 400 foreigners. Let us then take expenditure of £1 million as the datum level and say we reduce the number of foreigners relevant to this at rate of 20 per annum. Then after one year, number required per million pounds is 380 and for £5 million, 1900, but should expenditure rise to £6 million after 2 years then the number of foreigners permissible under the formula would be 2160 or [400-(20x2)] x 6. This is one curve and your next curve should be production in millions of tons relative to foreign employees, and might then be possible by a combination of two curves to obtain a satisfactory relation between the three factors. The great advantage of this is whilst we do our best to reduce, and possibly succeed, our basic figure, our totals always bear an automatic relation to the programme. However, any reduction at all can as a rule only take place if competent Iranians available.

In practice, the company eliminated a political issue by making an explanation in economic and technical terms to legitimize and guarantee its long-term control over a racially and hierarchically organized labor regime. The technical device of the formula,

[# Foreigners – (Rate of Reduction of Foreigners × # of Years)]
    × Expenditure = # of Foreigners (permissible under the formula)

constructed an argument about the competency of the Iranian oil worker, producing the effect of a replacement of non-Iranian labor with Iranian labor in terms of three variables (number of foreigners required, production in tons, and expenditures). The limits built into the definition of the Iranian oil worker relied on various technologies of difference employed by the company as a device to keep technical and economic issues separate from political questions of national control and social concerns about the racial organization of living and working conditions (i.e., keeping British and Indian employees as supervisory personnel).[106] The oil worker was internal to oil operations for economic arguments about efficiency, production, and costs but external when considered for the replacement of British and Indian labor at higher grades or the improvement of housing conditions and treatment in the workplace.

This formula was a remarkable power grab, as it transformed a political issue into a purely technical-economic calculation, ensuring AIOC's total control over its recruitment policy. On the other hand, petroformulas organized heterogeneous actors, such as oil workers, who were increasingly involved in the reassembly of the state, creating a kind of uncertainty. The Iranian government, public opinion, and oil workers themselves sought, in different ways, to reconnect the so-called technical argument about training Iranians at higher levels to the political question of national control of the oil industry. Thus, labor strikes and the reassembly of the state toward national control were entangled in the terms of Article 16 and threatened to deconstruct the company's formulations.

Through the 1940s, both AIOC and the Iranian government responded to labor dissent by forming new institutions and disciplinary regimes to protect oil operations. But the formation and resurgence of new political parties calling for national control of the oil, such as the reformist National Front Party and the communist, Tudeh Party, threatened the Pahlavi shah's power. The national government and the oil company had a shared interest in blocking the possibility of a more militant alliance between oil workers, the Communist Party, and national control. Only by redirecting labor dissent into the more manageable framework of reformist nationalism might the company and government ensure profits for large-scale military and industrialization projects domestically while also protecting the international monopoly arrangements of the largest transnational oil corporations.

Article 16 (III) of the 1933 concession confirmed that the two parties had agreed to "study and prepare a general plan of yearly and progressive reduction of the non-Persian employees" with the aim of replacing them "in the shortest possible time and progressively with Persian nationals." The two parties signed the final agreement over the terms of this General Plan in 1936.[107] The terms of the plan (Part I (a), specifically) guaranteed a rate of reduction in foreign employees on the condition that the "highest degree of efficiency and economy is maintained." The Iranians had agreed to a peculiar formula, which was built into this statement and graphed (figure 3.1).[108]

No subsequent progress was made in the struggle to replace foreign employees with Iranian labor. New negotiations commenced in 1947 as the Iranian parliament responded to the lack of progress by passing the Single Article Law.[109] The law built on Article 16 of the 1933 concession, calling for increased worker benefits in health, housing, and education. The Iranian government pursued further negotiations from 1947 to 1951, when the reformist Mosaddiq government passed a law calling for nationalization of the British-controlled oil industry. In the meantime, more oil worker strikes erupted.

As its policy, AIOC rejected any possibility of a redefinition of terms and variables by rival actors. In 1948, the Iranian government continued to argue that employing more Iranians would help the company avoid the costs of expatriation and transportation for non-Iranian employees.[110] It proposed a formula (thus attempting to deconstruct the British formula) to reduce the number of non-Iranian employees at an increasing rate annually—that is, 150 for the first year, 200 for the second year, and 50 more for each subsequent year. Razmarra, the Iranian prime minister at the time, made a final attempt to deconstruct the British formulation for Persianization[111] by presenting his government's formula as a condition for securing Iranian parliamentary approval for the terms of the 1949 Supplemental Agreement, renegotiating Iran's oil royalties.[112] AIOC officials rejected the proposal.

The two parties disagreed on whether achieving a formula for Persianization should serve as a condition for the resolution of negotiations on the General Plan or whether the formula might be put aside. AIOC officials had hoped that an agreement might be reached based on the plan's other provisions, thus avoiding the resolution of the formula problem altogether.[113] Company strategy sought to avoid the resolution of Clause 16 (III), or to avoid transforming the political question of labor into numbers, as this risked inviting a discussion of pay scales.

In 1950, the company finally declared the impossibility of reaching a resolution with the ruling Iranian government as the controversy had spilled into national politics. The reassembly of the Iranian state toward the nationalization of its oil industry was finally defeated in an Anglo-American engineered coup d'état, which successfully replaced Mosaddegh's reformist-national government with a pro-West Pahlavi regime in 1953 (discussed in chapter 6). The coup was, in part, a dramatic attempt to block the national government and oil workers from deconstructing AIOC's labor formula, which linked the number of foreign employees in a dependent relationship to levels of oil production and expenditures. Formulas organized around oil and labor served as calculative equipment, helping AIOC build a world in which more democratic forms of oil production and politics were, for the moment, blocked.

Politics of Formulas and Formulation

The 1901 oil concession's articles on royalties, labor, and arbitration did not occur collectively as a legal contract between two contractually equivalent parties. When put to work in the course of the 1932–1933 concession crisis, the concession exhibited a flexibility enabling the various actors involved to frame the same issues in different ways (legally, technically, numerically, or mathematically) and according to multiple interests. The practical work of information exchange, calculation, international standardization, and decision making was connected to politics and was precisely what enabled both the oil industry and the national state to be conceived and built at this moment of crisis.[114]

Following the activities of the technical dimensions of mathematical formulas and legal formulation in this history has made visible their connections to the political management of oil workers, national control of the oil industry, and the stability of international oil markets. A social history of Iran's oil workers cannot account for the actual equipment— the technical work of formulas—and their political activity in defining what conceptions of worker competency were necessary to protect British control of oil operations and ultimately block nationalization.[115] As discussed in the next chapter, the more dangerous alliance of militant workers with national control threatened to disrupt the company's labor regime and its undemocratic forms of oil production by opening them up to alternative political arrangements of the energy system.

In the history presented here, the concession dispute unraveled at a critical moment in the interwar period when a new set of calculating technologies and devices were needed to limit the production and distribution of energy. These techniques and controls, such as the concession device and monopoly arrangements, shaped the transnational oil corporation and the oil-producing state. Thus, the concession contract worked alongside formulas, laws, and concerns attached to it by contributing to the production of the reality it sought to describe.[116] The oil company worked in southwest Iran to build a social world it could manage through the institution of the concession, but the various articles and formulas built into this device were also built into the oil environment. Their entanglement with physical and social forces did not appear to factor into the articles, but actually worked in reality by producing a situation of uncertainty. Such uncertainty provided opportunities for Iranian government officials, AIOC negotiators, national public opinion, and oil workers to frame issues advantageously.

Standard accounts of Iran's interwar concession crisis place the controversy in the geopolitical and economic context of the Great Depression and the worldwide drop in oil demand, the 1928 cartel agreements for limiting global oil production, and looming disputes with independent producer governments in Mexico and Venezuela. This chapter has argued somewhat differently that Iran's concession crisis was not a mere "context" to be evaluated as a backdrop or dress rehearsal for nationalization. Rather, the crisis occurred as a set of traceable connections between the local and the global (e.g., formulas, laws, and cartel arrangements), constituting the political machinery for managing disputes that would increasingly involve the question of national control. The resistance to such an explanation in the scholarship about oil in favor of the standard account may have to do with their apparent incommensurability, and the mixing of the social and technical worlds such an alternative account entails.[117] The nature of politics at work concentrated legal and accounting knowledge at new sites to guarantee a corporation's claim to expertise and authority embodied in the concession. Politics was also working to resolve the world into what seemed human calculation and company expertise inside oil operations and a passive nature and society outside.[118] But the world of oil operations and its borders was produced out of a set of political projects and entanglements traced here.

There are many different ways in which taking seriously the technical aspects of the interwar concession crisis alters one's understanding of the kind of social world the oil industry sought to build. For example, the company sought to transform a southwest corner of Iran into its laboratory and extend it through a chain of concession terms, monopoly arrangements, legal arguments, paperwork, lawyers, and accountants to centers such as the League of Nations. The traceability of Anglo-Iranian oil and the delineation of its properties in technical, legal, and mathematical terms ensured the stabilization of "facts" about oil operations. AIOC pursued this goal through a series of battles to keep associations and alliances among many different entities enrolled in the process of oil production, transport, refining, and marketing stable as it would help to eliminate further disruption.[119]

One of the company's central tactics to avoid controversy resided in its attempts to separate "technical questions" from "political" ones, or oil operations and monopoly arrangements from local political struggles, violence, and labor controversies. AIOC used this manufactured divide as a weapon, in part by enrolling the calculative equipment and services of a most unusual actor—a formula. A careful tracking of the technical problems that emerged in the crisis, however, has led to politics and the proliferation of agencies, organizational forms, and calculations constantly in play. Iran's political claims to sovereignty over oil produced excesses that could not be factored into the company's formulations and overflowed into national politics, rendering the sociotechnical world of Anglo-Iranian oil highly unpredictable.

4

What Kind of Worker Does an Oil Industry Require to Survive?

The workmen employed in the service of the Company shall be subjects of His Imperial Majesty the Shah, except the technical staff such as the managers, engineers, borers and foremen. (Article 12, D'Arcy Concession, 1901)

(II) ... *the Company* shall recruit its artisans as well as its technical and commercial staff from among Persian nationals to the extent that it shall find in Persia persons who possess the requisite competence and experience. It is likewise understood that unskilled staff shall be composed exclusively of Persian nationals. (III) The parties declare themselves in agreement to study and prepare a general plan of yearly and progressive reduction of the non-Persian employees with a view to replacing them in the shortest possible time and progressively by Persian nationals. (Article 16, 1933 Concession)

Between the 1920s and 1950s, oil workers helped transform the oilfields, pipeline, and refinery of southwest Iran into sites of intense political struggle. The struggle triggered one of the most dramatic political events of the mid-twentieth century, the Iranian government's decision to nationalize the oil industry in 1951. The interwar dispute between AIOC and the Iranian government over the terms of Iran's 1901 oil concession included the unresolved issue of "Persianization," the gradual replacement of foreign employees with Iranian workers at higher skill levels. This particular controversy, along with a series of oil worker strikes promoting national control of the oil, raised the question of the kind of worker the oil industry required to survive.

The Iranian government's call for Persianization of the British-controlled oil industry was, in part, a response to the first organized industrial action by Iranian oil workers in 1929. Indian labor went on strike in 1922, and more strikes, often allying Iranian, Indian, and Arab oil workers, followed in 1945–1946 and 1949–1951.[1] This chapter considers the kinds of social technologies that targeted striking oil workers to help stabilize a labor regime peculiar to the transnational oil corporation of the twentieth century.

Striking oil workers generated a kind of vulnerability, which threatened to disrupt the energy system at any moment. But studies on the development of Iran's labor movement exclude the legal, economic, and organizational content of labor controversies from the politics.[2] They suggest that social forces, interests, and resources are somehow separate from the technicalities of the battle. Worker disruption of oil infrastructure, such as pipelines and refinery processes, had political consequences for the powers of the transnational oil corporation and the national state. Disruptions to the flow of oil and thus the flow of profits to AIOC, and income to the Iranian government, shaped the emergence of certain kinds of political arrangements that favored British control.

This chapter follows oil workers as they built connections between politics and the control and distribution of oil. It considers the kinds of social technologies and practical work involved in organizing oil workers in the locations of housing and work. The chapter argues that AIOC's organizational techniques of intervention and control constituted a political project that worked by enrolling diverse actors to build divides in terms of racial-technical difference in the oilfields. Labor disputes marked decisive moments when the company attempted to devise and implement a scheme to address the question of employing more Iranian labor at higher skill levels (discussed in chapter 3). The controversies that attached to this recruitment scheme were the outcome of a peculiar process of bifurcation in which the company attempted to exclude the local from oil operations by transforming political questions of labor into technical and economic issues.

AIOC's formative years of constructing an oil labor regime in Khuzistan, Iran, coincided with the first episodes of industrial labor action in the 1920s. As one of their management techniques, mining and oil firms in the first half of the twentieth century resorted to paternalism in their efforts to defeat union building and worker dissent in the mines of America's Southwest and in the oilfields of the Middle East.[3] Paternalism involved the provision of benefits and the construction of housing and recreational facilities for a small segment of employees as a means of securing loyalty and thus stability in oil operations. Similar to the other global oil firms of the twentieth century, AIOC relied on a peculiar combination of racial and technical ordering and coercion as a strategy to battle union formation and stabilize oil operations.[4] These practices were not new to the transnational oil corporation and had their origins in other industries located in other parts of the world. Robert Vitalis has tied the portrayal of the Arabian-American Oil Company, Aramco, as a benevolent force to a larger American corporate history of mining in the Southwest of the United States. AIOC's company managers and technologists had a different colonial past, however, connected to the oil operations of Burma and colonial administrative apparatuses of India.[5] Dominant Iranian political groups, such as the communist Tudeh Party, played an equally significant role by portraying themselves as the national spokespersons for the oil workers to redirect oil worker dissent toward alternative political possibilities.

This chapter traces shifting conceptions of the oil worker through AIOC's mobilization of benevolent and paternalistic practices, in response to the most pivotal oil worker strikes. By doing this, it pinpoints the critical moments in which the company decided to respond (or not) to the oil workers' demands by building divides between technical and economic issues (inside oil operations) and political issues of labor (outside oil operations). Technologies of racialization and other kinds of difference were often introduced and legitimized in technological terms—according to skill and wage structures to justify the use of the workforce—and then limited by delineating the suitability of certain jobs over others according to race. Article 12 of the 1901 concession, quoted above, designated all "workmen" as Iranian subjects, "except the technical staff such as the managers, engineers, borers and foremen." The division of labor in terms of race was marked in terms of a technical difference between unskilled (Iranian) and skilled (British) labor. As the scale and demands of striking Iranian workers intensified, AIOC managers and accountants devised inventive ways of delineating the unsuitability of skilled jobs according to race by explaining that the Iranians did not possess the requisite training and experience. The last section considers the extent to which technologies of constructing a segregated labor regime flowed into national politics, especially during ongoing battles over how to make Anglo-Iranian oil governable, and in coordination with other oil corporations to manage labor internationally. The chapter ends with a consideration of how a careful examination of the social and calculating technologies involved in building an oil labor regime with a specific kind of worker alters our understanding of the history of nationalism, the role of the subaltern, and the so-called emergence of a labor movement in the formation of a national state.

Points of Vulnerability within the Energy System

Oilfields, pipelines, and refineries became the sites of powerful political battles throughout the Middle East in the twentieth century. However, as Timothy Mitchell explains, organizing the control and distribution of oil did not offer oil workers the same power as the triple alliance of coal, railway, and dockworkers did in building more democratic forms of energy production.[6] Because oil comes out of the ground under its own pressure, it requires a smaller workforce than coal.[7] Oil's unique physical and chemical properties demand that each category of work—drilling, pipeline construction, well maintenance, transportation, and refining—utilizes specific kinds of skilled and unskilled laborers such as drillers, pipeline fitters, engineers, geologists, and chemists. The layout and design of oil infrastructure, namely, that it has an enclave character and requires oil wells, a pipeline, and a refinery to transform the oil into marketable products, result in distinct methods of monitoring and surveillance of workers. The oil workers' capacity to form unions and "engage in strike activity" is drastically reduced, especially when considering that other sources of oil can be relied on and tankers can be rerouted to replace a sudden loss of oil elsewhere.[8] Thus, one reason oil companies have succeeded in making enormous profits has been "their ability to contain labor militancy."[9] Where labor militancy has occurred, it has generally been concentrated in refinery operations where there are large concentrations of skilled workers who occupy strategic positions to disrupt the economies of both oil-exporting and oil-consuming countries. Over time, pumping stations and pipelines replaced railways as the main means of transporting a liquid form of energy, rather than a solid, from the site of production to refineries and tankers for shipping abroad. This meant the infrastructure of oil operations was vulnerable but not as easy to incapacitate through strike actions as were railways that carried coal, for example.

These points of vulnerability on the technical side of oil operations extended to, and were reinforced by, the segregated layout of residential areas according to race. As operations expanded after World War I, AIOC built an almost completely segregated populace through housing accommodations and the use of buses, clubs, and cinemas. It was comparable to the racial system built into Aramco's organization of Saudi Arabia's oil labor regime.[10] Worker skills were divided along racial lines and these were translated into the organization of housing, transport, leisure, and work. By 1922, Indian workers were living separately in "tents and mud huts in the barrack-like 'coolie lines' located to the southwest" of the refinery (figure 4.1).[11] Iranian recruits lived in separate quarters, either in sun-baked mud houses in the old village, or in structures made of sticks or bamboo and covered with palm leaves.[12]

Mark Crinson has shown that AIOC's development of housing and facilities in Abadan "heavily favoured the small European section of its population and indeed its policy towards Abadan as a whole was largely to treat the town as a place divided by race."[13] After the discovery of oil in 1908 and the formation of AIOC in 1909, all building resources and facilities were imported from abroad and an area was laid out for the construction of bungalows for European staff.[14] AIOC built its first "pucka bungalow constructed in the local style" with a mat and "chandle roof." A chandle roof is constructed of poles placed close together and overlaid with mats made from date palm leaves covered with earth. The bungalow form was symptomatic of sociospatial divisions of labor within colonial urban development, particularly in colonial India.[15] The

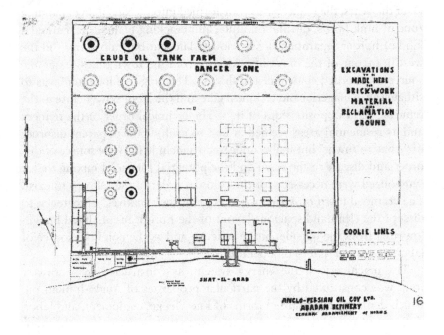

Figure 4.1
Right, "Coolie Lines." Map of Abadan refinery in 1910. *Source:* Mark Crinson, "Abadan: Planning and Architecture under the Anglo-Iranian Oil Company" *Planning Perspectives* 12, no. 3 (1997): 341–359, esp. 343. Reproduced with the permission of the BP Archive.

"bungalow area," known as "Braim," was built exclusively by company engineers for European employees in 1912. It also consisted of buildings and a pattern of roads including "specialist bachelor barracks" known as "Slidevalve" and "Sunshine," built in 1923. The buildings had thick walls, shutters, and arcaded verandas to block out the heat, which could reach 125 degrees Fahrenheit in the summer. Communal buildings such as the "Gymkhana Club" as well as many gardens were constructed in the vicinity. AIOC's transformation of Braim into a "green oasis" for the European workers was a major undertaking. Its construction involved the transportation of materials and extensive labor for irrigation and planting as well as the employment of professional gardeners with work experience at Kew and in New Dehli.

The Abadan refinery itself was located between Braim and the town of Abadan. Located on the Shatt al-Arab waterway at the end of the 130-mile-long pipeline, it would soon become the largest refinery in the world, pooling the liquid and transporting it through plants for all stages of refining before pumping it onto tankers to be transported abroad.[16] Over the years, the refinery was transformed into the site of an expanding zone of tank farms, distillation units, and cracking plants.[17] It marked a kind of border separating the spaciously laid out bungalow area "in the west in favour of the prevailing winds" from the town that became the source of disorder, epidemic, and disease. The "logic of location" was to situate the managerial and technical elite and the labor power close to the refinery, "if at opposite sides of it."[18] The technical order of the refinery and its residential areas seemed at first to exclude the apparent disorder of "Abadan town," but it also appeared to include it.[19] The source of disorder and disease in the town and the potential for racial mixing that it symbolized were necessary for the colonial order of the refinery to exist. The cramped town of Abadan rioters and other "natives" threatened to disrupt the clean and spacious layout of the European-inhabited bungalow area from the outside, but in practice and as the potential source of labor, it was *internal* to oil operations.

The machinery of the energy system, as constructed in southwest Iran, was constituted by the particular properties of Anglo-Iranian oil, which demanded specialized forms of knowledge, equipment, and labor, skilled and unskilled. The problem was that, from the start, the company chose to organize and manage oil operations by fixing the skill set of the managerial and technical elite and the labor power to race, in locations of housing and work. As discussed below, this ordering along racial-technical lines produced situations of immense uncertainty and vulnerability, particularly during moments of labor unrest, because they put into question the kind of worker AIOC required to expand operations and keep profits high.

Beneficence and Violence

As of January 1921, 4,942 Indian workers out of a total of 20,000 workers were employed in AIOC's oil operations and of these, 3,816 were concentrated at the Abadan refinery in the Persian Gulf.[20] Indian workers were recruited either through an agency from India or transferred from the Rangoon Refinery in Burma, through the mediation of the Burmah Oil Company.[21] AIOC had commenced its policy of recruiting Indian labor for the expansion of operations during World War I.[22] The British government helped the company by intervening to suspend the Indian Emigration Act of 1883, which placed restrictions on the migration of Indian labor to certain destinations. The suspension was considered a "war measure in respect of recruitment of skilled labour required by the Company."[23] Company officials believed it essential to have the "greatest possible freedom to recruit labour from India."[24] But the specificity of Indian laborers' technical knowledge—they were mainly employed as artisans and clerks but also in construction work at the refinery—gave them a kind of power to disrupt oil operations at any moment. In the period between World War I and the early 1920s, Indian workers at the Abadan refinery disrupted refining processes by going on strike on multiple occasions over pay rates for different classifications of labor.[25] In each instance, they built connections between their technical knowledge of oil refining and their economic and social demands about wages and housing.

In 1922, the immediate threat came from the "coolie lines" to the southwest of the refinery. The complaints of maltreatment and the miserable conditions of Indian workers employed in AIOC operations were expressed in Indian newspapers such as the *Bombay Chronicle*.[26] In terms of pay, the position of the worker went from bad to worse.[27] In a letter from the General Committee of the (Indian) Workmen to AIOC's joint works manager at Abadan, the workers declared their intent to go on strike. They expressed their dissatisfaction regarding the company policy of discriminating between "workman and a cooly." Several demands were listed concerning improvements in wages as well as housing and working conditions.

Led by the General Workers Committee, approximately 2,000 Indian workers went on strike for eleven days in March 1922. In response to the articles published in Indian newspapers as well as the demands submitted by the workers, AIOC challenged the notion that the accommodations provided for Indian workers were poor. On the contrary, company officials argued, the newest "clerks' quarters" were in excellent condition.[28] They consisted of a "large airy bed-room with bathrooms attached for each clerk and a dining room for every two clerks."[29] An alleyway separated the living rooms from the latrines and cookhouses, but "periodical congestion" was evident, the official admitted. The constructed quarters were not quite finished, and cookhouses and latrines, which were "nearing completion," were "unexpectedly demolished by a heavy storm of rain."[30]

The situation of the 1922 strike at the Abadan refinery improved with the arrival of Shaykh Khaz'al from Kuwait. The shaykh had an interest in securing the stability of AIOC operations. As a claimant to property in the oil regions (discussed in chapter 1), he was receiving rents through a secret financial deal with the company that granted access to the lands where the refinery and part of the pipeline were built, as well as the provision of security to ensure that local nomadic groups did not disrupt company operations.[31] The shaykh posed a threat to the Iranian government, as his power over Khuzistan Province persisted outside the control of the central state.[32] According to the British political resident in the Gulf, the shaykh had "practically scotched [put an abrupt end to] the strike of Arab and Persian labour."[33] There were 1,500 Chittagonian laborers (Sunni Muslims originating in the northeast region of India known as Bengal) who had continued to work under the guarantee of a pay increase of around 15 percent, but they were wavering.[34] "It was very important," the British political resident explained, "to keep these men loyal because if they struck all the benches where oil is refined would have had to close and the entire refinery would have been at a standstill."[35] In the view of AIOC and the British government, the Abadan refinery constituted one of the most vulnerable points in the energy system where large concentrations of skilled workers could easily disrupt international oil markets.

To avoid addressing the social and economic demands of the Indian workers, the company made distinctions between economic issues and politics by putting the blame on a specific group of "agitators." The British political resident alleged that the motivations behind the strike were not solely economic but "largely political," and that it was engineered by "the Sikh element."[36] As a result, AIOC repatriated approximately 2,000 male workers to India, while the remaining Indians, Arabs, and Iranians resumed work and new laborers were recruited.[37] In this early crisis of industrial action, the British government, operating on behalf of AIOC, felt the company was completely "justified in using force if necessary to compel men to leave as they have no right to insist on staying."[38] The company argued that the Indian strikers had no "reasonable grounds for going on strike and for intimidating others and local labour into joining."[39] In terms of pay, there were no legitimate grievances because the "company's scale is very liberal and is ... higher than that given by other firms."[40] The company portrayed itself as the most liberal and beneficent of the international oil firms. However, it simultaneously resorted to forms of coercion such as deportation and violence to eliminate dissent and place the blame on the political agitations of a particular group of workers.[41]

Kaveh Bayat has highlighted the strikes and grievances expressed by Indian and some Arab workers in the 1920s to argue that the Iranian workers were "not yet ready to take an active part in these actions or to organize their own."[42] But it is equally important to follow the more complex forms of coercion at play—for example, the technical procedures and terms, calculative equipment, infrastructures, and coordination among diverse actors that helped render flows of oil and people governable.[43] Whether the Iranian workers were ready or not, strikes were critical moments in which assemblages of workers, company officials, and the British and Iranian governments struggled to frame controversies in advantageous ways to block opponents.

Indian workers performed specific tasks at the refinery without which the set of interconnected mechanisms involved in producing, transporting, and refining oil was rendered vulnerable, especially at these moments of political uncertainty. The company feared that Indian workers would encourage local laborers to join and open up new political possibilities in oil production that threatened to weaken British control. To counter this, AIOC worked hard to dismiss grievances with respect to pay by comparing its wage scale to that of other firms and workers in Iran. In these early years, the success of Persianization was explicitly linked to the reduction of Indian labor in a way that took into account Iranian contract labor not directly employed by the company.[44] The Indian worker was, at the same time, necessary to company operations, which allowed AIOC officials to make technical and economic arguments about a lack of Iranian labor with the requisite efficiency and experience.[45] Company interests were "obliged to rely on India not only for unskilled but for skilled labour as none is obtainable in Persia."[46] On the other hand, political possibilities were narrowed as striking workers could always be deported through the exercise of force, as in 1922.[47]

The 1922 strike by Indian workers marked an early moment when the company did not have to address worker demands by setting up institutions or mechanisms for the management of labor. It could easily resort to violence and deportation, an option not available to them with regard to Iranian workers. Most importantly, the foreign Indian worker equipped AIOC officials with the "evidence" it needed to develop a working definition of the Iranian oil worker in technical and economic terms as too inefficient and inexperienced to replace foreign labor, Indian or British. In what followed, however, the oil labor regime did not stabilize so easily. Additional equipment and forms of expertise were necessary to manage AIOC's expanding oil operations because it was becoming increasingly difficult to deal with ongoing labor crises and the looming threat of national control using the techniques of coercion framed as corporate beneficence.

The 1929 Strike and the Question of National Control

The strike by Indian laborers in 1922 coincided with the Iranian government's pressure on AIOC headquarters to decrease the amount of Indian labor. In response, AIOC managers argued that the company was hindered from executing such demands at Abadan on account of "the very large percentage of Indians employed who resisted every effort to introduce local labour to their own exclusion."[48] Resistance among Indian workers was acceptable in the company's working definition of the oil worker as long as it reinforced the company's arguments about imposing limits on the possibility of employing more Iranian laborers at higher skill levels.

Abadan in the late 1920s was an overcrowded "township" with 60,000 residents, a large number of whom were living in "squalid and unsanitary dwellings with no public services like clean drinking water."[49] In 1928, the company employed 16,382 workers, but as Bayat argues, the presence of the company in southwest Iran was "constraining Iranian sovereignty and other issues like the appalling condition of Iranian workers." Although the town was nominally under local municipal control, Abadan was practically a company town.[50]

The company's initial plans for constructing housing and work facilities at the site of oil operations were internal to the British imperial
project to bring a peculiar kind of order to its colonies around the world.
AIOC assigned James Mollison Wilson as the architect in charge of developing town-planning schemes and designing large numbers of buildings
in Abadan.[51] Wilson received commissions from AIOC starting in 1927,
but his career experience and training included work in the mandated
territory of Iraq, first organizing the Public Works Department and then
serving as Director of Public Works from 1920 to 1926. He had previously worked in New Dehli from 1913 to 1916 as an assistant to Sir
Edwin Landseer Lutyens, the British imperial architect of the company's
future headquarters in London.[52] Led by Wilson, the work of designing individual buildings and planning large residential areas, especially
in Abadan but also in other company areas in Iran, Iraq, and Kuwait,
mushroomed in the 1930s.

By 1929, the degree of planning for the organization of working and
living conditions, as well as the political groups involved, had changed.
The Iranian Communist Party, known as the Tudeh, launched a campaign
in early 1928 to reorganize the "Iranian working class."[53] Moscow's decision to "Bolshevize" the international communist movement in the 1920s
encouraged communist groups such as the Tudeh to take a more radical
line of action in their respective countries. The party sent trained agents
to the Khuzistan oilfields to take advantage of the growing anti-British
sentiment among oil workers, communist sympathizers residing in the
area, and other "nationalists," who together contributed to the formation of what Bayat has called "the semblances of a trade union."[54] Work-
ingmen's clubs were social clubs that served as an important vehicle of
worker organization and education.[55] The miserable conditions faced by
the oil workers, the discrimination between Iranian and foreign workers,
the poor living conditions and low wages, and the official campaign to
force AIOC to revise its 1901 D'Arcy Oil Concession were among the
many grievances voiced at these social gatherings. Meanwhile, the new
British-backed ruler of Iran, Reza Shah, was working with the company
to centralize and consolidate his power over semiautonomous political groups in the provinces, such as the Bakhtiyari khans and Shaykh
Khaz'al, to secure the stability of oil operations and keep oil profits
flowing to the state.[56]

The Iranian oil workers finally went on strike May 1–6, 1929, disrupting operations at multiple points of vulnerability, including the main oilfields, pipeline, and refinery.[57] AIOC's general manager, E. H. O. Elkington,

hoped to avoid difficulties by securing the "fullest support" of the Iranian minister of court, Abdulhusayn Teymourtash, for the governor-general of Khuzistan. The governor-general must be "empowered to deport such agitators as he may deem fit, irrespective of their nationality, in the interests of law and order."[58] The governor-general responded swiftly to the minister's call to "strong action" by arresting forty-five "ringleaders."[59] On May 6, strikers attempted to prevent laborers from returning to work in the refinery. The disruption was "quelled by a detachment of soldiers" summoned from a neighboring town, Mohammerah, after it became evident that the Iranian police force was inadequate in size to cope with the situation.[60] In total, "two to three thousand workmen" were reported to have demonstrated, causing shops and bazaars to close, but other studies suggest up to 9,000 workers protested at the most vulnerable point in the energy system, the Abadan refinery.[61] No oil processes were successfully shut down, however, and the arrest of oil workers led to an escalation of violence, more arrests, and the dispersal of "mobs" by the military. The company made arrangements to have "reserve men in the works … in case of emergency."[62]

The strike was disruptive and yet ineffective. The Iranian government responded in a novel way by collaborating with the company to end the strike while making Reza Shah's upcoming visit to oil operations coincide with the announcement of an increase in wages. The strike ended on May 7 with the resumption of work. "Mainly owing to the loyalty of Indian and Persian labour," AIOC successfully maintained all refining processes throughout the period of disturbance.[63] The strike was disruptive, however, and forced the Iranian government to suspend all geological surveys conducted by the company. An oil operations manager confided to AIOC's chairman, John Cadman, that the company suffered a "great loss" by not being able to "employ its Geological staff, as also the valuable time which was being lost securing evidence for future development."[64] The technical infrastructure of oil operations, whose upkeep demanded constant geological exploration and knowledge gathering to support future expansion (discussed in chapter 2), was vulnerable to striking oil workers and easily disrupted at moments like this.

AIOC responded to the striking oil workers by disseminating information in certain newspapers accusing the oil workers of conspiring to strike and burn the refinery under the instigation of communist influences.[65] Reacting to a meeting between AIOC and the Iranian government on the status of workers in the oil regions, Ali Dashti, editor of *Shafagh-i Sorkh*, dismissed attempts by the company to paint an alternative picture of the oil workers. In reality, the demands of the Iranian oil workers were about securing better wages.

The Iranian government conducted an investigation into the causes of the strike and concluded that "Bolshevik" instigation was an inadequate explanation.[66] Economic grievances and concerns about racial discrimination provided a more feasible motivation. Indian workers had gone on strike to demand higher wages prior to the arrival of Bolshevik influence in the Persian Gulf region, and they were receiving higher wages than Iranian laborers with a higher literacy and skill level. In fact, the government's investigation claimed that the company fired Iranian workers who were close to learning technical skills and specialization. Iranian workers suffered from a lack of housing and lower wages compared to foreign workers. It was this kind of racial segregation, concluded the government, that was the cause of dissatisfaction, and ultimately led Iranians to organize themselves like the Indian workforce.

The investigative report claimed that the British company fabricated a link to the Bolsheviks to frame the strike as a political issue and dismiss the employees' economic demands, despite the success of the communists in organizing the workers to assert such demands in the first place. The workers who had intended to protest their meager wages were arrested at the instigation of the company. This was a tactic that the Iranian government took to quell the strike. The group consisted of thousands of people from which "hundreds" attempted to break away and occupy company installations.[67] But the editor of *Habl al-Matin*, a Calcutta newspaper, explained that while communist influence was shaping troubles in Khuzistan, the Iranian oil workers had legitimate economic demands. They were not receiving the same treatment as the "Indians and the Iraqis."[68] The Calcutta paper called on the workers to establish a union to protect their rights. It declared that to survive, the company would have to separate itself from political concerns and preserve its identity as a "commercial institution only."[69]

On a national scale, the 1929 strike opened up alternative political possibilities connected to questions of political sovereignty and national control of the oil. For the first time in the company's history, industrial action triggered a reorganization of AIOC's labor management in order to "increase contact" between laborers and company managers.[70] Likewise, AIOC's efforts to quell oil workers marked a period of heightened involvement in labor issues by the new British-backed ruler, Reza Shah. Various government ministries worked with the company to repress the strike and impose martial law, while simultaneously subduing the workers by announcing a wage increase. Indian workers were useful as a company strategy for replacing agitating Iranian oil workers in various technical tasks, but this was a short-term solution. Oil worker dissent along with the call for Persianization in the 1930s triggered new efforts by the company to provide all Iranian employees with access to education, transportation, health benefits, leisure facilities, and even their own traffic police.

The increasing gains made by populist politics in both Iran and Iraq forced AIOC to expand housing facilities to accommodate Iranian laborers and their families, but also to institute more layers of segregation, monitoring, and surveillance. Physical spaces were designed according to what Ehsani has called the "authoritarian spatial design of a company town," to impose "time-discipline" and a "hierarchy, distinguishing laborers, supervisors, managers, engineers, white collar staff, and the unemployed."[71] Oil towns became the "first modern industrial towns" in Iran designed in a hierarchical and segregated form that would later serve as a model for state-owned industrial urbanization projects.[72]

Social technologies of racial and technical difference, mapped here, were built into the very design of housing and work sites. Guaranteed by the authority of the oil concession, the company dismissed the economic demands of its workers by maintaining a boundary between technical operations and questions of labor, violence, and increasing calls for national control of the oil industry. This was best exemplified in the company's attempts to construct a labor formula to manage the rate of Persianization during the 1932–1933 concession revision negotiations, discussed in the previous chapter. Calculating formulas for managing the rate of Persianization were entangled with the company's development of housing, work, and training facilities. These technologies equipped workers with forms of agency, informing their decisions to hold strikes and make demands in increasingly nationalist terms. The construction of these social and calculating technologies exemplifies one of the peculiar ways the oil corporation operated and its political agency was constituted by delineating the kind of Iranian oil worker, preferably unskilled, that was necessary for oil operations to succeed.

Newly involved political groups such as Iran's communist Tudeh Party acted on behalf of the oil workers by attaching themselves to their cause as local and national spokespersons. Iranian public opinion, government agencies, and political groups aligned themselves with the oil workers as proponents of national control, but with different interests at stake. Benevolence and violence on the part of AIOC managers were no longer as effective interventions as they had been in response to the strike in 1922 by Indian workers. Shifts in strategy included new efforts by the company to provide Iranian employees with access to social services and more training, as well as better coordination with the Iranian government in quelling strikes to keep the threat of union formation, communist influence, and more militant politics out. New responses were worked out precisely at moments of crisis, when technical concerns about developing and expanding oil operations were increasingly intertwined with political questions of sovereignty and national control, producing important albeit inadequate concessions to the oil workers' demands.

New "Conception of the Rights of the Persian" after World War II

The oil worker was now calculable and manageable in new ways that had not been possible before. The resolution of the labor crisis in 1929 and the concession crisis in 1932–1933 triggered the introduction of new institutions of labor and disciplinary regimes to block workers from the threat of nationalism and militant forms of politics.[73] The events in 1929 and 1932–1933 marked a political moment in which the exercise of direct force by a colonial power on a subject population justified in terms of racial differentiation and "civilization" was no longer possible. Populist politics were gaining ground across the Middle East and especially in the oilfields of Iran, Iraq, and Saudi Arabia. If a foreign oil corporation wanted to operate locally in these countries, it would have to develop new forms of social control to manage militant forms of worker dissent, while also accommodating domestic laborers and their families by providing them with social services and access to training at higher skill levels.

After the Second World War, this emerging politics of national sovereignty could not be overlooked. AIOC mobilized paternalism—that is, the provision of benefits, housing, and recreational facilities for a small segment of employees—as a means of securing loyalty and managing the increasing threat of nationalism. These practices were legitimized, on the one hand, in terms of economic costs, and on the other, in terms of the modernization and moral uplift the company believed would come from new living quarters and leisure activities. The strategy necessarily included the impossibility of relinquishing British control over oil. This had consequences for political possibilities tied to national control of the industry. As Vitalis says, firms began disguising the "supremacist origins and resonances of their labor regimes" after World War II and through the Cold War. It was not a question of racism but of "skill levels" that the international oil firm would begin to insist on in expanded public relations campaigns, which AIOC commenced as far back as the 1930s in Iran.[74] Overcoming the vulnerability of the energy system at particular nodes in southwest Iran, such as with the oil wells, at the refinery, and in segregated living spaces, required new kinds of arguments justified in terms of skill levels rather than race.

By the end of the Second World War, there were 65,641 AIOC employees in Abadan alone, of whom 2,357 were British.[75] The company was expanding exponentially, drawing labor directly and through subcontractors from southwest Iran, the Persian Gulf, and India.[76] Company resources continued to be channeled in favor of the construction of housing and other facilities for its senior European staff. Housing for non-European junior staff and wage-earning laborers, especially the contract laborers the company increasingly relied on, however, was "left as a matter for the market or the municipality."[77] They simply lived in "shanty towns on the edges of the Company and municipal areas." AIOC did not regard the large numbers of contract laborers arriving in the 1940s as its responsibility.

AIOC's deployment of paternalism, which favored only a small group of elite workers, did not solve the problem of "a very great and widespread spirit of Nationalism," according to Wilson, the company architect.[78] He pointed to disparities in housing as contributing the "most to the dangerous divide between Iranian and British employees."[79] As a solution, Wilson proposed to create a new residential area known as Bawarda "as a kind of manifesto of racial mixing, an experiment in non-segregation."[80] Situated on the other side of the bazaar from the refinery, the town was designed by Wilson as a "showcase vision of company paternalism."[81] Iranians living in Bawarda would soon abandon the traditional "purdah system," the practice of veiling and secluding women within the home, and future homes would be designed in the European style, encouraging the Iranian employee to "desire British conventions of domestic life."[82] Wilson's project for harmonious racial mixing failed in the face of "potential violence."[83] Racial mixing was likely to breed violence. The "very spaciousness of the plots in Bawarda and its generous ... road provision could only be provocative to the Iranians in 'the town.'" In subsequent years, Wilson built offices, bungalows, and "dormitory estates." New industrial estates were laid out all over Abadan Island and maintained their connections through the introduction of company buses, which transported workers to the refinery.[84] These new clusters of small towns rather than large townships would help reduce political activism, it was thought, by establishing a distance between them and the "social disorder" of Abadan town. In effect, the racial-technical construction of housing and work within oil operations was helping the company exclude the local while maintaining a distance from politics.

In the spring of 1945, a representative of the British Ministry of Labor and National Service, A. Hudson Davies, accompanied by the Inspector of Labor Supply, J. B. English, traveled to the oil regions for three months to examine ways of addressing labor problems in AIOC's operations. The demands and costs of World War II, according to the report, had exacerbated the isolation of company management in Iran from "direct knowledge of industrial experience" in the country.[85] The oil company was a "highly technical industry in a backward, foreign, and remote country."[86] This meant that skilled British employees constituted "the backbone" of the company. The need for British employees might decline in the future, but there were such "deficiencies in quality, skill, and training of the native staff and labour that an increase of output" could only be achieved by increasing the number of British employees.[87] For the oil industry to survive, the report argued, the ratio of British to Iranian workers must favor the British employee of a higher skill level. Framed in technical and economic terms, this was not due to a lack of domestic labor, but to the Iranian worker's inherent inability to acquire the qualities, skills, and training necessary for increased economic productivity.

Compared to the oilfields, the refinery at Abadan was suffering the most, both in terms of skilled labor and in terms of adequate housing. "So short is housing that only a man with more than seven years of service can be joined permanently by his wife."[88] The lack of housing to accommodate spouses and families was affecting the attitude of the workers. Over half of new recruits were housed for the first few years in "emergency housing of a relatively poor standard," and this was encouraging the spread of discontent. Davies advised the company to pursue an "accelerated house building programme" in Abadan. The program would address labor problems, though it would not go far enough to solve the housing shortage. That all British staff were in specialist or supervisory posts could not be altered in proportion to the relatively low percentage of Iranians (7 percent) in these senior posts. The British inspectors explained away the problem in technical terms, arguing that it stemmed from a serious shortage of "natives capable and trained in supervision and of clerks."[89] The Abadan Technical Institute was built in 1938 to give Iranian apprentices basic technical skills, but by 1945, only 1,700

Iranians had received training.[90] The training covered a five-year appren-
ticeship and included short, intensive training courses in craftsmanship
and semiskilled occupations for adult workers together with special
courses for the upgrading of certain employees, night classes in languages
and other subjects, a four-year course for junior foremen, and courses
in clerical work and commerce. More advanced course topics included
mechanical and petroleum engineering, and during World War II, the
institute introduced a bachelor of science degree in petroleum technology.[91] However, Michael Dobe explains that, as a strategy, the company
"sought to minimize the number of Iranians sent for university training
and maximize the number sent for trade training" in the United Kingdom
as it would block the threat of returning superior-skilled Iranians from
stirring up trouble among the workers.[92]

Technical obstacles of inadequate training were adding to the burden
placed on British employees, who had to continue to "organize, supervise, and teach."[93] The "concessional obligation" to reduce the "British
component" and employ more Iranians had fallen into the background
during the war, but there were additional "limitations in all directions"
that "forbid a quick reduction" of the British-to-"native" worker ratio.
Davies warned that while trade unions had been illegal in Iran, a new
labor law was before the Majlis to make them legal. If the law passed,
it would be wise for the company to consider the possibility of "forestalling pressure from the Persians" by taking the initiative to set up "its
own plan for consultative machinery with the native staff and labour."[94]
After World War II, AIOC and the British government were finding it
increasingly difficult to maintain a boundary between social concerns
about labor, nationalist politics, and technical operations. To maintain
its precarious control over the energy system, AIOC would have to
resort to other kinds of explanations such as the threat of communism
and would have to make additional plans to implement paternalistic
welfare work.

Iran's 1936 labor law did not discuss the right to strike or form trade
unions, but after the 1941 Allied invasion of Iran, intended to protect oil
and supply routes from the Nazis, unions became strong and in 1942, the
communists established a Central Council of Trade Unions in Iran which
in 1944 became the United Central Council of the Unified Trade Unions
of Iranian workers (CCUTU).[95] In 1946, the CCUTU was affiliated with
the World Federation of Trade Unions and membership included 90,000
workers in the oilfields of Khuzistan. In this period, there were twenty-five
major stoppages and five separate regional general strikes of which the

two main centers were Tehran and the oilfields. The oil workers' main form of unrest was to strike.

On May Day 1946, over 50,000 workers demonstrated in Tehran. On the morning of May 5, Tudeh officials in the oilfields called on 350 AIOC workers employed in distillation and bitumen plants at the Abadan refinery to walk out without warning.[96] There were major strikes in the refinery and in six distinct production centers during the 1945–1946 period including a three-day general strike in July 1946, specifically over pay and working conditions.[97] The May strike was organized in sympathy with a strike that had occurred six days earlier in the company's locomotive shops in which fifty men were involved.[98] The effect of the walkout on production and refining in 1946 was significant, revealing the vulnerability of the technical side of the energy system at certain key points in the process.[99] On the first day of the strike, instant crude oil dispatches from the main oilfields, amounting to some ten million gallons per day, were shut off. As a result, oil production was shut down at Gach Saran, since the percentage of incoming oil to the Abadan refinery was too high for the maintenance of products. Due to picketing activity, the loading of tankers was also suspended.

An oil worker writing during the strike action of 1946 and appointed head of branch affairs for the CCUTU, claims that the British government had larger interests than breaking the strike because it was exerting a lot of effort in time and money to neutralize and break the Tudeh organization.[100] AIOC was allying itself with certain Arab "tribes" in the province to infiltrate the workers' union affiliated with the Tudeh-backed United Council. More skilled and trustworthy oil workers "of the first rank" needed to be recruited in leadership roles and each node in the energy network needed to come under a central Abadan administration.[101] A set of social and economic demands and future goals were listed in the report, including the election by workers of a representative for each oil region to address worker complaints, an increase in wages, and the prior approval by worker representatives of any form of punishment, such as deportation, pursued by the company.

In its report to the British government, the company took a different view from the Iranian oil workers by building connections between the causes of the strike and the rise of "Tudeh infiltration" in Abadan at the end of November 1945.[102] Leaders included drivers, fitters, and plant attendants. Strikers resumed work on May 6 at Abadan, but another group of workers at Agha Jari, one of the main production sites producing four million tons of oil per year, struck on May 13 to protest poor housing conditions and general amenities.[103] The Tudeh Party was attaching itself to the cause of the oil workers by acting as their national spokesperson.[104] AIOC rejected their demands, and Iranian military forces were dispatched to maintain order. In response to the 1946 strike, company management had instructions to "expedite as far as possible" the construction of housing and expansion of medical and other welfare facilities, to review wages in relation to the cost of living, and to encourage representation among workers, which might be developed into union organization.[105] The company knew that it must be prepared to deal with Tudeh leaders so long as they remained in control of the situation at Abadan.

A delegation of British members of parliament paid a visit to the oilfields in June 1946 to investigate the causes of the militant Iranian labor movement within AIOC's oil operations. The delegation advised that the "conception of rights of the Persian, defined twenty to thirty years ago by the Company," must undergo "a complete and fundamental change."[106] The British secretary of state for foreign affairs, Earnest Bevin, had warned that around the world, "the sense of equality is rapidly developing."[107] The way to tackle the problem, advised the report, was through a concerted effort by the company to develop its social program and to engage in "greater consultation with their workpeople."[108] The background of anti–trade union organization by companies such as AIOC had "inevitably brought the present situation to a head."[109] The company's management was composed of men inexperienced in negotiations with unions. This needed to change for management to be better equipped to deal directly with the internal representatives of oil workers rather than members of Iran's Communist Party. Such a strategy would work to "separate the political aims of the Tudeh Party from the economic desires of employees."[110]

Iran's pending national labor legislation was making it difficult for the company to avoid addressing social and welfare activities, with the most important consequence being the recognition of an oil workers' union. Iran's Council of Ministers had passed the national labor law back on May 18, 1946.[111] The labor law made it possible to enroll new institutions and standards of work to help manage oil worker dissent in terms of reformist nationalism. In particular, the law called for the establishment of a National Ministry of Labor and Department General of Labor charged with executing the various regulations in conjunction with the Ministry of Commerce and Industry.[112] Provisions of the law stipulated that work must not exceed forty-eight hours per week.[113] The main labor office was opened in Abadan, with subsidiary labor offices scheduled for opening in each of the oilfields. The Iranian government welcomed a labor representative from the United Kingdom to help formulate procedures for union activity under the new law.[114]

A committee was subsequently formed to meet in the same period as the British delegation's visit, consisting of representatives of AIOC and the Abadan Workers' Union. Under pressure from the Iranian Ministry of Commerce and Industry, the Ministry of Justice, and the prime minister, the company agreed to provide full pay for the period during which workers were absent at the oil-producing region at Agha Jari.[115] This was further induced by the technical calculation that production losses had already amounted to some 30,000 tons of oil, losses limited by the efforts of British workers to operate wells and production plants throughout the strike.[116] While the industrial actions posed a significant threat, the technical knowledge of British workers enabled a flexibility that prevented the total disruption of operations, particularly at certain points of vulnerability such as wells and production plants.

Having spent two weeks in the oilfields, the British parliamentary delegation concluded that a labor crisis was "inevitable."[117] The British delegates met with leaders of the Abadan Workers' Union, who reported on the intolerable living and working conditions.[118] The delegates observed the status of housing conditions in Abadan:

On one hand we are able to see splendidly built modern houses with air-conditioning, ice-boxes, etc., passing through a variety of stages right down to small individually built shacks in some cases the only roof being empty paper cement bags and old pieces of matting which some of the Persian labor had tried to create shelter from. Never in the whole of my experience indeed in any other country which I have had the privilege of visiting, did I see so close together such extremes in Housing Accommodation.[119]

The oil regions of southwest Iran did not appear to have changed much since the first "coolie lines" were built outside the Abadan refinery in the early 1920s. The delegates concluded that "the place looks like a penal settlement in the desert" and "houses we visited little better than pig-styes."[120] AIOC faced mounting trouble due to the strike actions, reported the delegates. Its "previous policy had been one of paternalism," and they had depended on their "undoubted good relations with their staff" to solve labor problems as they arose.[121] The disappearance of a "means of discussion" for handling workplace issues and disputes needed to be redressed as an "essential feature of future policy."[122] Nevertheless, the limitations remained. It was "of course, humanly impossible," the British delegates confessed, "to expect even a Persian to advance his knowledge technically in the interests of British Oil Production and at the same time expect any individual Persian, or Persians, not to acquire knowledge which creates within them a desire for a much better existence."[123] On the one hand, arguments about Iranian oil workers were factored into calculations when making arguments about protecting the British-to-"native" ratio due to a "deficiency" in technical skills for adequate economic output. On the other hand, arguments about the acquisition of technical knowledge were impossible to factor in because they would create within that same Iranian worker "a desire" for more equitable forms of treatment, causing the ratio to fall apart.

According to one former AIOC worker with forty-six years of experience fixing drinking-water problems in Abadan, the company's aim had always been to subordinate the Iranian laborers. H. Gholami-Ghanavati has recalled the specific ways the hierarchical and racial system of labor was organized after the Allied invasion of 1941.[124] He started working for AIOC at the age of fourteen as an unskilled laborer, although he managed to study English at the company night school for seven years. He remembers that when he was first hired, he worked under the supervision of the English and Indian employees: "They were never interested in teaching us technical skills."[125] For example, "when they wanted to repair equipment or pumps they either sent us to work on something else or covered up what they were doing, so that we would not learn the particular skill involved." Gholami-Ghanavati explains that on other occasions unskilled laborers were expected to lubricate machines and equipment, and in doing so, secretly learned the mechanisms involved in their operation. When Gholami-Ghanavati realized that he had acquired as much technical skill as many of the Indian workers, he applied for work at the higher skill level, "but the English were severely against this." Their refusal was "natural," according to Gholami-Ghanavati, because their aim had never been to promote the Iranian laborers.

The defense of the new rights of the Iranian oil workers was at stake. The Tudeh Party, the Iranian government, and AIOC battled to attach themselves to the interests of the oil workers and serve as their national spokesperson. The Tudeh leaders contributed to disruptions in the oilfields by picketing all petrol-distributing stations in Tehran and preventing any vehicle from refueling on certain days in June. The company feared that a cessation of "oil distribution would entail break down of public transport, electric light supply and closing all bakeries."[126] Cutting off Abadan's supplies of oil would have "disastrous economic consequences in the UK and throughout the Empire."[127] The British Foreign Office viewed the labor unrest less as reflecting "genuine concern for the welfare of Persian workmen than as a political campaign against British interests." Acknowledging the strength of the connection between the legitimate demands of militant oil workers and the Iranian Communist Party was impossible if the British side wanted to maintain control of oil operations and keep profits high. On the other hand, using arguments about workers in political terms was possible when it threatened the stability of oil operations, framed here as a political campaign against British interests.

Both the British and Iranian governments viewed the creation of new labor offices in the oilfields and refinery area as the best mechanism for managing the crisis and avoiding the more militant alliance of oil workers with the Iranian communist movement. The goal was to "wean AIOC employees from the Tudeh by persuading them to submit complaints to the Government Labour Office."[128] Through the establishment of new labor institutions and laws, the Iranian government sought to position itself as the oil workers' spokesperson.[129] By doing so, the Iranian government could collaborate simultaneously with the British government and the oil company to redirect labor militancy into the more manageable frame of national labor reform. This was a strategy to secure the loyalty of a small segment of employees and to create enough stability to prevent the more militant alliance of communism, the control of oil, and nationalism from coming together.[130]

The Iranian government, acting as national spokesperson and policeman, made official statements to the press that minimum wages, including Friday pay, were under negotiation. Pending the outcome of the Work Commission's inquiry, the government deemed any future strikes illegal.[131] The 1946 strike that started on May Day finally ended on July 17. It drew some concessions from AIOC, but ultimately led to the mass arrest of strike leaders and the shutting down of the oil workers' union and the Tudeh Party in Khuzistan.[132] Laborers gradually returned to work, but there was still a "hostile feeling between Iranian labor and the British." Halliday argues that "this action in which workers won most of their demands ... demonstrated how a small but strategically placed working class can play a major role in an economy like Iran."[133] Although the government imposed martial law and stationed troops at the oilfields and refinery,[134] the strike was decisive in terms of the disruption of particular tasks and operations involved in the energy system, and also in terms of the scale of violence and involvement of concerned groups. The

Iranian prime minister, Ahmad Qavam al-Saltaneh, eventually stepped in and encouraged AIOC to compensate the oil workers in national-legal terms.[135] The politics of a highly technical operation that the company had hoped to preserve as an isolated economic enclave was overflowing into national debates and laws.

There appeared here to be a tension in defining the limits and possibilities affecting the competencies of the oil worker, which was triggering battles over who would win as national spokesperson for the oil workers. There was no question that the design and operation of AIOC's oil labor regime were working to the disadvantage of local and non-European labor. At the same time, the British delegates expressed anxiety that the necessary and rapid improvement in housing, work, and training would instill in the Iranian worker a desire to acquire much more, namely, national control of the oil industry.

The company defended its housing policy, arguing that it was "geared to produce a good general quality of housing rather than a rapid production of quantity."[136] The Iranian government was collaborating with the company by imposing martial law and authorizing troops to suppress the strikes, while also redirecting the militancy of the oil workers into the more manageable frame of reformist nationalism. The company continued to view itself as a benevolent social force. This was justified by comparing its housing projects to the housing available in the rest of Khuzistan Province, beyond the borders of oil operations. At the same time, there was a heightened awareness that the company must address the economic and social demands of the oil workers, but that it lacked experience negotiating with union organizations to do this.

By 1951, only 18.5 percent of the labor force was accommodated in AIOC quarters.[137] AIOC housing was allocated largely by seniority rather than longevity of service or basic rate of pay.[138] Over the course of two decades, AIOC had done little to improve the living and working conditions of its non-European employees, but instead was pressured by the British government to respond with more institutions and layers of monitoring to weaken unions and reduce the threat of future disruptions. Iran's government worked hand in hand with the company's efforts to defeat the communist Tudeh Party and rechannel workers' interests into the more manageable outlet of reformist nationalism, as noted earlier. The impact of the 1946 general strike coincided with the rapid construction and expansion of government institutions, laws, and political parties with a stake in representing the oil workers as their national spokesperson. Suddenly, the mobilization of benevolence, economic and technical arguments about worker competency, and various forms of monitoring, surveillance, and coercion were inadequate. These techniques and controls flowed into national and international debates about the rights of the maltreated workers and who had the right to control the oil of a sovereign country.

Coordinating Labor Nationally and Internationally

Labor crises opened up new political possibilities involving the nationalization of the British-controlled oil industry that might include the confiscation of company property, oil infrastructure, and the replacement of all foreign labor with Iranian labor. Labor crises also threatened to disrupt the coordination of international monopoly arrangements among the largest oil corporations over the control of production, profits, and now labor. In 1947, the Iranian government set up a judicial committee that ruled that the General Plan of 1936 concerning Persianization was invalid because its provisions were contrary to Article 16 of the 1933 concession (discussed in chapter 3).[139] Iranian government ministers did not endorse the General Plan for Persianization. An alternative had been discussed between AIOC officials and the Iranian ministers for salaries and working conditions, in the "widest sense of all staff and labor," to be the joint responsibility of the government and company. The Iranian government was unconvinced by AIOC's efforts to improve living and working conditions for the laborers.[140] There was strong pressure coming from the Iranian government for the company to seek a formula whereby Iranian staff would receive regular and automatic promotion to senior and management posts or to serve on committees, which would be directed toward the same ends and on which there would be government representation.[141]

AIOC was refusing to do numerical reductions and continued to insist on a formulaic reduction in relation to the "scale of operations," discussed in the previous chapter. The government was also demanding financial assistance for the construction of additional schools in Khuzistan, maternity facilities in Abadan, and the immediate reduction of Indian personnel.[142] Husayn Pirnia, director of the Petroleum Department, insisted that the Iranian government's interpretation of the phrase "annual and progressive reductions" in the text of the 1933 concession could only be satisfied by the execution of an annual numerical reduction. Pirnia also declined to recognize a joint proposal for the implementation of Article 16.

In a telling interview conducted in 1948 by AIOC, Pirnia expressed the prime minister's grave disappointment with the company's attitude toward the discussion of a general plan and its failure to reduce its foreign personnel.[143] The company's willingness to help was "more theoretical than practical" and showed an utter disregard for the repeated explanations offered by the Iranian government and public opinion. The Iranian public was "riveted on the attitude of the Company towards the country," as evidenced by numerous speeches in the Majlis.[144] An announcement would soon be made to the press that general ongoing discussions surrounding the question of Persianization had broken down. Pirnia warned that the government was prepared for AIOC to restrict its operations in Iran in order to keep within its obligations, rather than expanding operations and going beyond the "limit, which the Government could acquiesce in."[145]

Acting as national spokesperson for the oil workers, the Iranian government decided to enlist the help of the International Labour Organization (ILO), originally created by Article 23 of the Covenant of the League of Nations in 1918, and transfer the dispute to a new international site in Geneva in November 1948, just as the British side had done at the League of Nations in 1932–1933.[146] After a three-week stay in the oil regions of southwest Iran, the ILO published a report that was presented at a subsequent ILO meeting in November 1950. The delegates continued to legitimize oil operations by defining Iranian oil workers in the industry in economic and technical terms. The report argued that there remained a "shortage of workers with the required skills" and qualifications, which the "oil industry needs."[147] The report also relied on arguments about inadequate modernization such that the "minds" of local labor seemed "firmly set in traditional ways," and that this constituted "one of the big problems of the industry."[148] Increasing the rate at which Iranian nationals were recruited for employment in the higher categories of wage earners and members of supervisory staff would therefore be difficult. As international spokesperson for workers worldwide, the ILO mission confirmed what the AIOC had always argued, namely that positions were open to all who "acquire the necessary qualifications and experience." To the dismay of the Iranian government, the report presented a positive view of the company's treatment of its laborers, while admitting the urgency of constructing better housing facilities.

Shifting the labor controversy to the international arena failed to strengthen the Iranian government's bargaining position and that of its oil workers. AIOC won the battle by receiving a positive assessment from the ILO mission, which was urged by the British government not to "worsen" relations between the Iranian government and the company by emphasizing labor problems.[149] The controversy had been transformed into an international labor issue, but in favor of corporate interests, which continued to operate through a hierarchical and racially organized regime of labor. The ILO advised that it remained in the interests of both sides to pursue a cooperative strategy. The time had come to "elevate a Persian of proved merit and temperament for work ... to be the third link in the vitally important advisory chain."[150] Mostafah Fateh, AIOC's most senior Iranian employee, was appointed to the post of managing employee relations.[151] The industrial labor regime must be controlled, an official explained to the US government's labor attaché in Tehran, but still appear to produce a "collective relationship between employers and workers in Persia."[152]

The stability of relations between politics and international oil markets was at stake. Achieving this stability necessitated the coordination of labor organization nationally, between AIOC and the Iranian government, and internationally. In a meeting held at the company headquarters, Britannic House, in London, representatives of the largest oil companies gathered to reconsider the whole question of standards of employment for oil companies operating in the Middle East.[153] Representatives of AIOC, the Kuwait Oil Company Limited (KOC), the Middle East Pipeline Company (MEPL), the Trans-Arabian Pipeline Company (TAPLINE), Standard Oil of New Jersey, Shell Petroleum Company Limited, IPC, and United Overseas Petroleum Company agreed that "tremendous sums of money" would be poured into the Middle East in upcoming years.[154] This spending would have consequences for labor matters such as salaries, wages, and other terms of employment.

AIOC representatives discussed the increased pressure to provide accommodations for families of British staff. In response to suggestions by other company representatives, Elkington insisted that concessions could never be implemented for British staff in Iran without extending them to indigenous employees. AIOC firmly held the view that family allowances were the responsibility of the state and not of the employer. It therefore discouraged KOC and MEPL from granting family and separation allowances, as it would be impossible to resist the pressure to extend this practice to Iran, "with embarrassing results."[155] Also, AIOC's European employees worked a forty-two hour week, while local labor worked a forty-eight hour week, but new companies were well advised to adopt the longer week for all shifts. With regard to "local labor problems,"

Elkington argued that industrial relations officers were a useful remedy along with more extensive contact between management and local labor. Industrial relations officers from one's home country could be helpful in advising on labor legislation toward a "practical and workable form suited to the conditions of a country."[156] This was the case in formulating the labor law in Iran. In response to inquiries from David Rockefeller Sr., Elkington went on to discuss the improved standards of technical and other education as implemented in Iran.

AIOC's expertise in managing oil workers and addressing the question of Persianization in Iran was circulated among the major oil companies to determine the best strategies for managing labor and welfare in the Middle East as a whole. AIOC officials urged that steps be taken to "present a united front" by developing a system of wage policy and grading labor, "which will be common to all companies."[157] The coordination of production, prices, and now labor formed part of a system of information exchanges among companies that acted individually but coordinated through associations. Transnational oil companies operating in the Middle East were highly conscious of the need to develop standardized strategies for avoiding labor dissent and inhibiting strong labor organization. They achieved this not only by collaborating with national host governments, but also with other companies internationally. AIOC officials treated the subsoil as one geological unit in order to maintain larger production and pricing arrangements with the largest transnational oil companies. In the same fashion, the oil workers on the surface needed to be managed through a unified labor policy among the largest international oil companies, defying the goals of national politics. Only this kind of coordination and information exchange could successfully protect international production, pricing, and patent arrangements among the world's largest oil producers.

Defusing the Oil Worker's Power

The Majlis finally approved the nationalization of the oil industry on the recommendation of a parliamentary committee on March 20, 1951. The parliament appointed Muhammad Mosaddiq as prime minister on April 28, replacing the shah's choice, Husayn Ala.[158] This monumental decision coincided with the most decisive confrontation between oil workers and the Iranian government, in March and April 1951, when the local Iranian governor attacked the strikers and arrested their leaders.[159] The reformist government, represented by the Iranian provincial governor, attempted to manage the oil workers' power by arguing that mass action against the company would provoke British military intervention and undermine the oil nationalization campaign.

There were a series of work stoppages in the oilfields between 1949 and 1951.[160] The communist movement was revived in 1949, but the government banned the Tudeh Party and the CCUTU after an attempt on the shah's life in February 1949.[161] At the end of 1949, approximately 33,000 of the 38,000 employees at Abadan and 15,000 of 17,000 workers in the oilfields were involved in construction, maintenance, transportation, loading, and pipeline work. A large proportion of the workers were unskilled and the upper level of the managerial and engineering staff remained, for the most part, British. By 1951, an additional 15,000 Iranians worked as contract laborers for the oil company.[162] The status of the Iranian oil workers had not changed significantly since the 1920s. To survive, the Anglo-Iranian oil industry appeared to require unskilled Iranian oil workers, described in the terms of Article 12 of the 1901 oil concession, ensuring that all skilled technical staff would remain under British control for the foreseeable future.

Oil workers at the Bandar Mashur oilfield went on strike on March 24, 1951, to protest new-scale allowances.[163] As a result of loading stoppages at Bandar Mashur, oil production from Agha Jari was halted. Extra troops from the Iranian military were also drafted into the area and the Ministry of Labor issued a circular promising negotiations with AIOC's general management on bathhouses and other amenities.[164] Workmen in the garage and workshops at Agha Jari had also struck in solidarity with Bandar Mashur, and by midday, three-quarters of the laborers had ceased work.

The strike at Agha Jari quickly spread to the main workshops, the welding operations, and the main stores, followed by the electrical departments.[165] Workers submitted complaints about the inadequacy of water, housing, and other amenities, but managers agreed to meet with worker representatives only after the men returned to work. AIOC officials were well aware of the urgency of providing permanent housing and amenities in the oilfields, but they were also concerned about the possibility of the strike spreading nationally as a result of "Tudeh encouragement."[166] On the other hand, working in alliance with the Iranian government to repress the more militant forms of labor protests provided the company with a degree of security. As a means of restoring order, the government ignored the workers' complaints and declared martial law in the strike areas.

Demonstrations by apprentices outside the Abadan Technical Institute commenced the following day, March 26.[167] Workers at Masjid Suleiman struck on March 28 when those employed in the garage, workshops, stores, and electrical and other departments did not report for work and gathered outside the main company gates.[168] Work stoppages at Naft Safid and Lali followed, with workers demanding the restoration of certain allowances, as well as improved promotion rates and a minimum wage.[169] In March, the oil workers immediately formed a strike committee representing the different striking areas and presented a series of grievances concerning pay and amenities.[170] The Iranian government declared martial law on March 27, and AIOC imposed a curfew on the oil workers at Agha Jari. By the start of April, bungalow servants had left their places of employment, and there was a falling off in attendance of "Production Labor." British staff were forced to take over and operate the "Production Unit" and "Flow Tank."

In the meantime, the Iranian government sent a "Commission of Three" to investigate oil workers' claims and meet with six striker representatives along with AIOC management. The director of labor also arrived to distribute notices describing the legal machinery embodied in the 1946 labor law, through which disputes must be solved collectively rather than by strike action, which the company had declared illegal.[171] Reza Dinashi of the garage workshops at Masjid Suleiman dispatched a telegram to the Majlis stating that "since the question of nationalization of the Southern oilfields had been deliberated, the Co. had increased its pressure on the workers, hence the reason for their strike."[172] The strike ended at Agha Jari on April 11 with the announcement that the company would offset strike pay against leave entitlement—that is, in place of receiving payments to help meet their basic needs while on strike, the workers could make use of their legal right to take leave or be away from work. According to company estimates, 95 percent of the workers returned to work the following day. As in 1946, the 1951 strike occurred as an assemblage of diverse actors, organizational forms, and coercion in relation to the flow of oil, but the central issue was now national control of an entire oil industry.

The 1946 strike was different, however, in that it was almost entirely "industrial in its origin," while the 1951 strike action, according to company officials, was "clearly political."[173] AIOC managers were alarmed by the degree of support offered to the oil workers by the Iranian government, the Majlis, and public opinion. The Tudeh appeared to be exploiting the situation for "its own ends," and officials admitted the need to "eradicat[e] the Tudeh element."[174] AIOC managers hoped that the Iranian government would "capitulate completely" with regard to increased pay, amenities, and workers' strike pay. British managers were also alarmed about oil production, recalling that it took nearly a month to restore refinery operations after its last shutdown during the 1946 general strike.[175] Tudeh agitators and labor were making the political origins of the strike quite clear. According to Elkington, "in the same [that] way you will recall how whole cities were brought out on strike in India during the Congress struggle for home rule," many Iranians possessed "no illusion whatever in regard to Nationalization," but wished a greater share in the profits and administration of the oil industry, which occupied a dominant place in the country's economy.[176]

The British government urgently needed to decide future oil policy based on a very real loss of production at Abadan due to the strike.[177] The main issue was whether the company should move to rely on refineries outside of Iran (e.g., UK refineries at Llandarcy or Grangemouth or even Trinidad). The problem was that the quality of the oil would change, altering the kinds of products that could be produced.[178] The shutting down of a refinery constituted a major point of vulnerability where labor movements could organize and apply pressure.[179] Podobnik explains that refiners have a more strategic position versus drillers and oilfield laborers in that refining requires the application of heat, pressure, and chemical agents to unprocessed oil to extract impurities and produce standardized categories of fuel.[180] To the extent that labor militancy has emerged in the oil sectors of specific countries, it has "focused on these refinery operations." This was the case in southwest Iran, where the oil was not as easily replaceable as company experts had hoped, and adjustments in refining operations abroad would have to be implemented to accommodate the loss of Iran's oil and its replacement with a comparable, salable product.

The 1951 strike succeeded in creating a situation of immense political uncertainty in the national arena. Ayatollah Abolghassem Kashani, a prominent religious leader, member of the Majlis, and member of Mosaddiq's National Front coalition, suddenly emerged as another national spokesperson, issuing a communiqué urging strikers to return to work on April 17. Kashani argued that the strike action provided a pretext for disturbances to the "enemies of the country." The workers pressed on. The representative for the Abadan strikers issued a second list of demands to AIOC's Industrial Relations Department arguing that the minimum wage needed to be fixed as soon as possible. Until strike pay was paid, the workers were not prepared to "cease the strike, our last legal weapon."[181]

The strikers were making their demands in nationalist terms proclaiming the defeat of the British oil company by the Iranian nationalist movement.[182] The Tudeh Party was viewed as a legitimate spokesperson for the oil workers and demanded that the Iranian government declare the cancellation of the 1933 concession. The proclamation declared that the oil concessions of 1901 and 1933, "which had been exchanged between the internal and international traitors for dividing our resources of wealth," were "nothing but scraps of paper." The oil workers were exhibiting new kinds of power, building connections between economic and social demands as well as national demands about control over the oil industry. In a message to workers at the Abadan refinery, the strikers at Bandar Mashur denounced the legitimacy of the commission sent by the government to address their demands. The message expressed "the Iranian nation's" desire that the Majlis announce the cancellation of the British-controlled oil company and its nationalization.[183]

As with the strikes that preceded it, the 1951 strike constituted a shift in methods of managing concerns as well as in the degree of violence exercised to reinforce a particular organization of labor. A violent confrontation had emerged between the Mosaddiq government and the oil workers, leading to the arrest of the strike leaders. Both British and Iranian employees were killed in the violence. The government imprisoned strike leaders and extended martial law through the course of the strike.[184] Whereas AIOC sought to maintain secrecy about the impact of the strike by withholding all data on the loss of production,[185] the Tudeh Party, like the Communist Party of Iraq, sought to expose and disrupt the most vulnerable points in the technical structures of oil production.[186] Coordinated decisions by the company and the Iranian government to direct the oil workers into a manageable union framework, and ultimately reformist nationalism, were most evident as they battled to repress striking oil workers.

As in Mexico in 1937, a reformist government in Iran had attempted to "defuse the oil workers' power by nationalizing the country's oil industry ... on terms more favorable to the foreign oil company than those demanded by the union and the communist party."[187] As the following chapters reveal, the international oil companies, including AIOC, would ultimately refuse to accommodate the national organization of Iran's oil industry. In 1953, a CIA-organized coup successfully reestablished foreign control. Each controversy over the control of oil in Iran was shaped by the very real threat of "Persian nationalism" and its connections to the motives of the Iranian government in relation to national control of the industry.

Reconfiguring the History of (Oil) Nationalism

To survive, AIOC required new definitions of the oil worker in technical, economic, and legal terms into which was built a peculiar limitation. The limitation expressed the impossibility of employing a skilled Iranian worker demanding a better life and more equitable forms of oil distribution and profits. As the company continued to make promises to redress the plight of the Iranian oil worker, it also promoted a process of exclusion along racial-technical lines. However, this process became less effective after World War II as nationalist and populist politics spread across oil infrastructures around the world. The excesses of AIOC's efforts to police the boundary between unskilled and skilled Iranian oil workers had the unintended consequence of transforming the political possibility of nationalization of the entire industry into a reality.

A sociotechnical approach to a history of oil workers demands closer attention to the machineries of building an oil labor regime, which were simultaneously sites of politics. By following the ways the building of a racial-technical oil labor regime leads to politics, we see that the history of nationalism, or the nationalization of the oil industry, was about the opening up and narrowing down of different political arrangements involving Anglo-Iranian oil and its infrastructure.

Following Geoffrey Bowker's study of a French oilfield services company, this process of racial-technical exclusion was not so much due to the workings of a company state within a state as it was "a series of filaments (roads, pipelines, oil wells) operating within a different social time than the old state, and with minimal spatial interference."[188] This process of inclusive exclusion helped constitute the energy system's form. New kinds of definitions of the oil worker were possible that were impossible before, particularly at oil operation sites in which the locals from Abadan town and the whole of Khuzistan Province did not have access. Instead of unskilled Iranian labor, the company now needed skilled Iranian labor to justify its recruitment policy at a controlled rate that would not disrupt its Iranian-to-British worker ratio. As with oil workers around the world, to arrive at the enclave of the oilfields, unskilled Iranians would ultimately have to go abroad to acquire adequate (Western) training and technical knowledge.[189]

AIOC mobilized paternalistic practices in response to a series of labor crises over the legitimacy of British control of oil operations. The sophistication of the social technologies traced here, and of the calculating technologies traced in the previous chapter, reveals the inventiveness of AIOC and other oil companies that transformed political questions of labor into technoeconomic issues and cooperated in the management of labor in the Middle East and Latin America. In a corner of southwest Iran, political possibilities were worked out in the course of each strike, with consequences for the organizational design of an industry that did not respect borders marking the national state as distinct from a foreign oil corporation. These local events were directly related to the ways in which the flow of petroleum energy was organized at various points in production, distribution, and refining. Each of the strikes constituted a battle over the connectivity of issues that AIOC fought hard to keep separate from political concerns. By 1951, the mechanism of the strike remained the only legal weapon to achieve economic demands and defy the company strategy of inhibiting independent union organization. Through each strike action, the oil workers revealed new powers as they built connections between their economic demands, technical knowledge, and efforts to win national control of the oil industry.

The company and the Iranian government also sought to maintain a border between oil operations and the question of violence and security. The strategy entailed policing a peculiar regime of labor organized according to racial difference but justified in terms of the technical (in)competencies of the workers. The strikes were power struggles in which workers mixed the technical and the social in order to make their impact on the various points of vulnerability in the energy system decisive, particularly at the Abadan refinery. Disputes over the question of Persianization, which lingered throughout the nationalization crisis, served as the occasion for the production of expert knowledge. This technoscience would then shape which aspects of the labor question acquired political significance and which aspects remained strictly technical. Such an approach, it was thought, would stabilize the energy system by eliminating any controversy over the definition of the oil worker.

Rather than viewing the politics of oil nationalism in terms of large-scale political events, institutional (racial and paternalistic) ideologies, and actors rescued as historical agents in the making of their own history,[190] this chapter has located the most important political battles in the fields of technicality pertaining to the working definition of the

Iranian oil worker. The oil workers helped generate vulnerabilities in which the powers of subaltern dissent as well as the control of energy were taken up simultaneously, distributed, and transformed toward militant nationalism. The excesses of violence and failures of the company to suppress labor unrest through the intervention of alternative organizational machineries exposed the kind of labor regime that AIOC, soon to be BP, needed to survive and the very shape of the national state. This was most evident in the ways a few dominant spokespersons—the Communist Party and a reformist nationalist government—would claim to speak on behalf of the many silent workers in the oilfields.

5
Assembling Intractability: Managing Nationalism, Combating Nationalization

Any differences between the parties of any nature whatever and in particular any differences arising out of the interpretation of this Agreement and of the rights and obligations therein contained as well as any differences of opinion which may arise relative to questions for the settlement of which, by the terms of this Agreement, the agreement of both parties is necessary, shall be settled by arbitration.

If one of the parties does not appoint its arbitrator or does not advise the other party of its appointment, within sixty days of having received notification of the request for arbitration, the other party shall have the right to request the President of the Permanent Court of International Justice ... to nominate a sole arbitrator ... the difference shall be settled by this sole arbitrator. (Articles 22(A) and (D) of the 1933 Concession)

Before the date of the 31st December 1993 [date of expiry] this Concession can only come to an end in the case the Company should surrender the Concession (Article 25) or in the case that the Arbitration Court should declare the Concession annulled as a consequence of default of the Company in the performance of the present Agreement. (Article 26 of the 1933 Concession)

The post–World War II petroleum order witnessed a series of governments, including Venezuela, Saudi Arabia, and Iraq, taking steps to nationalize their foreign-controlled oil industries. On March 8, 1951, Iran's parliamentary oil committee approved a resolution recommending nationalization of the oil industry and requested that the Majlis allow the committee two months to study how to put nationalization into effect. The parliament's resolution did not mark the cause of the dispute, however, but was merely the outcome of multiple controversies concerning the authority and terms of the oil concession, in its various forms, proliferating over the course of five decades.

The origins and transformation of Iran's oil nationalization dispute emerged from many sites between 1948 and 1952, when the validity of the oil concession, the calculation and control of profits, and the replacement

of British managerial and technical staff with Iranian labor were scruti-
nized. These disputes reopened as controversies susceptible to alternative
arrangements with political ramifications larger than ever. This chapter
explores the conditions and factors contributing to the success and failure of oil industry nationalization. This battle was fought locally in the
oilfields of Khuzistan, nationally in Tehran, and on an international scale
between the leading and up-and-coming oil producers of the world and
the major oil corporations, attempting to protect the terms of their concessionary relations with them.

The extensive scholarship on the oil nationalization crisis ignores
the machinery of nationalizing an oil industry. Starting with the act of
nationalization, much focus is placed on the diplomacy of the Tehran-
London-Washington connection, which is disrupted by the Anglo-
American engineered coup that overthrew the Mosaddiq government in
1953 and reinstalled Mohammad Reza as shah and General Zahedi as
premier.[1] This dominant narrative of failed oil diplomacy ending in a
coup d'état overlooks the specificity of technical battles and the organizational varieties that emerged in the working out of a highly successful
consortium arrangement among the major oil companies.[2] Other studies treat governmental and nongovernmental entities, such as AIOC or
the British and Iranian governments, as separate groups and thus take
their organizational forms for granted.[3] This is insufficient because the
focus is placed on the impact of powerful private and commercial interests on official government policy or the reverse.[4] In all of these studies,
questions of diplomacy, law, public and private interests, the coup d'état,
and the working out of the dispute in the oilfields are treated as separate events and entities with their own set of historical actors. In what
follows, I leave these relations open and follow how diverse actors are
enrolled in disputes over nationalization. Certain organizational forms,
such as a national oil industry independent of foreign control, are closed
off while others, such as the consortium arrangement reestablishing foreign control, are left open. I show how an understanding of the most
important events and actors in this crisis changes by working through
Iran's oil nationalization dispute in terms of its technical dimensions or
the machinery of the process.

In response to the Iranian government's ratification of a series of
nationalization laws,[5] the British government and AIOC mobilized economic and legal arguments and organizational entities to defend their
right to control the oil. For example, AIOC built connections between
flows of Anglo-Iranian oil and the International Court of Justice (ICJ) at

The Hague (the PCIJ, prior to 1946) in order to block Iran's attempt at nationalization and restore the terms of the 1933 concession, not due to expire until 1993. Additionally, and in coordination with the American and British governments and oil firms, AIOC pursued financial restrictions through the use of economic sanctions, and technical blockages through the use of the oil boycott, to preserve a politics of oil production according to international monopoly arrangements. With what kinds of arguments, procedures, and calculative equipment was this connectivity achieved?

As in the concessionary dispute of 1932–1933, the nationalization law rejected the oil concession as an illegal document. The act of nationalization marked the breakdown of normal apparatuses of control that AIOC managers had set in place for the calculation and control of profits and oil operations. The decision to nationalize the oil industry was an open issue that generated multiple technical proposals and diplomatic missions. The mechanism of international law was enlisted to manage the dispute in terms of a kind of intractability, diversion, and delay, which was also a political weapon. I investigate the political consequences for the role of the Iranian state as a sovereign entity, and the role of the oil corporation battling against nationalizing oil producers in the Middle East. The successes and failures of each kind of technical arrangement are tracked to demonstrate the ways intractability was constructed to impede the flow of oil, and thus the flow of income to the Iranian government, helping to stabilize a peculiar global oil economy based on an artificial scarcity of supply.

Strategies of intractability worked in concert with other kinds of technologies and interests. For example, in the first section I examine the politics of profit sharing such as the 50–50 arrangement, which was constructed to manage the mid-twentieth-century oil order against the threat of militant nationalism and communism, as well as to protect the profits of the increasingly powerful American oil companies. This strategy involved an unusual mixing of oil contracts and currency flows, US tax law, and Iranian popular opinion. I then follow the dispute, again, to the international legal arena where an extended mechanism of temporization framed national sovereignty in legal terms rather than political ones in order to block the implementation of any real nationalization. Next, I follow how Western technical consultants and government officials arranged Iran's oil nationalization in terms of reformist nationalism and the power to determine markets. The strategy worked, not by ensuring the flow of oil, as they so often claimed, but by imposing limits through an insistence on the economic facts about oil. Oil consultants argued that the only way oil could flow was within the apparatus of the international oil industry, as separate from political questions of national sovereignty.

Nationalizing Iran's oilfields involved reworking the power of AIOC and the national state locally in southwest Iran through the elimination of British technical staff. The peculiar way oil operations were managed along racial-technical lines worked in tandem with sanctions and a boycott to help the British and American allies constrain more democratic forms of oil production. I place collectives of legal, diplomatic, and economic machinery of the nationalization crisis at the center of the analysis. By doing so, I identify the political role long-distance machinery played in the maintenance of a particular oil economy in the Middle East.

Technologies of Profit Sharing and the Threat of Militant Nationalism

The Majlis took one of its first steps toward nationalization in 1947 when it declined to ratify a draft for a future agreement granting an oil concession to the Soviet Union for the development of Iran's northern oilfields. Qavam-al-Saltaneh, the prime minister, had promised Stalin an oil concession in return for the withdrawal of Soviet troops during the Azerbaijan crisis of 1946.[6] The same law that declared the draft null and void carried a provision that when the rights of the Iranian people and the economic wealth of the country had been infringed on, whether in the matter of subsoil resources or otherwise and especially in regard to AIOC's oil concession, the Iranian government must take action toward the reestablishment of these rights and inform the Majlis of the results obtained.[7]

Prior to the call for nationalization of its oil industry, the Iranian government made a series of attempts to renegotiate the terms of the 1933 concession with AIOC. Iranian officials initiated formal discussions with AIOC in the autumn of 1948, presenting the company with a memorandum that put forward twenty-five points for discussion, concerning the calculation of profits and treatment of Iranian labor.[8] Iran was demanding the same profit-sharing terms as Venezuela, which received 50 percent of the profits of Creole Oil (a subsidiary of Standard Oil of New Jersey) under its concession. Iran had received £7 million in royalties in 1947, but following the Venezuela model, should have received royalties of approximately £22 million. Other sticking points included Iran's demand that it be exempted from British taxation on its share of AIOC's annual profits, that it acquire a share of profits from the sale of Anglo-Iranian oil overseas, and that it inspect AIOC's books to ascertain whether the Iranian government was receiving its fair share of royalties. Finally, the Iranian government pointed out that the company had ignored its demands to improve the working conditions of the Iranian workforce and train Iranian staff to replace foreign employees in skilled jobs.

In contrast to demands made in earlier disputes, the Iranian government was now proposing an equal sharing of income based on the profits obtained by AIOC from all of its activities, whether inside or outside of Iran. Negotiations continued through 1949 with AIOC showing indifference to the points raised. On January 20, 1949, the Majlis submitted a bill calling for the cancellation of AIOC's concession, which triggered new negotiations. When these discussions also failed, William Fraser, who took over from Cadman as chairman of AIOC in 1941,[9] traveled to Tehran and presented Mohammad Sa'ed, the prime minister, with his final offer: a draft of a "supplemental oil agreement," framed as a supplement to the 1933 concession.[10] The offer proposed a mix of payments to Iran, which included royalties, taxes, and Iran's 20 percent share in dividends and reserves.

AIOC was eager to make a deal prior to an anticipated devaluation of sterling by the British government. The deal would ensure that Iran would not be in a position to make demands later. Company officials expressed concern for the general economic state of Britain, but also that of Iran in light of the sterling's planned devaluation.[11] It is important to note here that Iranian oil operations supplied Great Britain with twenty-two million tons of oil products and seven million tons of oil per year.[12] It also generated £100 million annually in foreign exchange, which Heiss says "the British sorely needed."[13] In terms of profits and income tax alone, AIOC, not counting its numerous subsidiaries, paid approximately £15 million to the British Treasury in 1948.[14] The oil currency issue was also connected to Iran's domestic plans for economic development. Abolhassan Ebtehaj, head of the National Bank (Bank *Melli*), had requested £10 million from AIOC to stabilize the exchange rate between Iranian rial, sterling, and dollars. This occurred in addition to a separate request from the British ambassador for £10 million worth of goods for the purpose of increasing their flow and raising additional revenue for Iran's Seven Year Plan to implement large-scale projects of modernization and industrialization throughout the country.

The Supplemental Agreement was signed on July 17, 1949, awarding Iran 32–37.5 percent of total net profits. When the agreement was presented to the Majlis in the form of a bill on July 23, however, the opposition, headed by Hossein Maki, deferred the decision to the next Majlis session. The Bank of England devalued the British pound on September 18, 1949, causing the pound to fall against the dollar by 30.5 percent.[15] The Iranians continued to complain that the British government was, through various tax laws and regulations, making far higher revenues from its oil than the Iranian government as owner of the natural resource.[16] For example, in 1945, Britain made in taxation almost triple the amount that Iran received in taxation and royalties on the sale of its oil and products.[17]

Mosaddiq and his allies in the National Front (*Jebhe Melli*), the political opposition party founded by Mosaddiq, considered the oil bill on the Supplemental Agreement a symbol of Iran's subjugation to foreign interests.[18] Members of the Majlis had turned against the bill after the finance minister, Abbas Golshayan, presented a fifty-page report commissioned by Gilbert Gidel, a professor of international law at the University of Paris, that "documented the accounting tricks by which Anglo-Iranian was cheating Iran out of huge sums of money."[19] Anxious to see their oil agreement approved, British officials encouraged the shah to replace Ali Mansur, the prime minister, whom they viewed as too weak to push the oil bill through, for a more authoritarian figure. On June 26, 1950, the shah appointed Ali Razmara, a military general, to replace Mansur.[20] The following day, Razmara asked the Majlis for a vote of confidence and, against the wishes of the National Front minority, was endorsed as prime minister.[21]

Razmara promised AIOC officials and Sir Francis Shepherd, the British ambassador to Iran, that he would help get the Supplemental Agreement passed but only if a number of points, not unlike the previous twenty-five-point proposal, were improved.[22] As before, Razmara failed to receive a positive response from Britain on any of the issues. He then turned to Henry Grady, the US ambassador, for help. Grady, in turn, notified Dean Acheson, US secretary of state, who advised Ernest Bevin, the British foreign secretary, that Britain's failure to take positive action against AIOC's "intransigence" was inappropriate to the current situation in Iran and world conditions. As Vitalis has highlighted, US officials blamed the British for clinging to their "outmoded imperialist ways and blocking reforms that might check communism's advance."[23] The American government viewed nationalization as an act of communism that threatened the security of their national interests at home and international oil markets abroad.

In a top-secret summary of US views on the nationalization question, American officials speculated that a general 50–50 profit-sharing for-mula would have a "useful stabilizing effect in the Middle East."[24] In this arrangement, the Iranian government would have a contract with the oil company, which would retain control of operations, have its capital costs reimbursed, receive 50 percent of the profits, and continue to "determine markets." British and American interests, in the format of the transna-tional oil corporation, would never give up their power to "determine markets." Losing this power meant increasing the risk of nationalization in terms of a loss of control of oil production, as well as the chances of Iran controlling pricing and marketing.

"Determining the oil markets" gave Britain and the United States the power to manage the global flow of Iranian oil and impose limits to artificially keep prices high. Under the Truman administration's Point IV Program, the United States placed a priority on providing all possible sup-port to the Iranian government and the shah in the form of economic and technical assistance as well as military equipment and training. It favored 50–50 profit sharing as an alternative for managing the economic order of moderate nationalism in the Middle East. Moreover, profit sharing would avoid the more dangerous alliance of militant nationalism, communism, and national control of oil operations. In September 1950, the US State Department held a series of discussions with British officials within the wider constellation of Anglo-American oil interests in the Middle East.[25] According to Richard Funkhouser, the State Department's oil economist, the 50–50 profit-sharing arrangement between Venezuela and the Creole Oil Company was well known in Tehran. Iran and other Middle East oil producers could not be prevented from demanding the same terms. For forty years, AIOC had operated with production costs under 10 cents per barrel and a sale price between $1 and $3. This long-term arrangement had guaranteed much higher profits to the company than to the Iranian government. Funkhouser advised George McGhee, assistant secretary of state for Near Eastern affairs, to encourage the British Foreign Office to accept Razmara's demands before Aramco finalized its agreement with Saudi Arabia on a 50–50 basis.

The conventional arrangement for the control and sale of oil prior to the 1940s had been for a host country to receive a fixed royalty rate from the concession companies per ton of oil produced.[26] The govern-ment of Venezuela, however, successfully challenged these terms in 1943 and 1947, resulting in an approximately equal or 50–50 division of prof-its with the transnational oil companies.[27] In 1943, the Hydrocarbon

Law was enacted in Venezuela. Under its provisions, royalty rates and exploration, development, and surface taxes were increased significantly. The Venezuelan government expected that these increases together with a general income tax would equalize its income with the profits earned by the oil companies. It later became evident that this was not achieved and in the next several years, the government levied further compulsory payments on oil company profits to bring about the equalization. In 1948, it became evident, again, that high production and prices were bringing profits to the oil companies that exceeded the income accruing to the government. As a result, an amendment was added in November 1948 to Venezuela's "Income Tax Law" in the form of an "Additional Tax," applicable only to mining and petroleum companies. This tax was imposed at the rate of 50 percent on the amount by which a company's net income exceeded all royalties and other taxes accruing to the government. This tax ensured that the government's revenues would at least equal the net profits of each oil company. However, if royalties and normal taxes received by the government exceeded company earnings, no downward adjustments in the government's share were provided for. By the end of 1948, Venezuela had achieved its objective of an approximately equal sharing of oil company profits.

The US oil firms assented more easily to such an arrangement because the additional rents to Venezuela were in the form of "taxes" that allowed the oil companies to deduct the payments from those owed to the US Treasury. American oil corporations, with the encouragement of the State Department and US Treasury, were exploiting an enormous loophole in US tax law. The loophole allowed them to evade antitrust laws by filing the amount paid as a higher royalty to the Saudi government as a foreign income tax that could be deducted from the company's regular taxes to the US government.[28] The American oil corporation's preference for this new technology of profit sharing, the 50–50 arrangement, was due to favorable tax laws and pressure from US government agencies to rely on the US taxpayer to subsidize the higher royalty paid to Saudi Arabia. As negotiations began in 1950, AIOC refused to agree to similar revisions when pressed by Tehran, where a Venezuelan delegation was visiting.[29] British law did not grant firms the same capacity to rely on the British taxpayer to offset tax payments.[30]

Negotiating generous dealings with host governments in lieu of nationalization, Heiss explains, motivated US oil giants to negotiate 50–50 profit-sharing deals with Venezuela in 1948 and Saudi Arabia in 1950.[31] US oil officials believed that AIOC should follow this sort of strategy in

Iran. In a September 1950 meeting with officials from the State Department, US oil executives warned that dealing generously with the Iranians would indeed be "a sound commercial proposition." Failure might lead to disaster not only for AIOC (and the British government), but to all the Anglo-American oil majors.[32] However, Heiss's analysis overlooks the specific ways in which the mechanism of 50–50 profit sharing worked as a new technology of management to protect oil profits and manage political outcomes. The United States was using 50–50 as a way of blocking the coming together of militant nationalism with national control of oil, which would destabilize relations between politics and international oil markets.

The dispute over royalties and profit sharing marked the beginning of the drive in the Iranian parliament toward nationalization of AIOC's holdings. It also marked a shift toward accommodation on the part of Aramco's owners in Saudi Arabia, who signed their own 50–50 agreement in December 1950. While the first 50–50 agreement was signed in Saudi Arabia, the first 50–50 negotiations in the Middle East actually took place in Iran in connection with calls for the revision of the 1933 concession. But there was a "world of difference," psychologically, in AIOC's offer to increase *royalties* up to 50 percent and Aramco's 50–50 *profit sharing*, even if the division of revenue was roughly the same.[33]

McGhee's US mission to London in 1950 to encourage the British Foreign Office to accept Razmara's offer proved unsuccessful after the British blocked a $25 million credit request from Iran that the Export-Import Bank had authorized the same year.[34] Sterling and dollar flows, 50–50 arrangements, US tax law, American and British missions, and the demands of the Iranian public and parliament to take control of the oil industry were now entangled in Iran's oil nationalization dispute with AIOC. The participation of American officials and profit-sharing technologies were also bound up in larger connections that sought to stabilize national oil economies, manage the threat of militant nationalism, and protect the profits of the largest American oil companies.

In these early moments of national and potentially international crisis, the Majlis, led by Mosaddiq and his National Front allies, served as a political obstacle against the shah and British interests. In October 1950, Mosaddiq argued the legality of the concession, claiming the 1933 concession was null and void because it was enacted during a dictatorship when Majlis deputies were not the real representatives of the people. The Supplemental Agreement was framed as the affirmation of a basically invalid concession, representing Iran's continued subjugation to unjust terms.

The deputies unanimously passed a resolution on November 25, 1950, opposing the agreement for offering Iran inadequate rights.[35] Gholam Hussein Forouhar, the finance minister of Iran, was forced to withdraw the bill on the Supplemental Agreement and resigned after the December 30 announcement of Aramco's signing of a 50–50 agreement with Saudi Arabia. The Majlis instructed its oil committee to formulate a new policy that the government could pursue in securing Iran's oil rights.[36]

A different kind of moment marked this mid-twentieth-century political order in which new technologies of profit sharing were open to oil-producing countries. The nature of Anglo-American interests was also shifting, with oil majors taking a decisive role in building a particular relationship among oil currency flows, military, financial, and technical aid, and the battle against communism. Negotiations surrounding the Supplemental Agreement were very much a rehearsal for the calls to nationalize Iran's oil industry looming in the background and in light of experiences in Venezuela and then Saudi Arabia. Unwilling to follow the American lead by pursuing 50–50 profit sharing, AIOC hoped to maintain the arrangement it had worked hard to stabilize over the years, insisting on the concessionary relation as a basis for accumulating profits and controlling Iranian oil in world distribution and marketing arrangements. The uncertainty remained, however, as Iranian public opinion, reformist nationalists, and striking oil workers threatening more militant forms of nationalism were increasingly unwilling to turn back to more undemocratic forms of oil production.

Return to International Law and National Sovereignty

An alliance of oil workers, religious political groups, the reformist-nationalist National Front, and the now clandestine, communist Tudeh Party came together in support of nationalization (figure 5.1). This collective of conflicting political groups and oil workers as well as national public opinion and the law framed the oil issue as one about Iran's national sovereignty and, therefore, its legitimate claim to a natural resource. In a speech to the Majlis, Mosaddiq justified why AIOC had not fulfilled the terms of the concession in good faith. He referred to Article 16(III) of the 1933 concession calling for the progressive replacement of non-Iranian employees by Iranian nationals in "the shortest possible time"; Article 14(B) with regard to the duty of the company to place at the disposal of the Iranian government "the whole of its records relative to scientific and technical data"; and Article 10(V), the gold clause, which was intended to protect the Iranian government against losses that might result from fluctuations in the value of the British pound sterling.[37] Majlis deputies submitted a total of sixty-seven proposals ranging from a reversion to the D'Arcy Concession of 1901 to a 50–50 arrangement with Iranian control. But, as Elm explains, it was the proposal drafted by Mosaddiq and his National Front allies in January 11, 1951, calling for nationalization of Iran's oil industry, "that set the future course of the Majlis and the government."[38] Soon thereafter, Ayatollah Kashani, the prominent religious leader and member of the Majlis, joined in backing the National Front's calls for nationalization. AIOC then attempted to reopen negotiations on a 50–50 arrangement, but these attempts were blocked on February 19, 1951, when Mosaddiq presented the Special Oil Committee of the Majlis with a formal resolution for nationalization.[39]

On March 15, 1951, the Majlis passed the "Single Article" bill, nationalizing the Iranian oil industry. In April, a more detailed bill of nine articles was prepared and passed by both the Majlis and the Senate, and then approved by the shah on May 1 and 2, 1951.[40] By approving both the Single Article and the Law of Nine Articles, the Iranians transformed the question of oil nationalization into a legal-national issue to increase their bargaining power. The passage of the articles included the formation of a Mixed Parliamentary Board consisting of eleven members who worked on implementation of the nationalization law.[41] The nationalization law and its law of implementation had three effects in legal terms:[42] (1) they canceled the 1933 oil concession; (2) they expropriated all the property of AIOC in Iran insofar as such property related to the oil industry; (3) they vested the expropriated oil industry in the new political actor, the National Iranian Oil Company.

Following the passage of Iran's nationalization laws, AIOC responded in legal terms by requesting arbitration according to the provisions of Articles 22 and 26 of the 1933 concession, quoted above.[43] In the midst of an imminent crisis in Tehran and in the southwest oilfields where workers were preparing to go on strike, the British Foreign Office decided that it was vitally concerned with AIOC's policies in Iran and proceeded to take an active part in efforts to prevent the oil industry's nationalization.[44] Before the Iranian government had a chance to reply to the company's request for arbitration, Shepherd, the British ambassador, presented the Iranian government with an aide-mémoire, which set forth "in strong language" the British government's view on the legal position of the company.[45] If Iran rejected the company's request for arbitration, it specifically reserved the right to transform the dispute into an international

Figure 5.1
Scene of Tehran supporters hauling down AIOC sign on June 24, 1951. The company captions the photo as "An angry mob, incited and supported by the Islamic groups and the National Front, tore down the signs above the Anglo-Iranian Oil Company's store yard, Information Centre and Central Office and paraded them upside down in lorries around the town." *Source*: INP P 277 Special 53961, 1951, 78148, BP Archive. Reproduced with the permission of the BP Archive.

issue, just as it had done in 1932–1933, by taking the case to the ICJ at The Hague. The aide-mémoire also reiterated the suggestion that the dispute be solved by negotiation.

For the second time in its history, the British-controlled oil company was threatening to transform a concession dispute into international legal terms as a means of protecting its concession rights and rendering Iran's oil nationalization illegitimate. On May 15, the British government took more drastic steps, announcing that paratroopers were being held in readiness in the United Kingdom to protect its nationals and prevent the "illegal seizure of the property of AIOC."[46] In response, Iran's minister of finance, M. Ali Varashteh, wrote to the AIOC representative in Tehran that his government rejected AIOC's request for arbitration. The nationalization of the Iranian oil industry was not referable to arbitration and "no international authority was competent to deal with the matter" because it was entirely within the domain of the Iranian government.[47] However, AIOC representatives were invited to meet with the Majlis's Special Oil Committee to arrange the implementation of the nationalization laws.

In a speech to the foreign press, Mosaddiq spelled out Iran's reasons for refusing to arbitrate by framing the dispute in terms of national sovereignty. First, the sovereign right of the Iranian nation entitled it to nationalize its oil industry, but AIOC was arguing that the Iranian nation had annulled its contractual agreement.[48] Second, Mosaddiq argued that the 1933 concession was signed under duress, rendering it invalid. Third, the Iranian Majlis had not acted in a way that gave AIOC reason to refer the matter for arbitration. He agreed that if the question had been raised regarding the validity of the agreement, the need for arbitration would arise, but "neither the Majlis nor the Iranian government has raised any point regarding the agreement."[49] Therefore, the oil company could not invoke the arbitration clause. The only remedy for a situation in which a government exercised its sovereign rights, causing loss to a private corporation, was to claim compensation from that government. The law, Mosaddiq explained, already provided for such compensation, but again, the 1933 concession was not the subject of the Iranian government's discussion and any reference to it was misplaced.[50]

AIOC responded that as a consequence of the Iranian government's refusal to rely on arbitration, the company was applying to the president of the ICJ to appoint a sole arbitrator in accordance with Article 22, paragraph D of the 1933 concession. The following day, the British embassy in Tehran notified the Iranian government that the British government, as a separate party from AIOC, had brought the oil dispute before the ICJ. The Iranian government had declared the concession null and void, yet AIOC continued to draw on its articles for leverage. The Iranian government and AIOC battled to frame the dispute in legal terms, one group on the basis of its sovereign rights and the other on the basis of its oil concession. Now the British government stepped in, on behalf of a private corporate interest in which it had a controlling stake, to transform the dispute into international legal terms by taking the case to the ICJ. An entire legal-metrological order was being invoked for the second time.

Eric Beckett, legal advisor to the British Foreign Office, submitted the "Application" instituting proceedings at the ICJ. The Application requested that the Court declare the execution of the Iranian nationalization laws a violation of international law insofar as it claimed the unilateral annulment of the terms of the 1933 concession. By rejecting the company's request for arbitration, the Iranian government denied the company the exclusive legal remedy provided for in the agreement, which constituted a denial of justice contrary to international law.[51] The British government made a series of claims to the ICJ to justify the authority of the concession and ensure its stability.[52] If the Iranian government persisted in rejecting this legal remedy, it would be responsible for "a denial of justice against a British national."[53] In its conduct, the Iranian government had "treated a British national in a manner not in accordance with ... international law and have in consequence, committed an international wrong against the Government of the United Kingdom."[54]

As with the proceedings at the League of Nations in 1932–1933, the British government made a case on behalf of a private company, framed as a British national that had not been treated in accordance with the principles of international law. As Ford explains, there were two basic questions that would determine the legal validity of the Iranian nationalization laws from the perspective of the British government: the actions taken in the course of their execution, and the international responsibility for such actions.[55] The first question concerned the propriety and implications of Britain's exercise of diplomatic protection. The second question concerned the international law governing the right of a state to expropriate the private property of aliens, such as a foreign firm, located within its borders.

Contrary to Mosaddiq's view, the British government argued that the ICJ had the jurisdiction to resolve the dispute because it fell under the terms submitted with the League of Nations in September of 1932, when the Iranian government "accepted jurisdiction of the Permanent Court of International Justice in conformity with Article 36 (2) of the Statute of that Court."[56] Thus, according to the British, Iran was bound to accept "on the basis of reciprocity vis à vis any other government" the jurisdiction of the ICJ.[57] The legal question of whether a state had committed a breach of international obligation could not be decided exclusively within the domestic jurisdiction of the state. The Application referred to various treaties and other agreements accepted by Iran, which indirectly obliged it to accord to "British nationals the same treatment as that accorded to nationals of the most favoured nation."[58] The British government concluded that Iran had breached both "the rules of customary international law" and "the treaty obligations accepted by that

Government" in accordance with the terms of the declaration it signed at the League of Nations.[59]

By asking the ICJ to acknowledge its Application, the British government was requesting that it oblige the Iranian government to submit its dispute with AIOC to arbitration, or alternatively, to declare that the Iranian Oil Nationalization Act of May 1951 was contrary to international law. The British government also reserved the right to request that the Court indicate provisional measures that might be taken to protect the British government and "their national," AIOC, allowing it to enjoy the rights to which it was entitled under the concession.

How could the British government make its case for diplomatic protection on behalf of a private corporation based within its territories by framing it as a "British national"? According to Ford, "it is an elementary principle of international law that a state has the right to protect its nationals when they have been injured by the internationally illegal conduct of another state."[60] Further, if the state, as with the British in this instance, "takes up the case of its injured national," through diplomatic channels or by pursuing international judicial proceedings, "the fact that the dispute originated in an injury to a private person or interest is irrelevant, since the state is asserting its own right."[61] The injury to the national was an injury to the state, and internationally, the state is the sole claimant. Furthermore, "a state can interpose on behalf of a corporation incorporated under its own laws, the nationality of the corporation being derived from the place of incorporation."[62] Thus, in the present case, "it's clear that AIOC is a British national on whose behalf the British government would be entitled to interpose," if the Court establishes that AIOC "suffered injury as a result of the illegal conduct of another state."[63] The intervention of the British government on behalf of AIOC generated new controversies in international law about the juridical character of the concession agreement and whether the parties to the agreement, one a sovereign state and the other a private entity, were equal subjects under international law.[64]

The Iranian minister of foreign affairs informed the ICJ that his government did not recognize the jurisdiction of the Court to deal with the matters in the dispute.[65] The Iranian government announced that while it would agree to discuss its oil requirements (not AIOC's) with the British government, it refused to consider the British government a party to the oil dispute, which was a domestic matter between itself and a private company. The Iranian government and AIOC still needed each other. Iran needed the company's expertise and facilities to keep oil and profits

flowing within a national framing and the company needed Iran to stabilize its concessionary rule, allowing for "some form of nationalization."[66]
AIOC's chief representative in Tehran, Richard Seddon, argued that while
the British government could not accept the Iranian government's right to
repudiate contracts, the company was prepared to consider a settlement
"which would involve some form of nationalization, provided ... it were
satisfactory in other respects."[67] In a meeting with Varashteh, the finance
minister, the Iranian government expressed its desire to draw on the company's experience to implement the nationalization laws, and thus invited
AIOC to submit proposals on this. AIOC headquarters responded with
an announcement of the names of four directors who would go to Iran to
hold the planned discussions.[68]

The Iranian delegation rejected AIOC's proposals on the grounds that
they were inconsistent with the laws of oil nationalization.[69] AIOC's proposals offered a £10 million advance against any sum that would be due
to the Iranian government as a result of an eventual agreement and based
on the understanding that Iran would not interfere with the company's
operations while discussions continued. The company also proposed that
the Iranian assets of the company be vested in a "National Oil Company," while the use of such assets be granted to another company established by AIOC, which would have a number of Iranian directors on its
board. Such concessions, Ford suggests, were very small "with no real
changes."[70] Negotiations ended the following day, and the British government returned to the ICJ to submit its "Request for Interim Measures
of Protection."[71] Proposals and counterproposals multiplied as AIOC,
the British government, and the Iranian government battled to frame the
dispute in advantageous ways to either block or enable nationalization
of the oil industry. While the boundaries between AIOC and the British
government were blurred, differences in their interests and approaches to
the dispute would soon emerge.

With the AIOC offer rejected, the British government submitted its
request for interim measures of protection on the grounds that if the ICJ
were to decide in favor of the claims made by the British government,
the decision could not be executed because the "gravest damage" would
already have been done to AIOC operations in Iran.[72] According to the
British government, the ICJ must indicate interim measures of protection to "prevent any step that might be taken to aggravate or extend
the dispute."[73] The Iranian government was using "the inflammation of
national feeling" through broadcasts and propaganda to unjustly vilify

the company, illustrated in the death of three British personnel during the strike at Abadan in April 1951.

The interim measures of protection, argued the British government, must oblige the Iranian government to permit AIOC to continue with oil operations as before.[74] Such provisional measures were another kind of technical procedure the British government used to intervene in the dispute. It was a technique used to impede the flow of oil (Iran wanted to increase production) and frame the issues in favor of the British bargaining position, which would delay the Iranian government in executing nationalization by pursuing more meetings and legal proceedings at the ICJ. This process of temporization did not work alone. Rather, it was connected in part to impeding the flow of oil income accrued by the Iranian government by allowing time for other kinds of technical and financial blockages, such as economic sanctions and an oil boycott, discussed below, to take effect.

In subsequent Oral Proceedings, Sir Frank Soskice argued on behalf of the British government that the ICJ should accept the request for interim measures prior to the determination of its jurisdiction in the case.[75] The British government was itself invoking an entire legal-metrological order, thus extending the dispute's scope globally, rather than keeping it a matter to be resolved by the Iranian government, domestically, with a private company. Effectively erasing the political question of nationalizing an oil industry, Soskice concluded that the Iranian government should permit the company to execute the terms of the concession before any further damage occurred. [76]

Kazemi, the Iranian foreign minister, replied to the British government's request for interim measures with a message consisting of a series of grievances challenging ICJ jurisdiction and the validity of the 1933 concession.[77] The message stated that the Iranian government hoped the ICJ would declare the case as "not within its jurisdiction because of the legal incompetence of the complaint and because of the fact that the exercise of sovereignty is not subject to complaint."[78] The claim to national sovereignty over a natural resource was a political tool that the Iranian government relied on to generate a disruption in international law, which needed to account for the emergence of the transnational corporation as a new kind of "nonstate" actor. International lawyers writing in the 1950s would refer to the AIOC case and other arbitrations arising out of disputes between Middle Eastern states and international oil corporations to devise a novel legal doctrine, combining domestic, private, international, and public law for the regulation of these "new realities."[79]

The ICJ finally ruled in early July that "pending its final decision in the proceedings instituted on May 26, 1951 (the Application)," a series of provisional measures would apply "on the basis of reciprocal observance."[80] The ICJ indicated measures proposed in the British government's "Request." Two judges, Winiarski and Badawi Pasha, submitted dissenting opinions to the ICJ's order. They argued that the question of interim measures and the question of jurisdiction were necessarily linked: the ICJ had the power to indicate interim measures "only if it holds, should it be only provisionally, that it is competent to hear the case on its merits."[81] While in municipal law, "there is always some tribunal which has jurisdiction," international law required the "consent of the parties, which confers jurisdiction on the Court."[82] Thus, "the Court has jurisdiction only in so far as that jurisdiction has been accepted by the parties."[83] Without this, there is no jurisdiction for indicating measures of protection. Such measures, they insisted, were in fact "exceptional in character in international law and to an even greater extent in municipal law."[84] The measures might be "easily considered a scarcely tolerable interference in the affairs of a sovereign State."[85] Because Iran did not accept the jurisdiction of the ICJ, the Court should be compelled to withhold its jurisdiction and under such circumstances, "interim measures of protection should not have been indicated."[86] The dissenting opinions by Winiarski and Pasha, ruling in favor of Iran's national sovereignty, did not hold, however, and the British foreign secretary accepted the ICJ's decision.

The Iranian government took immediate steps to return the dispute to a domestic-legal venue by sending a telegram to another international regulatory body, the United Nations Security Council. Addressed to the secretary general of the United Nations, the Iranian government gave notice of abrogating its declaration of September 19, 1932, recognizing the compulsory jurisdiction of the ICJ, and one of the bases on which the British government maintained that the Iranian government must respect the Court's authority.[87] The Iranian government regarded the ICJ's order as unenforceable and would not carry out the provisional measures indicated by the Court because the ICJ had acted without jurisdiction, contrary to the provisions of the Statute of the Court and in violation of the United Nations Charter. It was not until July 22, 1952, over a year after the British government's Application, that the ICJ would make a final ruling regarding its jurisdiction in the Anglo-Iranian nationalization dispute (discussed in chapter 6).

As I argued in chapter 3, transposing the dispute into the framework of international law and moving it to various sites in the international arena worked by associating the oil with international law, and keeping certain interests and associations tied together on behalf of the British government and AIOC.[88] Made possible by a proliferation of legal procedures, "internationalization" was a mechanism by which the British stabilized the control and distribution of Iranian oil. The mechanism transported the oil from local to international settings and back and won allies in a series of trials of strength, helping to eliminate any controversy. The technical and procedural work of law entailed the translation of actors' goals to the advantage of both AIOC and the British government's bargaining positions.[89] It also enabled the Iranian side to push for a reformulation of the law in terms of national sovereignty over oil, something never argued before in international legal terms. More meetings and technical argumentation took place on the domestic front in Iran as various political forces sought to resolve the dispute. The collective work of this legal disputation and machinery oscillated on the one hand between the authority of international law and on the other, of national sovereignty. The outcome was to help render Iran's oil nationalization intractable just as other technologies of intractability, in the format of economic sanctions and an oil boycott, got underway.

Financial Restrictions and Technical Blockages

Anglo-Iranian oil stopped flowing on June 1, 1951, and the Abadan refinery was shut down on July 31, 1951. The stoppage occurred as the British government was submitting its complaint to the ICJ at The Hague. In late summer and early fall, the British government and AIOC imposed the machinery of economic sanctions and an oil boycott to control and limit the distribution of Anglo-Iranian oil, working hand in hand with other technologies of intractability at sites around the world. These kinds of financial and technical blockages, according to the British government, were a useful technique of control that would help eliminate the possibility of nationalization by preventing Iran from operating its oilfields and having access to its financial reserves. Even if other foreign technicians were brought in, the lack of tankers would eliminate any outlet for oil exports from Iran. There would be mass unemployment in oil-producing areas, and Iran would face "economic chaos."[90]

The British government first discussed the consequences of economic sanctions against Iran in a secret document drafted by the Treasury.[91]

Through the summer of 1951, it instituted a series of economic measures designed to punish Iran for abrogating the AIOC concession. These included freezing Iran's sterling balances in London, withdrawing Iran's previous right to freely convert sterling into dollars, and halting exports to Iran of scarce commodities such as sugar and steel.[92] In the meantime, the government supported AIOC in taking measures in the form of an oil boycott. This involved AIOC's refusal to pay Iran any royalties, including advances or receipts with respect to the oil. The boycott prohibited AIOC tankers from loading oil at Abadan and ensured that any other company loading oil there was unable to dispose of it. Finally, AIOC withdrew the services of all British technicians working in oil operations.

The sudden cessation of refined products from Abadan, formerly contributing twenty million tons per annum of products, was a serious matter, but Anglo-Iranian oil did not possess special characteristics that made it indispensable. Refineries operating on these supplies could easily switch to utilize other sources of oil.[93] There was a continuous increase in consumer demand for petroleum products, however, such as bitumen and aviation fuel, with Abadan accounting for 45 percent of the total requirements of the latter product in the Eastern Hemisphere. Thus, the lack of refinery capacity to match this demand was creating a problem. Furthermore, Iranian oil was supplied for sterling. This meant that the replacement of Iranian products would necessarily involve "calling on dollar sources," intensifying existing problems with the British imbalance of trade in dollar areas.[94]

On the one hand, imposing economic sanctions on the Mosaddiq government would help the British government conserve dollars and preserve supplies of goods that could be sold for dollars, since the currency was now needed to pay for oil imports.[95] AIOC made the decision to impose an oil boycott for much the same reason and to show Iran that the company would not allow its property to be confiscated illegally and without adequate compensation. On the other hand, sanctions and a boycott would negatively impact the United Kingdom as well. As Britain's largest overseas investment, the critical issue in view of the need to protect its balance of payments and save dollars was to keep the refined oil from Abadan under British control.[96] The Treasury estimated in May 1951 that the loss of Iranian products would cost $504 million to replace. Refined products would be lost from closing down the world's largest refinery at Abadan, which would be impossible to replace except after a long interval. There might even be restrictions on the consumption of petrol products in the sterling area. The risks would be higher if the Iranians succeeded in maintaining oilfield and refinery operations as well as the distribution and payment for oil overseas by hiring foreign experts and tanker companies, the majority of whom would have to be American. It was unlikely, however, that enough foreign tankers would be available to carry the oil away.[97] From the start of the boycott, AIOC and the British government "worked hand in glove to ensure its success."[98] Their cooperation, argues Heiss, marked their policy on the boycott but did not necessarily carry over into other areas of the dispute.

The British government and oil company were also relying on the participation of another set of actors, whose cooperation was essential for the oil boycott to work—the American oil majors. US oil companies had no real reason to purchase Iranian oil because the world oil industry was experiencing a "buyer's market."[99] Despite increased consumption in the decade after the Second World War, production grew significantly as did a number of giant oilfields in the Middle East and the discovery of new fields. This fueled a scramble among the oil majors for concession agreements with producer countries. In light of AIOC's monopoly control in Iran, the US majors made arrangements with Kuwait (50 percent AIOC controlled) and Saudi Arabia. Buying oil from Iran in the absence of a dramatic increase in total worldwide demand would force production cutbacks in these countries and ultimately decrease the royalties they received from their oil operations. This could potentially hurt the US majors' concessions by triggering nationalization efforts elsewhere. Purchasing Iranian oil at substantial discounts might also upset the price structure of the Middle East petroleum industry, something the US companies strongly opposed. Having "no real need to buy Iranian oil and every apparent reason not to," the US majors helped impose limits on Iran's flow of oil and thus income to the Iranian government by supporting AIOC's boycott. The Truman administration took the same stance as these so-called private concerns, warning smaller companies prone to independent purchases of Iranian oil that official US policy supported the boycott. Independent purchasers would be on their own if AIOC prosecuted them for attempting to break the boycott.[100] According to Walden, US antitrust law says that "group boycotts" or "concerted refusals by traders to deal with other traders" have long been put in the "forbidden category."[101] US antitrust law deemed oil boycotts illegal.

Deploying the boycott arrangement posed risks in compensating for the loss of Iran's oil supply. In a memorandum, Walter Levy, an American oil consultant, warned that execution of the program would require strict coordination through US government regulation to avoid shortages in the

United States and would require a highly regimented petroleum industry, "a sort of partial war mobilization." In reality, a number of "specific supply crises" would result from a loss of oil from Abadan.[102] For example, the failure of aviation gasoline would result in the shutdown of commercial air operations between Europe and Asia. Practically all areas east of the Suez would run out of oil by August 1951. Major bunkering stations in the Red Sea and Indian Ocean area would also run out of supplies. Oil production and refining capacity would therefore have to be stepped up around the world. Lastly, normal trading and world oil prices would be totally disrupted as a result of the replacement of crude oil deliveries with the costlier shipping of refined products from abroad.

American oil companies and the US government aimed, like their allies in Britain, to minimize the damage they experienced as a result of the collapse of the Iranian oil industry while keeping world price levels high and production quotas limited. In July 1951, the US Petroleum Administration for Defense (PAD) organized representatives of nineteen of the nation's oil companies into the Foreign Petroleum Supply Committee.[103] The agency, headed by Stewart P. Coleman, a director at the Standard Oil Company of New Jersey and Aramco, was originally responsible for managing America's oil supplies during the Korean War. Known as "Plan of Action No. 1," the aim was to enroll the oil companies in the effort to replace 660,000 barrels of oil per day formerly supplied by Iran. This was equal to one-third of total Middle East production and a quarter of all refined products outside the Western Hemisphere. The loss of this oil was compensated by the parties of Plan of Action No. 1, who took steps to reorganize the world energy system by increasing production and refining in other countries, realigning imports and exports, and allocating markets.[104] The American oil companies involved were conveniently granted immunity from antitrust prosecution under the Defense Production Act of 1950. Ostensibly, the American oil companies signed the agreement to "prevent, eliminate, or alleviate shortages in oil supplies" of "friendly foreign nations" that might "threaten the defense interests or programs of the United States."[105] In reality, private and public American interests were coordinating to determine markets and replace the potential loss of Anglo-Iranian oil with other, American-controlled sources.

American oil majors supported AIOC's oil boycott not to secure a steady supply but to place limits on the flow of Iranian oil. The coordination of a different kind of cartel arrangement mixing the so-called public and private interests of the American and British governments and oil companies worked hand in hand to block Iranian oil nationalization and fuel a war effort. This framing device in terms of protecting the oil supplies of friendly foreign nations served as another kind of antimarket arrangement, coordinated with the economic boycott, to avoid supply dislocations in other concession areas and keep profits high.

Technical blockages and financial restrictions enabled the British government to reach a negotiating position with Iran with the aim of ensuring both that the government's oil supply needs were guaranteed and that AIOC maintained control of operations. The sanctions might have little effect if the Iranians "called our bluff by obtaining assurances of help from the Americans."[106] British government and oil company interests were internal to each other and aimed to use economic sanctions as a technique of intervention to frame interests advantageously and impede the flow of income to the Iranian government. Sanctions would strengthen the British negotiating position and help block the political possibility of oil nationalization. The success or failure of combined sanctions and a boycott would not only depend on cooperation between so-called public and private interests in the United Kingdom, but also on an alliance with American oil companies to ensure that Iran did not access alternative sources of expertise and technology for the distribution and sale of its oil.

The intertwining of public and private interests was crucial to the boycott's success, in spite of the antimarket and extralegal mechanisms involved.[107] If the US majors had broken these connections, the boycott would have failed. The relation between governments and transnational corporations is often understood in terms of an external relationship between two separate entities, but the clarity of the apparent divide between private and public was not so obvious as to delineate one actor as more powerful or as determining the role of the other. The Anglo-Iranian nationalization dispute involved assembling a set of technologies of intractability to help impose financial restrictions, technical blockages, and mechanisms of temporization to arrange oil nationalization advantageously by strengthening the British and increasingly American bargaining positions.

Economic Facts of Oil

The opportunity for the American government to play a mediating role in the crisis first came when Mosaddiq sent a message to Truman expressing his government's wish to maintain the flow of oil as before and to keep AIOC's foreign experts in the service of the oil industry, leaving the company's organization intact.[108] As the British request for interim measures at the ICJ was granted, Truman replied to Mosaddiq, offering to send Averell Harriman to Tehran on a diplomatic mission to mediate the dispute. Iran's Joint Parliamentary Oil Committee and cabinet approved the offer on July 11, 1951.[109] The impact of the Anglo-American economic boycott on Iran's dollar and sterling balances, ICJ court orders, and now an American diplomatic mission were coming together to organize the intractability of the dispute.

Iranian members of parliament had concluded that nationalization was the best resolution to the crisis for redirecting the demands of Iran's diverse political groups into the more manageable frame of reformist nationalism. But how could nationalization work without the participation of the British oil company? For advice, Iranian officials participated in a series of consultations with American oil business consultant, Walter Levy.[110] Levy previously served as head of the Petroleum Branch of the Economic Cooperation Administration (ECA), set up by the American government as part of the Marshall Plan to help subsidize the conversion of Europe from a dependence on coal to oil. He accumulated years of experience as an oil consultant for both the US government and the major American oil companies such as the Socony-Vacuum Oil Company.[111] Levy was eventually put "on loan" by the ECA to the State Department for Averell Harriman's diplomatic mission to Iran. The aim was to break the impasse between AIOC and the Iranian government by educating the latter about the "economic facts" of oil.

Presenting himself as an independent oil expert, Levy framed his role to both British and Iranian officials as presenting "the technical aspects of the problem," whereas "the political side was being handled entirely by Harriman."[112] Harriman felt it essential to "maintain a neutral position" so as to avoid the impression that he endorsed any particular plan.[113] Using technical and economic arguments, Levy and Harriman aimed to persuade the Iranians that they could not expect a financial return greater than that of other countries under comparable conditions.[114] They argued that oil operations in Iran must be run on an efficient basis. This could only be accomplished through a "foreign owned operating company" with freedom in daily management though acting under principles established by a government or national oil company. Oil diplomacy involved persuading the Iranian government to establish a peculiar connection between the "principles" of a nationally controlled oil company, only in name, while protecting daily management of operations under foreign control.

Iranian negotiators warned Levy that any arrangement that tried to nullify nationalization or render it ineffective was "doomed to failure because it will hurt national pride and be rejected by public opinion in Iran."[115] The practicable solution was to form a national oil company that would take over British and foreign technicians as well as a significant number of existing executives. The new company would make a long-term "Agency Agreement" for at least fifteen to twenty years with AIOC, ensuring the sale of as large a quantity of oil as possible. The proceeds of a portion of production would then be allocated toward the payment of the compensation agreed on for expropriation.[116]

During the course of the Harriman mission, Levy participated in secret talks with Kazem Hassibi, an engineer and undersecretary at the Ministry of Finance, and Allahyar Saleh, chairman of the National Oil Commission.[117] Levy framed his role in the dispute as concerning the "commercial and economic problems of the world oil industry," even though Saleh insisted that the problem was also political.[118] Saleh complained that the income his government accrued from the most profitable operation of the company was utterly inadequate. The company was "corrupting the political life of the country, controlling the radio, newspapers, and other communication facilities, and has paid no attention to the welfare of its workers."[119] The nationalization laws, Hassibi explained, would solve problems by enabling the government to obtain the necessary income from oil operations and the foreign exchange necessary to finance the country's import program. To do this, the country must export eight to ten million tons of oil in the first year.

Levy responded by framing his argument to suggest that oil sales could only be made within the framework of the international oil industry. He explained that Iran possessed large underground oil reserves producing, at the time, about 6 percent of the world's oil production and less than 5 percent of the world's refining output. This production was shipped and exported to many countries in the Indian Ocean area, where customers distributed it to various final consumers in importing countries. "To achieve this translation," Levy argued, "from a 'valueless' oil in the grounds of Iran to a commercial product in the consuming country" required the construction of a highly complex organization.[120] Coupled with production, transport, and distribution, the organization must also establish the "financial machinery" necessary to provide for the financing from production to the final distribution points.[121] This complex organization of oil infrastructure was already in place.

Levy laid out the arrangement of the world oil industry as a commercial-technical unit with the capacity to develop oilfields in producing countries for the establishment and maintenance of which "millions and millions of dollars are needed."[122] Only a long-distance machinery could procure the necessary technical expertise to provide transportation and distribution facilities to support worldwide operations. Oil companies ensured the smooth running of this machinery through their control of oil production, refining, and transportation "by sea and by land and by their network of distribution facilities which is spread over nearly every country of the world."[123] AIOC had spent the past fifty years building the petroleum energy system in collaboration with the largest Anglo-American oil corporations, helping to constitute the political agency of this actor in the twentieth century. According to Levy's expertise, this long-distance machinery was the only possible arrangement, a kind of obligatory passage point,[124] in which oil sales could occur on a large-scale. If the Iranian government wanted to sell its oil in international oil markets, it needed to know how to do this, following the model outlined by Levy, and to recognize that its alliance with Anglo-American oil interests around these econotechnical facts would benefit them, even with nationalization achieved only in principle.

Nearly all previous customers for Anglo-Iranian oil were subsidiaries of the Iranian oil company, companies closely associated either with AIOC or with other major British and American oil companies: "these facts they [the Iranian government] must face," explained Levy.[125] The Iranians would not be able to sell eight to ten million tons of oil, even at discounted prices. The major oil companies were, after all, "the only producers in the Eastern Hemisphere," and refineries depended, in the majority of cases, on such companies for their sea transportation.[126] For the "oil industry to survive," explained Levy, a means must be found to achieve "confidence and cooperation between the Iranian oil industry and the rest of the world," something impossible unless that arrangement involved the oil majors' energy network.[127] If this particular arrangement of the energy system did not happen, outlets for Iranian oil in world markets were bound to "shrink and disappear."[128] The long-distance arrangement formed part of the manufactured reality of the international oil cartel that transnational oil corporations had spent four decades building using concession terms, oil infrastructure, international law, and the production of expert knowledge about oil.

Framing his arguments in terms of a highly technical mapping of the global oil industry, Levy needed to convince the Iranians that their oil was not irreplaceable and would not seriously disrupt the world oil industry requirements. Iranian officials must be persuaded not only that oil sales were the only thing that could save their country's economy, but also that those sales could only be made within the framework of the international oil industry.[129] In sum, Levy's arguments worked to diminish the Iranian government's position of strength.

During the course of the Harriman mission, Levy continued to insist that Iran must not "sit on its oil, even temporarily."[130] The trend in the world, argued Levy, was definitely toward the development of new and additional sources of power, such as atomic energy, and these sources would "replace oil and other sources" currently in use.[131] As part of the Marshall Plan's reconstruction, Europe would soon have an annual refinery output of sixty to seventy million tons, which it would obtain from "increased production in Kuwait, Iraq, Saudi Arabia, and Qatar," to replace oil from Iran (figure 5.2). The seven major oil companies had huge sums of capital to weather production and marketing problems, owned most of the non-American refining capacity, and of 1,319 T2 tankers, only 44 were not owned or chartered by the large oil companies and their subsidiaries.

Levy was attempting to persuade the Iranians that building an alternative arrangement to oil production, transport, refining, distribution, and marketing was impossible. Iran must not pursue its current course of action "implying no satisfactory settlement with AIOC," because the oil companies "would never agree to help Iran in any way," as such help would "unsettle the oil industry over the entire world."[132] By the end of the discussions, Hassibi conceded that his government needed a "method of convincing [Iranian] public opinion" that Iran's nationalization laws were being fully executed and that the oil industry was under government management.[133] The connection between foreign influence and oil must disappear. Then, "public opinion could be convinced to enter a more permanent arrangement and one in which management control could be turned over to foreign technicians."[134] The Iranian government appeared to agree with Levy that for nationalization to work, the government must persuade the Iranian public that continued foreign control framed as nationalization was a successful outcome. This arrangement could not be achieved, however, without the participation of the United States. As long as the appearance of nationalization was achieved, Hassibi conceded, other arrangements involving the largest oil corporations were possible.

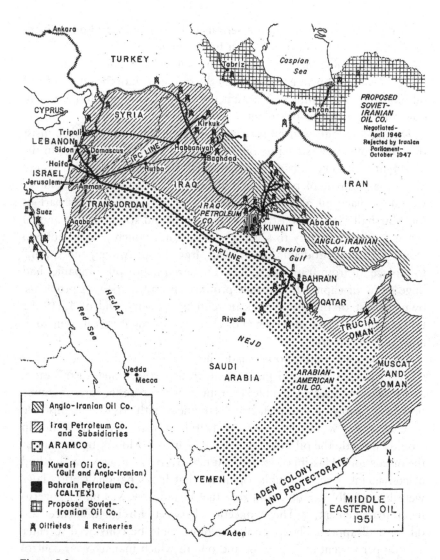

Figure 5.2
Map of Middle Eastern oil in 1951. *Source*: William Roger Louis, *The British Empire in the Middle East, 1945–1951: Arab Nationalism, the United States, and Postwar Imperialism* (Oxford: Clarendon Press, 1984), 690.

At Harriman's meeting with Iran's Joint Oil Committee, Levy reiterated the three major problems facing the Iranian oil industry.[135] A basis for friendly negotiations with AIOC must be pursued to solve the oil question, Harriman warned, "before arrangements are made by the world's oil industry to replace Iran's oil in world oil trade."[136] That Iran might choose to drop out of the international energy system was an unacceptable outcome and created instability. To avoid this outcome, Harriman acknowledged Iran's sovereign right to act on its own, but insisted that the government could not expect to "dictate oil operations outside the limits of its own frontiers."[137]

As a response to the American mission's efforts, the Iranian government, led by Mosaddiq, submitted a "definite formula" as its final view on the talks. In a report to the Majlis, the terms revealed the government's willingness to participate in negotiations with the British government on behalf of AIOC so long as the latter recognized Iran's "principle of nationalization."[138] Nationalization in "principle" indicated a novel framing that the Iranian government was persuaded to rely on in order to achieve the effect of nationalization without disturbing international oil markets and Iranian public opinion.

Conferences moved to London and Levy consulted secretly with both American oil company officials and British government officials.[139] They discussed the nature of the Aramco agreement in Saudi Arabia and the IPC plan, recently negotiated in Iraq. Levy noted that it would be "very difficult" for Iran to maintain a pricing arrangement under a profit-sharing plan like the one arranged in Saudi Arabia in which oil was priced at $1.43 per barrel because Iranian oil was being sold to other subsidiaries at the much higher price of $1.75 per barrel. Furthermore, while refinery profits from Ras Tanura were not included in the Saudi Arabian agreement, refinery profits from Abadan would have to be factored into any arrangement with Iran. The reason was that only 25 percent of Saudi Arabian oil production was refined locally, while over 80 percent of Anglo-Iranian oil was refined in Abadan. In another meeting at the Ministry of Fuel and Power, which included the US ambassador to London, Levy argued that the possibility of a future meeting between Mosaddiq and the British delegation was "doomed to fail," if AIOC representatives were included.[140] He also confessed the extreme importance of avoiding acute oil shortages, "during the course of negotiations in Tehran, particularly in India and Pakistan."[141] Such an event would "make the Iranians feel that the world was still greatly dependent on them for oil supplies," and translate into strengthened resistance to any agreement

with the British.[142] Finding a resolution to the oil nationalization controversy in Iran needed to occur in relation to pricing arrangements in Iraq
and Saudi Arabia, and oil supplies in India and Pakistan. Only through a
coordinated effort among oil experts, government officials, and company
managers would the economic facts of oil survive.

Levy concluded the meeting by outlining his ideas on the kind of
British position—a type of consortium arrangement—that might be
formulated and proposed to the Iranians. He emphasized the Iranians'
animosity toward AIOC such that it necessitated the creation of an operating company with additional representatives to introduce a "new character" to the operation.[143] A 50–50 proposal must be "eminently fair and
straight forward without any devious arrangements or unrealistic pricing
involved," and Harriman had authorized Levy to support such an idea.[144]
The idea was to obtain an arrangement that would be similar "in principle" to arrangements in Saudi Arabia and Iraq but "which might differ
considerably in detail."[145] Exposing his preferences, Levy revealed that
the "basic motive behind the idea was to protect British interests."[146] It
was clear, Victor Butler, of the British Ministry of Fuel and Power, and
Levy agreed, that the "current context" did not allow for a single company operating alone in a particular country, and that "there seemed to
be a much more stable situation where several different interests were
engaged."[147] Coinciding with the implementation of the Marshall Plan
to fuel the reconstruction of postwar Europe with cheap oil from the
Middle East, the consolidation of consortium arrangements emerged as
a new device to help restrict oil development in the Middle East, manage the rise of resource nationalism, and combat any real attempts at
nationalization.[148]

The two world wars had helped restrict the supply and movement of
oil, but between the wars a new set of devices were needed to limit the
production and distribution of energy.[149] During the interwar period, a
series of devices were developed to accomplish the task: the consortium
arrangement to restrict the development of new oil discoveries in the
Middle East, government quotas and price controls in the United States,
and cartel arrangements to govern the worldwide distribution and marketing of oil.[150] As discussed here, these techniques and controls shaped
the development of the transnational oil corporation, which emerged as
a long-distance machinery for maintaining limits to the supply of oil in a
period of abundance.

Though claiming independence and neutrality as a private oil consultant on purely technical and economic matters, Levy was mixing

econotechnical matters with politics. He was effectively coaching British and Iranian officials in a trial of strength that would close off the working possibilities for nationalization and open up alternative political arrangements for a kind of consortium arrangement, more amenable to the "current context," but by no means one in which strictly British control returned to Iran.[151] In practice, the "current context" was a set of connections and long-distance machinery that oil company managers, consultants, scientists, engineers, lawyers, government officials, and oil workers had battled to construct (and often deconstruct) over the course of half a century, starting in the oilfields of southwest Iran.

From the American perspective, the Harriman mission, conducted in its technical detail by Walter Levy, was another weapon to weaken the Iranian position, block nationalization of the oil industry, and thwart any return to British control. Harriman admitted to British officials that American oil companies were watching his activities, encouraging suspicions on the British side that Levy was undermining "Anglo-Persian" interests through the pursuit of a consortium arrangement.[152] Through a series of diplomatic trials of strength, the success of securing the consortium arrangement to restore foreign control over Iran's oil worked precisely by making arguments about the economic facts of oil a reality. Building a world in which such facts could survive meant pursuing economic sanctions, imposing an oil boycott, and persuading the Iranian government about the impossibility of producing, refining, distributing, and selling oil outside the global arrangements of the seven major oil companies.

Like its American counterpart, the British government expressed a willingness to enter negotiations in response to Mosaddiq's proposal for an "Iranian formula." Harriman had also played a role, persuading Mosaddiq to accept a British delegation led by a senior minister of the Labour government.[153] This paved the way for a new diplomatic mission to Tehran, known as the Stokes Mission, in which Richard Stokes, Lord Privy Seal, led a British delegation to put forward another set of proposals allegedly in line with the "principle of nationalization" stated in the Iranian law of March 20, 1951.[154] This occurred as a British destroyer joined a British cruiser in the Shatt al-Arab, the Abadan oil refinery was shut down, and the leading American oil companies cooperated in the establishment of the Foreign Supply Committee to readjust to the loss of Anglo-Iranian oil.

These diplomatic battles were disputes over framing the issues and building connections in relation to the flow of oil: Mosaddiq fought to make a national and political claim to a natural resource in terms of state sovereignty, while Levy insisted that national politics could not be factored in because their outcome would be disastrous to the functioning of the technical and economic mechanisms of oil operations, which no other actors could manage except the seven major oil companies. Stokes withdrew his eight-point proposal on August 21, claiming the Iranian government was reading intentions into it that were not there.[155] The proposal was no different from the proposals first submitted by AIOC to the Iranian government, which involved the vesting of Iranian assets in a national oil company while the use of the assets would be granted to an operating company established by AIOC. Harriman made one final attempt to resolve the dispute, asserting that the British proposals were within the Iranian formula.[156] He argued that British proposals would achieve Iranian national aspirations, including Iranian control over its oil industry *within* Iran and would provide income to enable extensive development of the country's economic potential.

The Stokes and Harriman missions, in addition to Levy's efforts at persuasion, proved futile after Mosaddiq immediately rejected the notion of an operating company or agency to operate the oil.[157] Many studies attribute the failure of these diplomatic missions, in part, to Mosaddiq's highly irrational temperament.[158] Less emphasis has been placed on the British desperation to achieve a "nationalization arrangement" that retained as much British control of Iranian oil as possible by relying on technical, legal, and economic argumentation to redirect the Iranian government and public opinion toward a new consortium arrangement. Stokes was fully aware that Iranian oil was the cheapest to produce in the world and that there was no reason to follow the American 50–50 formula if they could retain their access to oil by offering a more generous share of profits.[159] British officials from different ministries disagreed with Stokes's assessment, as did the Americans. The Iranian delegation submitted a further reply, which essentially argued that it was unwilling to sell any of its oil production to AIOC.[160] As the crisis escalated with the ordered evacuation of all British staff members, the diplomatic missions officially ceased in September 1951. Mosaddiq prepared to make direct sales to potential customers of Iranian oil.

Thanking Levy for his efforts, Harriman blamed "the lack of technical knowledge on the part of the Iranian government representatives about the oil situation"—this was, he believed, "one of the greatest obstacles" confronting the diplomatic mission.[161] Multiple possibilities for controlling Iran's oil had seemed feasible, but in the end American and British interests would pursue alternative ones, including the use of force.[162] This would ensure, on the one hand, that a precedent was not set in which every foreign government felt it could unilaterally repudiate contracts with British firms and seize British assets in their countries upon the payment of compensation. On the other hand, it would ensure that the new 50–50 profit-sharing principle pursued by the Americans as a mechanism for managing contracts in a new international order would not be disrupted in favor of the more dangerous alliance of national control with Soviet-inspired communism.

The Mosaddiq government rejected the proliferation of diplomatic terms proposed for a resolution of the crisis as disconnected from the political question of national sovereignty over a natural resource. It refused these terms not because of a lack of technical knowledge, but because the control of Iran's oil supply was a powerful political weapon that could be used to redirect the energies of a militant labor movement into the more manageable framework of reformist nationalism and win more profits. Although the diplomatic missions failed to resolve the dispute, their technical work succeeded in rendering oil nationalization intractable, effectively impeding the flow of oil and thus oil income to the Iranian government.

Nationalizing Labor(atory)

The work of the various technologies of intractability opened up the definition of what it meant to nationalize an oil industry and the kinds of arrangements that would be necessary to secure older agreements and build new ones to the advantage of the major transnational oil companies. As diverse and competing interests put these technologies to work in the international and domestic political arenas, Iran's Special Commission on Oil, set up on June 20, 1951, and headed by Amir-Alai, the minister of economy, set out to build a different kind of arrangement from before, executing the nationalization laws by traveling to southwest Iran to take over AIOC operations.

Iranian officials arrived at the oilfields to find that the oil labor regime, fixing Iranians at the lowest level of the skill and wage hierarchy, was completely intact. Eric Drake, AIOC's general manager in Iran and Iraq, led the members of the commission around the oilfields, including the workers' housing sections at Chadorabad, Halabiabad, and Hasseerabad (figures 5.3 and 5.4). Amir-Alai explains in his memoir that the standard of living particularly at these three locales was so strikingly poor that

Figure 5.3
Photo of the oil delegations with a "background of primitive dwellings at *Chadurabad*. In the forefront are veiled women who have to bring water from miles away." Accompanying the visitors on the tour was Eric Drake, general manager of AIOC [tall with dark felt hat], "who merely shrugged his shoulders when asked for an explanation of these conditions." *Source*: INP P 277 Special 54043, 78148, BP Archive. Reproduced with the permission of the BP Archive.

AIOC officials indicated the workers were not employees of the company. The workers confirmed their employment at AIOC by revealing their company identification cards.[163]

How would it be possible to nationalize the oil industry, if the social world of oil had been built in such a way as to deny this very possibility—the replacement of foreign engineers, technicians, and managers with Iranian ones? A report drawn up by the British Treasury on the impact of the withdrawal of British staff on operations in Iran mapped how the organization of labor would ensure the impossibility of continuing oil operations in the absence of British workers.[164] The technical coordination of the oilfields and refineries, the report explained, was made possible by British staff with a high degree of skill in technical management and long experience operating the various units in or involving the fields,

Figure 5.4
On June 26, 1951, the Tehran delegation, which carried out the nationalization, paid a
visit to some of the slum dwellings of the company's workers, "where people live in primitive conditions in huts of matting or paper with little of the amenities of civilized life." The
company captions the photo with the following: "At Kagazabad, Hussein Makki, leader
of the Persian Oil Committee, leads a Persian worker by the hand out of one of the tents
made of paper waste and tree branches, followed by other members of the delegation."
Source: INP P 287 Special 54045, 78148, BP Archive. Reproduced with the permission of
the BP Archive.

the pipelines, and the refineries. Relying on a familiar formula, the report
argued that the Iranians still did not have "the capability or experience
which would enable them to play a real part at this level."[165] The work of
British staff was "largely associated with starting up and shutting down"
oil operations.

Compared to a total of seventy British staff employed directly in production, there were only nine Iranians employed in production operations with experience approximately equivalent to that of "Junior British
Staff." Production was dependent on specific tasks in mechanical engineering, maintenance, and workshops. These services would halt with
the withdrawal of foreign staff. In all fields, production depended on
the operation of machinery and plants, which took the form of power

stations driving electric-driven pumps, or steam- or gas-turbine-driven pumps and automatic control equipment.

The plants required "skilled operation, and although there are a small number of Persians with sufficient skill," the report explained, their numbers were too small to fill all the positions at "full production."[166] The amount of production drawn from each well and zone of the four main oilfields (Masjid Suleiman, Haft Kel, Lali, and Naft Safid) was controlled by senior British technical management staff "who have no Persian counterparts." Improper control of individual producing wells, the report warned, would rapidly result in water being produced in the oil, which would have a "disastrous effect in the Refinery."[167] A steady stream of tankers transporting the products from Abadan was also necessary for a steady operation.

The important feature in the disruption of operations caused by Iran's oil nationalization was therefore not the removal of Indian workers, but the removal of British staff, who were responsible for management as well as "technical direction and planning."[168] If the British left, the report concluded, "at least an equal number of Iranians must be found to replace them, and this is impossible."[169] The peculiar oil labor regime built over the past forty years had succeeded in transforming the oilfield into its laboratory, arranging an energy system divided into specific tasks delineated along racial lines. Could AIOC's laboratory be nationalized? The design appeared to ensure that the industry could not run without British management and technical expertise. The British admitted privately, however, that oil operations in southwest Iran could not be worked without the "assistance of Persian workers."[170]

As AIOC shut down its Iranian operations and withdrew its personnel, Iranian officials blocked ships from leaving the port unless their captains signed new receipts acknowledging NIOC as the true owner of the oil. This convinced AIOC to halt its loadings from Iranian ports, which, along with the departure of AIOC personnel from Abadan in October 1951, solidified the boycott.[171] After the withdrawal of British staff, the Iranians were confident that they could easily rehire non-British technicians to operate the industry and quickly train their own nationals to replace them.[172] This was, however, not the case. As a result of the boycott, the United States, Sweden, Belgium, The Netherlands, Pakistan, and Germany all refused to make their technicians available toward the Iranian goal of nationalizing the British-controlled oil industry.

The technologies of difference embedded in the hierarchical segregation of labor maintained their work over every aspect of operations from pipelines and pumping stations to water treatment and automatic control equipment.[173] These operations worked in tandem with the boycott to impose restrictions and close the possibility of alternative, more democratic forms of oil production. The control that the international petroleum industry exerted over the world's tanker fleet did not help matters. AIOC controlled 30.5 percent of the world's tankers and US petroleum companies controlled 42.5 percent, constituting a total of three-quarters of the total tanker capacity of 1,500 T-2 equivalents.[174] Iran controlled none. Although it signed contracts for oil sales, Iran's nationalist government failed to make good on these because the oil could not be delivered.

The peculiar technoracial organization of labor in the oilfields was further strengthened by the traditional cartel arrangement among the world's largest oil companies—Standard Oil of New Jersey (Exxon), Standard Oil of New York (Mobil), Standard Oil of California (Socal), the Texas Company (Texaco), the Gulf Oil Company, Royal Dutch Shell, and AIOC. These companies coordinated production restrictions and price fixing in their international operations and complied with AIOC's boycott of Iran's oil. This collective of public-private interests worked hand in hand with the racial-technical organization of Iran's oil industry. The power of international oil companies (i.e. the company we know today as BP), the authority of the Iranian government, and the process of nationalizing an oil industry were worked out within a battlefield of technicality and political contestation, concentrated along global distribution networks and in a small region of southwest Iran.

Collectives of Intractability

Arranging the intractability of Iran's oil nationalization dispute was a strategy that the British used to impose delays and avoid reaching a constructive solution. Delay was a useful strategy not because it worked alone, but because it worked in connection with other forces, such as the flow of income to the Iranian government. Controlling or impeding this flow of funds was part of what was at stake due to its interconnectivity with the flow of oil, which the British and eventually the Americans aimed to slow down. These blockages and constrictions needed time to take effect, which is why the temporizing diversions to new sites like The Hague were so important. The British company and government collaborated to ensure AIOC's control of oil operations and simultaneously weaken Iran's bargaining position so as to place Anglo-American interests at an advantage. And yet the proliferation of organizational forms and technologies of intractability that emerged to manage the dispute left the outcome unstable.

The Harriman and Stokes missions failed to bring an end to the transformation of the dispute into an international legal and financial issue managed by the work of the International Court of Justice, economic sanctions, and an oil boycott. These were political weapons the British government and AIOC deployed to build alliances and enroll supporters (e.g., the American government, Walter Levy, and the seven major oil companies) against the Iranian government and oil workers battling for nationalized control over oil operations along the lines of successful attempts in Bolivia, Mexico, and the Soviet Union. The British government had just nationalized its coal industry five years prior to Mosaddiq's call for nationalization.[175] Anglo-American interests sought to build a new international order that entertained the wording of nationalization, but only as a means of preserving old and new contractual relations between national governments and foreign firms. If the terms of nationalization were not built into the world according to an appropriate translation of Anglo-American interests, the achievement of real national control— that is, the expropriation of foreign firms without compensation— would set dangerous precedents in the Middle East, particularly in Iraq and Egypt.

The nationalization crisis was worked out in terms of the different political possibilities and collectives of intractability that opened up and closed down in the process. By considering the technical, legal, and diplomatic machineries involved in the distribution and control of oil, this discussion has opened up the black box of oil nationalization and the techniques of political rearrangement at play. Technologies of intractability were deployed to organize actors and translate issues in ways that separated political questions of nationalization from econotechnical issues of protecting the international oil network.

As in 1932–1933, the law worked as a political weapon and a sociotechnical device to redirect the crisis by invoking an entire international legal order. The fundamental issue at stake was who would control the oil industry, the British or the Iranians. British government officials, encouraged by the Americans, realized that AIOC, renamed British Petroleum in 1954, could not return to Iran as a controlling company. An alternative emerged, in the shape of a consortium, alongside the construction of 50–50 profit-sharing arrangements in neighboring countries. The proposed international consortium put forward the facade of nationalization but entailed the mixing of Anglo-American oil companies and government interests to determine markets by limiting the flow of oil supplies and retaining control along the lines of older concession agreements. With the failure to enroll the Mosaddiq government, other increasingly violent political alternatives remained.

6
Long-Distance Machineries of Oil

The post–World War II petroleum order marked a period of oil abundance and an emergent machinery of global economic governance set up to manage this abundance in the name of "Third World development." The problem for the Americans and British was not about restoring the supply of Anglo-Iranian oil after nationalization, but about constricting the flow of oil and income to the Iranian government using strategies of delay and intractability. Intractability allowed time to put in place a new set of political arrangements that appeared necessary to restore foreign control and profits from Iran's nominally nationalized oilfields. Iran's nationalization did not mark the end of the oil dispute, or the execution of the nationalization process through the complete confiscation of company property, the automatic takeover of the British company, and the evacuation of its foreign employees. Instead, over the course of three years, controversies emerged to redirect the nationalization crisis in legal and financial terms to ensure that nationalization did not happen in practice. As AIOC and the Iranian government battled to resolve the crisis, Anglo-Iranian oil stopped flowing in June 1951, with the official halt of operations at the Abadan refinery taking place on July 31, 1951. In the following summer months, international legal proceedings, economic sanctions, and an oil boycott were the machinery developed to limit the distribution of Anglo-Iranian oil (discussed in chapter 5). British employees were evacuated in October and alternative arrangements were put in place to make up for the loss of Iran's oil by relying on oil supplies from Saudi Arabia, Kuwait, and Bahrain. The rearrangement involved a significant degree of coordination between the major Anglo-American oil corporations and their host governments. While the British no longer maintained absolute control over Iran's oil, it did still have the power to influence the global market, of which Iran's newly nationalized oil was part. Foreign control of oil was no longer acceptable in a world of nation-states claiming sovereign control over their resources. With AIOC unable to return to Iran on its own, Iranian, British, and American government officials and oil executives needed to devise novel techniques to transform the formerly British concessionary arrangement into an oil consortium, an association of the largest foreign oil companies.

Alongside the emergence of the transnational oil corporation as a new political actor of the mid-twentieth century, this period witnessed the rise of a series of international regimes of legal and economic governance and security such as the United Nations (UN) as a successor to the League of Nations, the International Court of Justice (ICJ), and the International Bank for Reconstruction and Development (IBRD, known today as the World Bank). Ostensibly set up as "neutral" organizations to aid in postwar reconstruction and so-called Third World development, each organization served as a kind of disciplinary regime, playing a direct role in managing the Anglo-Iranian oil nationalization dispute by proposing a set of economic and legal arrangements for a resolution in 1952. The effect was to further delay Iran's reformist nationalist government from implementing any concrete nationalization while allowing for more time to put in place a set of mechanisms for keeping less democratic forms of politics inside oil operations and others, such as striking oil workers demanding nationalization of their country's oil, out.

This chapter follows Iranian, British, and American government officials and oil executives as they fought over the design of a newly nationalized oil industry and the consequences for global oil markets. The procedural work of restoring foreign control of the nationalized oilfields involved renaming the company as the National Iranian Oil Company while reconfiguring the concession arrangement. It also involved the reengineering of connections between flows of oil, mechanisms of financial governance, laws, methods of accounting, and the mechanism of the consortium arrangement at multiple sites in Iran and abroad. The success of the new consortium in blocking national control and rearranging older forms of monopoly control gets underplayed in the scholarship on the topic.[1] Instead, the consortium negotiations are analyzed as a separate organizational issue from the nationalization crisis, March 1951, and the coup d'état, August 1953.[2]

As the crisis unfolded between 1951 and 1954, British and Iranian lawyers debated whether a contract had been signed between a private entity and a government, two governments, or both. Following the transportation of the dispute from the ICJ to a site of international security at the UN and back again, I reveal the ways the two sides clashed over the framing of issues. The central issue concerned the question of political sovereignty over a natural resource and the right to nationalize a foreign-controlled oil industry, only for the British side to witness, for the first time in its history, the failure of the association of oil with the authority of AIOC's concession and the successful upholding of the sovereign political rights of the Iranian nation. As in the past, the British side hoped to enlist international law to mark the difference between order/civilization and disorder/ignorance, and to create delays until an advantageous solution—the preservation of concessionary authority—was achieved. When this particular truth-making strategy failed, the British side imposed further delays at yet another site, the World Bank. Back in Tehran, the work of violence and a CIA-engineered coup d'état transformed the proposed format of the international consortium arrangement from possibility into reality. The task of the rest of the chapter is to highlight the work of two unusual and unexpected actors in this history, US national security and another formula known as the Aggregate Programmed Quantity or APQ—two post–World War II devices designed to place limits on oil supplies and keep profits high as a means of reconfiguring the older concessionary arrangement in a new consortium. Working hand in hand with Iran's parliament to restore foreign control of a nominally nationalized Iranian oil company, these tools and techniques formed the last set of connections available to finalize the consortium deal within an appropriate reformist-nationalist framing.

Law by Association: The "Double Character" of the Concession

As discussed in the previous chapter, Iran refused to abide by the ICJ's interim provisional measures, intended to preserve AIOC's oil operations until a final ruling was made on the Court's jurisdiction in Iran's oil nationalization case. Instead, the Iranian government took steps to transport the dispute back to a domestic legal arena by giving notice to the UN Security Council that it was abrogating the declaration of 1932 recognizing the compulsory jurisdiction of the ICJ. As the crisis escalated at multiple sites in Iran and abroad, the British government responded with its decision to translate an international legal issue into an international security issue. The failure of the Harriman and Stokes diplomatic missions had first triggered AIOC's decision to evacuate all British staff members by August 1951. The Iranian government also served notice to all British staff to sign individual employment contracts with the National Iranian Oil Company (NIOC) or leave the country.[3] Iranian troops seized the Abadan refinery in late September, refusing admittance to all but ten British technicians. This triggered the British government's decision to return the dispute to another international regulatory venue by asking the UN Security Council to intervene.[4]

The British government, claiming to work on behalf of its "injured national," AIOC, of which it claimed over a 50 percent stake, was heavily involved in the work of translation, black-boxing one legal network and reopening another in multiple international arenas.[5] With the oil nationalization case still pending at the ICJ, two important legal questions remained unresolved: What were the rules of international law, first with regard to the right of a state to nationalize industries within the country, and second with regard to the right of foreign states to intervene diplomatically on behalf of their nationals? Three legal issues were entangled in these questions: the claim to national sovereignty, the act of expropriation, and the cancellation of the oil concession.[6]

The British government's request to the UN Security Council argued that the Iranian order expelling British staff was a violation of the ICJ's order calling on both parties to comply with interim provisional measures.[7] Iran's position was that the Security Council, like the ICJ, was incompetent to consider the British complaint because Article 2, paragraph 7 of the UN Charter forbade the UN to intervene in matters "essentially within the domestic jurisdiction of any member nation."[8] Sir Gladwyn Jebb, the British government's legal representative, argued that the ICJ's call for interim measures showed clearly that the dispute was at least "prima facie justiciable and not a matter solely within the domestic jurisdiction of Iran."[9] The Iranian government was creating an "inflammatory situation," which was a potential threat to "peace and security." Jebb argued that the Security Council should adopt the British government's draft resolution calling on Iran to revoke its expulsion order and comply with the provisional measures indicated by the ICJ. This would ensure that "the role of law in international affairs is upheld, to say nothing of the prevalence of reason."[10] Jebb continued that "on behalf of intelligent against unintelligent ... the resolution will create a landmark in the vast process of peaceful adjustment between ancient East and the industrialized West," which constituted the "major problem of our generation."[11] Doctrines of international law could be put to work to manage and maintain relations between those countries anchored in the past (the ancient East), constituting a threat to peace and security, and those in the present (the industrialized West).

International law and security were an expression of civilization that enabled British legal experts to step outside local constraints, "it was thought, and thus acquire a universal vision and understanding."[12] The transformation of the oil controversy in international legal and security terms was the difference between order and disorder in the dispute. If the dispute remained a domestic issue for municipal courts, it would be uncivilized and disordered, even though Iran's arguments reflected "certain incontrovertible and classic principles regarding sovereignty and domestic jurisdiction."[13] The British government sought to exclude the local by framing its interests in terms of a peculiar battle, defining the universal struggle of an entire generation on behalf of the rule of law, reason, intelligence, and the industrialized West. This universality of the principles of international law and security "fixed its difference from what was exceptional and local"—the Iranian government, oil workers, and public opinion—most obviously in their failure to follow "principles true in every country."[14]

As Anghie explains, attempts by countries such as Iran to regain control over their natural resources "generated a number of complex debates about several doctrines of international law."[15] As in the 1933 concession crisis, the British sought to invoke this emergent international legal and security order to reestablish the authority of AIOC's oil concession over any other possible arrangement. But this discussion does not aim to open up the details of court proceedings in order to get at the technolegal content of the dispute. Rather, it tracks how new mid-twentieth-century mechanisms of economic and legal governance, set up to deal with a new social reality of an expanded international community of nation-states, worked as a technique of political power that equipped the British with the power to associate Iran's oil with foreign control and block national control of oil production. In this period, "developing states" such as Iran asserted their right to control and exploit their own resources by relying on the principle of permanent sovereignty over natural resources.[16] Such efforts were connected to other emerging issues including the right to nationalize, the right to economic development, and the right to self-determination. The problem was that these states had played no role in the formulation of international law, which originated in Europe and became "universally applicable" in the nineteenth century "as a consequence of colonial expansion."[17]

First, in legal terms, every state has the undisputed right to vest properties and industries "which it owns in a national board or company." This constitutes an exercise of national sovereignty. As Cheng explains, however, "this right has a corollary duty," the obligation to protect within its territory the rights of other states, and in particular, "the rights which each State may claim for its nationals in foreign territory."[18] In the exercise of its sovereign powers, therefore, a state is left by "international law with a great deal of discretion, which is the field of the State's domestic jurisdiction." But, to the extent that a state has duties under "international—customary or treaty—law, the matter ceases to be one which is exclusively within its prerogative."[19] While the Iranian government had the legal right to nationalize its oil resources within the field of its domestic juridical order, so too was it limited in these rights by having to abide by certain duties under international law. If the British government believed that Iran had failed to protect the rights of its national, a private oil company, the field of domestic jurisdiction in the exercise of Iran's sovereign rights was suddenly open for intervention and redirection in international law.

Second, a state has the sovereign right of expropriating private property within its territory, but is, at the same time, subject to international duties when the expropriation proceedings affect the property of aliens.[20] Cheng explains that in legal terms, a state may, in "certain exceptional circumstances, under the right of self-preservation," legitimately cancel a concession, "provided that compensation be paid to the concessionaire."[21] On the other hand, the case of cancellation does not entitle the "grantor Government unilaterally and arbitrarily to cancel the concession."[22] Such an instance must be deferred to an "impartial or disinterested tribunal." According to Article 26 of the 1933 concession, Iran was not granted the right to simply cancel the concession. Instead, both parties to the concession agreement needed to agree to an arbiter, as stated in Article 22, and if they failed to do so, must resort to international law to appoint one. At the same time, the Iranian state, in "exceptional circumstances," appeared to have the right of cancellation provided that compensation was paid to the British concessionaire.

Prior to the reconvening of the Security Council on October 15, 1951, the British government made such a case, filing a memorial with the ICJ, asking it to declare that Iran's annulment of the concession and refusal to arbitrate was a denial of justice, and thus a violation of international law.[23] The memorial makes the case, as Cheng suggests above, that state sovereignty is not absolute but may be limited by international customary law and by obligations of a treaty or contractual character.[24] According to the British position, a state is not entitled to nationalize a concession if, by an international "contractual obligation towards the government of the State of which the concessionaire is a national," or by a provision in the concession, "it has expressly divested itself of the right to do so."[25] In the legal view of the British memorial, the act of expropriation directed exclusively against foreigners constituted a "discriminatory" move exposing nationalization as a "disguise for confiscation."[26] Even if the Iranian government was entitled to terminate the concession of 1933 unilaterally, this right did not extend to Article 22 calling for arbitration. Thus, the British government sought to restore the certainty built into the concession's terms by referring to its articles on arbitration and insisting that the Iranian government's motives were hostile, having nothing to do with the political question of the right to national sovereignty over a natural resource.

The crux of the British argument, which would serve as the basis on which the ICJ would rest its final judgment, was that the 1933 concession had a "double character." "On the one hand it was a contract between two parties," one a state and the other, "not a State but a national" of the United Kingdom.[27] On the other hand, the concession implied an agreement between the government of the United Kingdom and the Iranian government. In 1901, a "prima facie international obligation" (a concession framed in international diplomacy terms) was established upon a state, Iran, to observe the terms of the concession granted to a "foreigner," William Knox D'Arcy, such that the obligation was extended toward the United Kingdom, of which the foreigner was a national. Thus, the "international responsibility" of the "grantor state," Iran, was engaged. This international obligation embodied a "contractual character," and therefore, according to the British government, "may be described as an implied treaty or convention between the two States concerned." The concession also had the character of an Iranian law, but this did not prevent it from having the character of a contract or treaty too.[28] These legal points would serve as the basis for transforming the law by "internationalizing" concession contracts to enable foreign corporations such as AIOC to take on a "quasi-sovereign status" and deal with non-Western states such as Iran on an international playing field rather than a local-national one.[29] Anghie further clarifies that "whether a quasi-treaty between a sovereign and a quasi-sovereign entity, or a contract between two private parties, what is common to both characterizations is the real reduction of the powers of the sovereign Third World state with respect to the Western corporation."[30] As discussed in previous chapters, the concession's terms also maintained an ambiguity that could be exploited by multiple sides with conflicting interests. Over the course of the twentieth century, the ambiguity led to situations of immense uncertainty and opened the door for pursuing alternative political arrangements of the energy system.

The British government's case rested on two grounds. First, there was an international dispute between the two governments arising from the fact that the British government, in exercise of its right of diplomatic protection of its nationals, had taken up the case of AIOC when Iran purported to cancel the D'Arcy Concession of 1901. Both governments accepted the new concession in the settlement of 1933 as solving the international dispute, the British side argued. The memorial explained that "when there has been an international dispute between two Governments … there arises under international law an obligation binding the two governments to observe the terms of the settlement."[31] This obligation arose even though the resolution was in the form of a concessionary contract between a state and a private company.

The second contention was that the 1933 dispute, presented by the two governments before the Council of the League of Nations, was removed from the Council's agenda when, "but not until," the concession had entered into force on its ratification by the Majlis and approval by the shah.[32] According to the British truth-making strategy, both disputing governments and the Council had agreed on the withdrawal of the first concession dispute from the League in 1933. In other words, the act was the "equivalent of a resolution of the League accepted by the two parties," declaring that the dispute should be settled by enforcing and observing the 1933 concession. It is important to note here that the Council's actual ruling never expressed an opinion on the legality of the cancellation, the role of jurisdiction for the case itself, whether constituting diplomatic protection or a question for Iranian municipal law, or the legality of claims concerning the role of the company in its treatment of Iranian labor and its social impact on Khuzistan. While the ruling acknowledged "the important questions of law" concerning the case, the Council had simply approved a provisional agreement with both parties agreeing to the suspension of further proceedings. The legal standpoint of each party, as stated before the Council, remained "entirely reserved."[33]

The British memorial made its case for the "double character" of the concession by framing its role in terms of a historical entanglement in both the signing of the 1901 concession contract and its revision in the 1933 concession. The British position relied on particular truth-making strategies to frame the history of the oil concession in legal terms that secured the British government's backing of AIOC in 1901, 1933, and now 1951. This strategy relied on an assemblage of contracts, legal arguments, memorandums, provisional measures, meetings and delays, diplomacy in Tehran and London, international boycotts, and institutions of international law and security.[34] Working through the technolegal details of the dispute, over the course of two decades, has exposed the transformative work of the law in action—that is, the building and extension of British contractual control from oil operations in southwest Iran to international sites abroad and back. International law and security worked as technologies of intervention and control, helping the British side strengthen its bargaining position by transforming the political question of national control into a battlefield of legal technicality.

Contrary to the British position, the Iranian government did not want to associate international law with the flow of income from oil profits. Allahyar Saleh presented the legal basis of Iran's case to the Security Council, but this time on the grounds of a nationalist truth-making strategy that sought to include the political question of national control of the oil. First, he argued that the Council did not have competence to deal with the dispute because the oil resources of Iran, "like its soil, its rivers and mountains, are the property of the people of Iran."[35] This ownership and authority constituted inalienable rights that rested on Iran's national sovereignty and equality among the other sovereign states of the community of nations and of the UN. He argued that the provision in the "Law Regulating Nationalization" on compensation to AIOC,[36] and the offer to employ British staff, together demonstrated Iran's exercise of its sovereign rights, which was not "injurious to others."[37]

Saleh associated the law with universal principles of sovereign rights. He argued that it was a settled principle of international law that in matters of domestic concern to which the dispute related, the exercise of sovereign rights "can neither be abridged nor interfered with by any foreign sovereign or international body."[38] Referring to Articles 1 and 2 of the UN Charter, he argued that their terms provided the basis for Iran's position that the Council was incompetent to intervene in the oil dispute.[39] Invoking articles from the UN Charter as a legal defense represented an early attempt among "developing states" to use the UN General Assembly "to create a different type of international law," one that would work favorably in their interests when dealing with the West, particularly Western corporations.[40] The Iranian government had previously argued at the League of Nations that the 1933 concession was a "private agreement between the AIOC and the Iranian government," which could not limit Iran's sovereign rights to dispose of its resources as it saw fit. The

British government had acted in violation of international law by seeking to "usurp Iran's sovereign rights in matters of domestic concern, by interfering in the internal affairs of Iran, by placing its armed forces" near Iran, and by its "abusive use" of the ICJ. The sovereignty of states, argued Saleh, rested within the principles and laws established in Iran.[41] Saleh rebutted the British argument that the Security Council was competent, but his country was at a disadvantage in terms of strength—its military and economy were weaker than those of the British government, whose warships had gathered in the vicinity of Iran.

How the Law Was Transformed by Oil

After a series of delays on both sides, the Iranian government made its "Preliminary Objection" to the British government's mémoire of October 10, 1951, addressing the competence of the ICJ, on February 4, 1952.[42] The "Iranian Declaration" limited the jurisdiction of the ICJ to disputes arising "after the ratification of the said Declaration" and similarly with regard to treaties or conventions.[43] The said "Declaration" confined itself to the Iranian government's undertaking to "respect in regard to British nationals the rules of general international law, the violation of which," it argued, was not invoked by the British government and therefore did not provide grounds for the institution of proceedings before the Court.[44] The Iranian government attempted to deconstruct the legal formulation put forward by the British by arguing that the concession did not possess the character of a treaty or convention because it was not made between two states, nor was it registered with the League of Nations as such. Thus, the Court lacked jurisdiction.

On July 22, 1952, over a year after the British government's "Application," the ICJ ruled that the Court had no jurisdiction in the Anglo-Iranian oil nationalization dispute.[45] Anghie, who is quoting Ian Brownlie, explains that before the Second World War, the notion that concession contracts might operate on the field of international law was "heretical."[46] Furthermore, in 1952, the ICJ had "declared in effect that an agreement between a state and a corporation was simply a concessionary agreement and could not be elevated to international law." On the one hand, the truth-making strategies pursued by the British government and AIOC had provided a technical means for successfully delaying a resolution of the crisis until connections to other circuits and agencies—such as the flow of oil and income to the Iranian government—could be stabilized in an advantageous way. On the other hand, this strategy of temporization and its connections to impeding flows of Iranian oil and revenues had failed to win the backing of international law.[47] The Court argued that Iran obviously had "special reasons" for drafting the "Iranian Declaration" in a restrictive manner and excluding earlier circumstances. At the time, Iran denounced all treaties with other states relating to the regime of capitulations, uncertain as to the legal impact of these unilateral acts. It was unlikely that Iran would have willingly agreed to submit to an international court disputes relating to all of its treaties. Earlier treaties did not apply and thus the United Kingdom could not rely on them.[48]

The Court ruled that the 1933 concession had not resulted in an agreement between two governments, which may be regarded in terms of its "double character." The concession did not possess a double character because the United Kingdom was "not a party to the contract," which did not constitute a link between two governments or regulate relations between them. Under the contract, Iran could not claim any rights from the United Kingdom that it may claim from AIOC, nor could it perform any obligations toward the United Kingdom that it was bound to perform to AIOC. This "juridical situation," argued the ICJ, "is not altered by the fact that the concessionary contract was negotiated through the League of Nations."[49] At the League, the United Kingdom had exercised its "right of diplomatic protection in favour of one of its nationals," and this had nothing to do with the contractual relation between Iran and AIOC.[50] Thus, the Court concluded that it lacked jurisdiction.

The ICJ's judgment appeared to transport the dispute back to the domestic arena in Iran, suggesting that Iran's unilateral nationalization of AIOC was legal. However, the ICJ ruling did not induce the British to relax the boycott of Iran's oil at all.[51] As Heiss explains, the British considered the court ruling to serve "simply as a recognition that it could not decide the dispute, not a ruling that AIOC claims were invalid."[52] The British government's decision to resort to the rules of international law and security was intended to mark the difference between order and violence in the dispute, bringing the truth and order of the principle of law to the disorder of striking workers in the oilfields and a nation lacking in technolegal knowledge to take over an industry. Following the circuitry built between the control of oil in southwest Iran and technical arguments concerning international law and security has revealed the British strategy to temporize and delay in order to achieve an advantageous resolution. But this strategy of diversion failed to achieve the desired domination from London and the strengthening of the British bargaining position. Instead, international law ruled in favor of Iranian sovereignty in what would become known as a landmark case called "the Anglo-Iranian Oil Company Case (United Kingdom v. Iran)," which set the precedent for future disputes between national governments and foreign firms over the control of natural resources.

The law was transformed by its encounter with Anglo-Iranian oil having to make adjustments, for the first time, to deal with a national government intervening in a foreign-controlled industry to take control of it in the name of political sovereignty over its natural resources. International law regarding contracts claimed a universality that was, in practice, "specifically devised to deal with a type of agreement to which only Third World states were parties"—that is, "economic development agreements" such as the concession contract. In an emerging mid-twentieth-century context of resource nationalism, it was no longer possible for Western powers to resort to international law in order to preserve a distinction between the order of laws and contracts and the question of political sovereignty over a nation's resources. The law would have to be transformed. To survive, the British could not acknowledge the truth of international law, namely that the ICJ had made any decision regarding the legality or validity of AIOC's concessionary role in Iran's oil operations. Competing formulations of the law and concession terms did succeed, however, as a temporizing strategy, but they did not work alone, nor were they entirely successful in attaching themselves to other kinds of circuits to impede the flow of oil and income accrued to the Iranian government. The next step in avoiding any further disruption to the British monopoly over Iran's oil was to enroll another set of circuitry between oil and a new institution of global economic governance, the World Bank.

"Masterly Inactivity" at the World Bank

As it battled to stabilize the authority of AIOC's concession, the British side had shown extreme inventiveness in transforming and transporting the dispute to multiple sites in the international arena. According to Louis, "British policy pursued the hallowed course of 'Masterly Inactivity,'" which meant putting forward "no constructive solution" to the oil dispute.[53] There were good reasons for this, he notes, "as any alternative to AIOC would mean the breaking of the British monopoly." One such occasion for protecting British monopoly interests was the enrollment of the International Bank for Reconstruction and Development (IBRD, also known as the World Bank) in the first half of 1952. With Anglo-Iranian negotiations deadlocked and no sign of economic assistance,[54] Robert

Garner, the vice president of the World Bank, took the opportunity to offer his bank's services in order to break the stalemate during Mosaddiq's visit to the United Nations.[55] The World Bank had been established in 1944 at the Bretton Woods Conference to aid postwar reconstruction and Third World economic development.[56] As successor to the Mandate System of the League of Nations, it represented one of a series of international regulatory organizations, such as the United Nations and the International Court of Justice, which developed during the mid-twentieth century to formulate "new techniques with which to bridge the difference" between "the developed" and "the developing" countries.[57] One of the first instances in which the bank's powers were built was by proposing a series of technical arrangements for a resolution of the Anglo-Iranian oil dispute.

World Bankers of the 1950s, such as Garner, had professional identities as "economic professionals," deploying their technical expertise to manage crises such as the Anglo-Iranian oil dispute in terms of an "artificial dichotomy," a kind of economic purification, between so-called "political" problems and "economic" ones.[58] The bankers sought to limit national politics erupting in Iran (and Egypt) because they threatened to undermine future prospects for economic development and participation in the world economy. With no single raw material as important to the global economy of the 1950s as oil, the World Bankers saw an opportunity to use the bank's international identity to achieve a solution and boost the world economy. This strategy was most important in light of the British government's failure to enroll international law and security in the resumption of AIOC's oil operations.

The World Bank sought to act as a mediator in the dispute because the crisis was having a negative effect on the world economy and especially on the British and Iranian governments, which were deprived of their oil profits.[59] As Staples explains, the World Bankers believed that if they solved the economic problems of the dispute and resumed Iranian oil operations, the so-called political aspects of the crisis would become less urgent, each side would become more flexible, and outstanding issues would ultimately be settled. If the situation worsened, however, Iran would suffer from economic recession; this could undermine the government and provoke a civil war among Iran's political groups. The British would also suffer from losing significant AIOC tax revenues and the most important source of the Royal Navy's fuel oil. Finally, "the West" might lose Iran to communist takeover. The bankers decided to act.

To ease the financial difficulties on both sides of the dispute and create a suitable environment for achieving a long-term settlement, Garner outlined a plan to Mosaddiq. The bank would provide the funds needed to resume oil operations and would then act as a trustee of the oilfields and the Abadan refinery, operating the properties and marketing the oil through established AIOC distribution channels.[60] Proceeds from the sale of oil, deducting the bank's operating costs, were to be held in a trust pending a final compensation settlement between Iran and AIOC. Mosaddiq encouraged Garner to make this same proposal to the British, who believed that the bank's scheme offered possibilities.[61] In his reply to Garner, which was circulated to AIOC managers, Mosaddiq said that any intervention on the part of the bank to exploit the oil resources of Iran should be regarded as a delegation of authority from the Iranian government. The bank must act on behalf of the Iranian government.[62] The controversy over the nationalization of Anglo-Iranian oil was now bound up in multiple legal and economic institutions of international governance. To survive, the Iranian side needed to strengthen its bargaining power on a global stage by attempting to set the terms by which a resolution would happen.

On the British side, "masterly inactivity" also meant "keeping the Americans in play" while retaining British control.[63] In a confidential meeting between AIOC manager Neville Gass, Garner, and US State Department officials, the latter expressed a preference for avoiding the threat of militant nationalism by abandoning interim arrangements and pursuing a longer-term settlement. State Department officials suggested a plan for Iranian management control.[64] They argued that although sums of money received under the proposed new agreement would not be sufficient to "keep communism at bay" in Iran, the conclusion of an agreement, whether in the short or long term, would enable Mosaddiq to redirect his attention from "action against the British to action against the [communist] Tudeh Party," a group they assumed "would oppose violently any settlement."

With tentative approval from both sides, the World Bank staff drafted a formal proposal intended to save the Iranian oil industry and appease Iranian nationalism. British, Iranians, and Americans all looked to the bank as the "best chance" for bringing a constructive end to the crisis.[65] To restore the flow of oil quickly and protect the legal rights of all parties, the World Bankers believed that they must exercise exclusive managerial control, including full discretion to hire and fire personnel.[66] Staples suggests that the bank's staff "mistakenly believed the Iranians would accept the reintroduction of British workers," as long as they were responsible to the World Bank and "therefore, implicitly apolitical."[67]

The bankers worked hard to avoid the entanglement of political questions with economic and technical ones by separating the issues, a kind of purification that framed the controversy advantageously to avoid a disruption of international oil markets. In contrast, the Iranians, led by Mosaddiq, insisted that any settlement, interim or final, was a matter of national sovereignty and prestige. Therefore, a final settlement "could not be reduced to technicalities."[68] The bank's focus on economic issues as largely mirroring the position of AIOC "failed to grasp the dimensions of Iranian nationalism."[69]

In the end, Mosaddiq broke off talks with the bank because of its insistence on discounted oil sales to Great Britain.[70] The negotiations also stalled due to Mosaddiq's insistence that "'fair compensation' would be based on the current value of the oil installations."[71] AIOC's understanding of "fair compensation" was "based not on current value but on projected profits ... into 1993." America's rejection of Mosaddiq's formula for compensation "became part and parcel of the destabilization strategy" that led to the coup in August 1953. The crisis escalated. Three days prior to the bank's mission to the oilfields, a riot broke out in Tehran protesting the intervention of the World Bank in the crisis. Iranian public opinion was against transforming the dispute into an international legal, security, and now financial issue. Mosaddiq viewed the bank as a tool of the British, unless Garner recognized Iran's full control of its oil industry and the bank itself as acting solely on behalf of the Iranian government. Any suggestion otherwise would violate the spirit of Iran's nationalization decree.[72] Attempting to ease Iranian officials' concerns, Garner insisted that the bank was acting as a "friendly, neutral facilitator," framing its strategy on the basis of a sound economic and business rationale.

The impasse between the World Bank's economic approach and Iranian nationalism continued as Garner worked with oil consultant, Walter Levy to make another attempt to reach an interim settlement. Levy spelled out a highly detailed interim arrangement that split the profits from oil production and refinery operations on a 50–50 basis, with the World Bank serving as interim manager over oil operations and marketing.[73] The arrangement still contained the stipulation that a necessary number of foreign technicians must be employed. But the "real question," advised Levy, "is always whether the repercussions of any arrangement

... are likely to affect our national interest in a worse manner than if we would have no settlement at all."[74]

Garner traveled to Iran in February 1952 to make a second attempt at an arrangement. With the assistance of Torkild Rieber, Hector Prud'homme, and the bank's legal and economic team, Garner submitted a memorandum of agreement that Staples says "concealed some very difficult problems," and perhaps was "deliberately designed to do so."[75] The arrangement gave Iran veto power over all marketing arrangements, and the two sides agreed to a two-way split of profits, with AIOC's share going into an account on which the company could draw after the disputed points in the crisis has been settled. The issue of British technicians and the pricing of Iranian oil remained unresolved. The bank refused to accept Iran's conditions that it was operating solely on behalf of Iran. In a final and desperate plea, Garner first asked the British government to withdraw its case from the ICJ, which he believed would appease Mosaddiq's government and enable the readmittance of British technicians. Second, Garner invited Hossein Maki, Mosaddiq's oil adviser, to visit the United States and consult with American oilmen at the bank's expense.[76] The World Bankers hoped this would "enlighten Iranians about the facts of the situation," making them realize that "technical, rather than political, criteria determined the bank's insistence on readmitting British nationals." The failure of the World Banker's attempt to mediate the oil dispute did not merely signify the bank's attachment to an international "apolitical identity." On the contrary, it was the outcome of a series of arrangements and strategies that, for decades, deliberately aimed to transform political issues into economic and technical ones to build connections that would keep international oil markets stable.

The World Bank's strategy of mediating the dispute worked as another technology of intervention and control, not unlike the oil boycott and international proceedings at the ICJ and UN. It had failed to achieve a real resolution to the nationalization controversy, but it succeeded as a temporizing strategy to undermine Iran's sovereignty by ensuring that nationalization did not happen according to the formulations proposed by the Mosaddiq government.[77] Examining the technical details of the bankers' work has revealed their connections to other interests and agencies. These included the British and American governments, oil industry experts, and a new international financial order concerned with managing the political threat of Iran's oil nationalization and avoiding its spread to other countries with national claims to their resources such as the Suez Canal in Egypt.[78]

Regime Change

The history of Iran's attempt to nationalize its British-controlled oil industry involved the failure of a series of technolegal and financial arrangements at multiple sites at The Hague, the UN Security Council, and the World Bank. The arrangements appeared to operate as potential solutions but worked in practice as political weapons by helping to separate economic and technical concerns from political ones, leaving open the possibility for other more undemocratic political arrangements of the energy system. The Mosaddiq government refused all of the formulations for a compromise put forward. A permanent settlement needed to be devised to keep AIOC's interests alive, but without placing the management of operations solely within the hands of either AIOC or a subsidiary.

British and AIOC officials considered more violent political outcomes that might bring about the downfall of Mosaddiq.[79] The Mosaddiq government was unlikely to agree to AIOC or a subsidiary of AIOC returning to operate in Iran for the foreseeable future, if at all. As British officials began to redirect their energies toward bringing AIOC in line with government policy, William Fraser, chairman of AIOC, refused to discuss any new scheme with American officials until it was clear "under what conditions Mossadeq falls," who would be his successor, and what line his successor would take.[80] Fraser preferred a final settlement, but he doubted whether this was a practical solution, in the event of the overthrow of the Mosaddiq government. Instead, some sort of interim arrangement must be set up as a short-term solution with a new Iranian government.

As I discussed in the previous chapter, American oil expert Walter Levy first developed the idea for the consortium arrangement in early discussions with Iranian government officials in 1951, but according to Louis, "the actual architect" of the reorganization of the Iranian oil industry on the British side was Peter Ramsbotham, then at the oil desk of the British Foreign Office.[81] Ramsbotham argued that whatever the "concession" amounted to before in Iran must be "replaced." He envisioned a "contractual arrangement" in which a new company would negotiate a 50–50 profit-sharing agreement with the Iranian government. AIOC would receive compensation for losses, possibly through arbitration. Ramsbotham's proposed arrangement involved a new managing company operating as a "façade that would enable the Iranians to save face." The new agreement would also include a guarantee to prevent the Iranians from interfering "in the company's day to day operations." A British company could not operate as the sole company in Iran because this would be "too transparent a restoration of British monopoly."[82] British government officials including Anthony Eden, the foreign secretary, did not like Ramsbotham's novel suggestion of bringing in American companies. Interests were shifting, however, and even Fraser was coming to agree in principle to a management company in which the major American oil companies might participate.[83]

Another political crisis ignited when the shah agreed to dismiss Mosaddiq and replace him with Qavam briefly in July 1952, coinciding with the ICJ ruling that it had no jurisdiction in the Anglo-Iranian oil dispute. Mosaddiq returned to power after only a few days in July 1952.[84] The Americans feared that the economic collapse of the Mosaddiq regime might open the door for communist rule. Thus, economic assistance was necessary to counter what they viewed as Tudeh exploitation of the chaos and poverty. The British, on the other hand, acknowledged the threat posed by the Tudeh, but did not believe Mosaddiq could be appeased, nor would economic assistance impact the political situation. The crisis of July 1952, however, aligned the British more closely with the American view that if an oil agreement was to be reached, AIOC could not return alone.[85]

Policy lines ultimately shifted toward Anglo-American unity in the dispute. Unbeknownst to AIOC officials, whom Louis says "played no part in either the origins of the intervention or its execution," the British government had put plans in place to overthrow Mosaddiq since 1951.[86] They regarded three possible candidates for the replacement of Mosaddiq: Sayyed Sia and Qavam al-Saltaneh, former premier during the Azerbaijan crisis of 1946, and Fazlollah Zahedi, the army general.[87] British support for Qavam stemmed from their belief that he would collaborate in the conclusion of an agreement "satisfactory to the AIOC."[88] In the company view, Fraser argued that there was no rush for AIOC to return to Iran straightaway, as the Kuwait oilfields had already more than recovered losses from Abadan.[89]

With Mosaddiq restored to power, British and American government officials embarked on a series of "joint proposals" to the Iranian government as a solution to the oil dispute.[90] Officials at the British Ministry of Fuel and Power believed that future discussions between AIOC and the Iranians must proceed on an "entirely new basis."[91] The result was the Truman-Churchill proposal of September 1952 whereby the amount of compensation to AIOC would be arbitrated.[92] AIOC would negotiate with the Iranian government for the resumption of oil production and the United States would grant $10 million in budgetary aid. Mosaddiq responded with a demand for £50 million as an advance against the oil. This, argues Louis, was another turning point because the Americans now moved closer to the British assumption that it was "impossible to do business with Mosaddegh." Thus, British policy now shifted from "inactivity" to engineering the overthrow of a foreign, national government.

The Mosaddiq government expelled the British diplomatic mission in October 1952, and all diplomatic ties between the British government and Iran were broken. This marked a critical period when British and American interests collaborated in covert operations to overthrow the Mosaddiq government. MI6 relinquished control of its intelligence network to the CIA, which marked a decisive shift from the period prior to October 1952, when the plan to remove the Mosaddiq government was solely "British in inspiration."[93]

The details of "Operation Ajax," as it was known in American circles, to overthrow Mosaddiq in August 1953, have been well assessed in the scholarship.[94] The British and Americans aimed to engineer the installation of a pro-Anglo-American government in order to reach an advantageous resolution of not only the oil question, but also the Cold War battle against communist interests.[95] This entailed the building of a series of alliances and the enrollment of concerned groups to ensure that the Iranian government would actively participate in the takeover of its oil industry by participating in an international oil consortium framed as national control. With no constructive solution in sight, Anglo-American governments and corporate oil interests turned to a more fundamental political arrangement—regime change—through which Iranians would help execute the foreign takeover of the oil industry.

The 1953 coup and the reengineering of Iran's national government effectively closed off the political possibility of more democratic forms of oil production as originally demanded by oil workers, backed by the communist party and national public opinion. The coup did not mark the start or the end of the nationalization dispute over oil. Rather, it was one connection among a series of connections, conflicting interests, and alliances that continued to transform Iran into a "clinic" for the stabilization of private-public Anglo-American oil interests and a recasting of their interests as necessary for securing "strategic concerns" of the Cold War battle against Soviet expansion.[96] It was politically impossible for AIOC to return as the sole owner and operator of Iran's oil operations. This shift in interests drastically weakened the company's bargaining position and left open the possibility for American oil companies to redirect goals and alliances in other directions, namely, the assembling of an Iranian oil consortium.

US National Security Apparatus

The working out of American interests in the resolution of an Iranian oil consortium commenced much earlier than the events of the 1953 coup. The American government had repeatedly claimed that if a solution to the nationalization crisis was not found, Iran's economy would collapse from a lack of oil revenues, producing political instability and inviting communist control over the oil.[97] On the heels of Iran's nationalization act in May 1951, US government officials first consulted with American oil executives running operations in the Middle East about the possibility of involving their operations in Iran.[98] What impact would this have on their concession areas, and what could be done to protect their position in different areas? Howard Page, representing Standard Oil of New Jersey, responded by arguing that it was inadvisable for American oil companies to work in Iran. American participation was sure to invite objections from the Department of Justice and Federal Trade Commission (FTC) regarding antitrust legislation.

American oil companies suddenly had a direct interest in controlling the amount of oil Iran was producing by participating in the joint venture of a consortium and avoiding the threat of competition. After World War II, massive newly proved oil reserves in Saudi Arabia and Kuwait were controlled by the noncartel American oil companies, Gulf Oil, and Standard Oil of California (Socal), and this control threatened to introduce competition into world markets. In consequence, the activities of the postwar oil cartel shifted from controlling market distribution to controlling the supply end of the petroleum system.[99] Anglo-American oil companies embarked on joint production and long-term supply arrangements in the Middle East, as a means of maintaining high oil prices and profits. First, the strategy was for companies to expand the number of interlocking, jointly owned production companies to unify their control of concessions and of oil output from Middle East sources.[100] Second, they established a system of long-term mutual supply contracts, under which they sold or reciprocally exchanged (i.e., bartered) enormous volumes (upward of a billion barrels) of oil and products among themselves at substantial mutual savings, while at the same time, quantitatively and geographically balancing surpluses and deficits. If supply could be controlled at the source, the elaborate market arrangements at the retail level throughout the entire world would be unnecessary.

The problem was that certain government agencies in the United States had an entirely different set of interests at stake concerning the antitrust activities of the largest American oil corporations and the interests of US national security in protecting the free flow of oil. Back in November 1951, the FTC wrote and circulated within government circles the *Report on the International Petroleum Cartel*, a study with damaging evidence that threatened to disrupt the stability of the energy system.[101] Oil companies' records indicated that their operations had retarded the development of oil in certain Middle East countries and purposely drilled dry holes to comply with the legal technicalities of their contracts. The report evidenced that during the 1920s and 1930s, the largest Anglo-American oil companies had endeavored, through company agreements, to control and divide the world oil markets.[102] According to State Department officials, publication of the report would seriously undermine the position of oil companies as well as the Anglo-American position in the Middle East. Opponents of the oil companies, advocates of nationalization, and the communists would have a "substantial amount of material in the report indicating that the oil companies had acted contrary to the best interests of the countries concerned." Also, the report was "highly embarrassing to the British" as it highlighted cartel activities through British operations in Iraq and IPC, in which AIOC was a shareholder.

Whereas the Federal Trade Commission identified itself as an "independent agency" with the authority to question the monopolistic activities of American corporations on behalf of the American consumer, the State Department recommended, in March 1952, that a request be put to the National Security Council (NSC) for a judgment on whether publication of the report would be "contrary to the national interest."[103] The State Department was attempting to put a gag order on the FTC. A battle over Anglo-Iranian oil was forming among American government agencies framed in legal, commercial, or foreign policy terms, depending on the interests at stake. These were the kinds of procedures and connections through which the resolution of Iran's oil nationalization controversy shaped the building of a different kind of international petroleum order focused solely on placing limits to world oil supplies.

The release of the FTC study was held up at the request of the State Department due to the "sensitive negotiations which were going on in an effort to straighten out the Iranian situation."[104] In a letter to the chair of the FTC, Dean Acheson, the secretary of state, made his case in terms of America's national foreign policy interests. He warned that the publication of the report would affect the "foreign policy aims of the United States in the Middle East" and could "seriously impair their attainment."[105] Mixing economic and political concerns, Acheson argued that the report would inevitably be "interpreted by the peoples of the region as a statement that, were it not for such agreements, they would be getting a higher return from their oil resources."[106] This would motivate movements such as the one in Iran for renegotiation of the present concession agreement and could give encouragement to groups pushing for nationalization. "Since the issues are not only economic but also political," explained Acheson, "the net effect will probably be to cause a decrease in the political stability of the region."[107]

State Department officials were connecting economic and technical issues of control and distribution to questions of national security, political stability in the Middle East, and the threat of militant nationalism. Publication of the report would "prejudice prospects for a settlement of the Iranian oil controversy" by "damaging, perhaps, irreparably, the status of the US as mediator between the UK and Iran."[108] Exposing details of the report would therefore contribute to economic deterioration in Iran, and increase opportunities for communist subversion to organize workers and gain control of the oil. Politics was factored into arguments about economic and technical concerns when it favored US interests and protected those of the oil companies.

The State Department connected the dispute over the FTC report to a peaceful resolution of the Iranian oil crisis, by framing its release as a problem of national security. The Department of Justice challenged this framing by transforming the problem into a legal-national issue. On July 17, 1952, the Department of Justice announced that a Federal Grand Jury would conduct an investigation into international oil cartel activities involving the five largest American oil majors, AIOC, and Royal Dutch Shell. Seeking to maintain a separation between the legal question of antitrust activities abroad and the political question of American national security in the Middle East, President Truman advised the FTC chairman, James Mead to declassify and release the report, "if the deletions and revisions suggested by you are made."[109] Furthermore, British officials pursued a similar line, insisting to the US attorney general that the Department of Justice could not expect to obtain the records of AIOC's overseas activities on the grounds of British national interest.[110] National Security Resolution 138 ruled that publication of the FTC report would have a "catastrophic effect" on American and British oil companies in

Venezuela and the Middle East, owing to the loss of oil sources to the "free world."[111] National security considerations required that nothing interfere with the free flow of petroleum and petroleum products from Venezuela and the Middle East. The Court conveniently annulled the subpoena for AIOC's records.[112]

NSC 138 helped protect the cartel arrangements of the largest Anglo-American oil companies by building a connection between the resolution of the Iranian oil dispute and the national security of the United States.[113] The US government was shifting from its role as mediator between the British government, AIOC, and the Iranian government to the leading spokesperson. The device of national security also helped Truman make the final gesture, opening the door for American oil companies to consider participation in controlling Iran's oil: he ordered the US attorney general, James McGranery, to transform the antitrust case from a criminal proceeding into a civil action suit.[114] This would help delay proceedings against the international cartel by four to eight years.[115] Working in the interests of the American oil corporation, national security operated as an apparatus to enforce delays and construct layers of secrecy by blocking the public disclosure of evidence concerning the operations of American oil companies abroad with "national security implications"— that is, the loss of oil sources to the so-called free world.[116]

An attorney-at-law who worked in the Antitrust Division of the Department of Justice recollected that by 1953 "the Iranian crisis caused decision making to be transferred from the Justice Department to the State Department and the National Security Council."[117] The NSC ultimately issued policy directives that, "for all practical purposes, gutted the oil cartel case." National security directives precluded the Justice Department from challenging the legality of joint production, joint refining, joint storage, and joint transportation ventures among the seven largest oil companies. These are the very joint arrangements that were the most significant features of the postwar supply cartel. Left open for prosecution were the older market cartel arrangements, which by then had become "relatively incidental to the basic joint venture, supply-control system."[118]

Government regulatory regimes and national security were some of the techniques that transformed postwar energy abundance into a system of limited supplies.[119] The system of limited supplies emerged through connections between US national security, American oil companies, and mechanisms for managing the nationalization of Iran's oil. The connections put in place a new consortium arrangement, which was constructed long before the American-led coup events of August 1953.

The US president and officials at the NSC and State Department worked hard to keep certain actors and interests in (i.e. the American oil companies), and other rivals out (i.e., Department of Justice, FTC, militant nationalism, and communism). The history of these sociotechnical and national-legal assemblages exposes the co-construction of the power of government and nongovernment (corporate) entities with certain mutual interests at stake. These interests were represented as issues confined to the public-government domain of "national security," foreign policy, and the fight against communism and militant nationalism.

Concession to Consortium: American Oil Majors Play "Dress Up"

The American oil companies and government needed to work together to achieve a solution in Iran sanctioned by US national law. Officials in the Truman administration advised the incoming Eisenhower administration to continue with this policy of achieving a coordinated solution. The new president must respect commitments on immunity from antitrust laws made by the outgoing Truman administration to employees of the US oil companies.[120] Eisenhower secured the approach by separating certain issues from others in discussions with the NSC. He argued that from a national security point of view, US oil companies must participate in an international consortium to purchase Iranian oil.[121] The NSC must ensure that "the problems of the cartel suit and the new consortium be kept wholly separate and distinct."[122]

Government officials, oil executives, and national laws came together as a matter of "patriotic duty" to achieve a solution to the Iranian oil problem. John Scott of the Socony-Vacuum Company and Harold Linder, assistant secretary of state for economic affairs, together agreed to "put AIOC and the other majors in the same frame of mind."[123] State Department officials invited oil company representatives to discuss participation in managing Iran's oil in December 1952. Government officials emphasized the importance of Iran to peacetime national security, the importance of winning the consent of the British government for US participation, and the necessity of reaching an agreement for compensation to AIOC.[124]

The oil majors had no particular desire to move Iran's oil as they had all taken steps to increase their production in other countries when Iranian oil stopped flowing. Difficulties would result with countries in which they had concessions, such as Saudi Arabia, which would be forced to cut back production in order to market Iranian oil. To maintain the concessionary arrangement, which the Iranians would "never agree to," an alternative arrangement might be put in place in the form of a management company to operate properties under a "management contract which had [the] same elements of [the] concession."[125] The concessionary arrangement was "politically impossible," argued Eugene Holman of the Standard Oil Company of New Jersey, but a "contract could be worked out to have the same effect" under a different name.[126]

The American government pursued discussions to coordinate with the oil companies in a new kind of management contract while at the same time assuming a position of neutrality in confronting the British and Iranian governments. Prior to the Anglo-American engineered overthrow of the Mosaddiq government in August 1953, the US government was unwilling to pursue any explicit change in its position as an "intermediary ... seeking a settlement of the Anglo-Iranian dispute."[127] Having reached a technical and economic impasse, British and American officials attempted to resolve the crisis with a political move: the overthrow of the intransigent Mosaddiq and his replacement with a more pliable shah and prime minister, General Zahedi.[128] The events of the coup itself helped manage nationalization by enabling a shift from concessionary control to a consortium. In the face of anti-imperialist public opinion, government officials and oil executives worked together to enroll the new Iranian government in helping to restore foreign control of the nationalized oilfields.

"Ways must be found of 'dressing up' the operating contract" to make it acceptable in Iran.[129] The first step was to employ a spokesperson to resolve the details of percentage participation, production quotas, compensation, and pricing among the oil companies prior to negotiations with the Iranian government. The American government chose Herbert Hoover Jr., son of the former US president and himself former president of the Consolidated Engineering Corporation with years of experience in the oil business, as just the man to serve as the new consultant.[130] Hoover agreed to take the lead on the condition that the government would guarantee the cooperation of the major oil companies in making room on the market for Iranian oil and that the Department of Justice would cooperate by not pressing the cartel suit.[131]

To block the threat of "extreme nationalization," American and British public-private interests were aligned with the position that the "negation of the Iranian nationalization proposal was inherent" in any proposal that would be put forward.[132] The arrangement was laid out in Hoover's "proposed Iranian consortium plan."[133] The so-called formula working elsewhere in the Middle East and South America must also work in Iran. Iran's parliamentary approval was "unquestionably required," however, before full-scale operations could commence.

Accepting the reality that a British company could not return to Iran as the sole operator, British officials listened to Hoover's explanations about the difference between a concession and the consortium arrangement.[134] In the "old concession type of enterprise the concessionaire had the implied ownership and possession of the subsurface resources." On the other hand, under the "more modern type of 'contract' agreement, the operator acted as the agent for production, refining and marketing," whereas the national government "retained title to all subsurface resources until they reached the well-head."[135]

Hoover used a different approach in his discussions with the shah and the Iranian prime minister. The argument worked in strictly technical terms along the lines of Levy's earlier strategy. If Iran were to dispose of any "appreciable quantity" of oil and its byproducts, it must use the same distribution channels controlled by the seven major oil companies moving the rest of Middle East oil.[136] This was the only way to maintain a system of limited oil supplies in a period of abundance, when oil supplies from Saudi Arabia, Iraq, and Kuwait could easily make up for any loss in Iran.[137] In 1974, at the congressional hearings prosecuting the international oil cartel, Howard Page, Middle East Coordinator for Exxon (Standard Oil of New Jersey), denied that the oil companies coerced the shah into dealing with the oil majors rather than with independent oil companies. To the contrary, Page explained, "the point is that we had the outlets for the oil."[138] Over a half century after the first oilfields were developed in Pennsylvania, the machinery of the transnational oil corporation had become an undeniable reality. Oil executives such as Page had intended to convey in no uncertain terms that the possibility of pursuing alternative arrangements of the energy system, by doing business with competitive and independent oil companies, was closed off.

Before embarking on negotiations between consortium members and the Iranian government, Hoover pursued one final set of connections to ensure the stability of public-private American interests in Iran. The American oil majors were reluctant to conduct any concrete discussions with either British oil companies or the Iranian government for fear of being charged with violations of the Sherman Antitrust Act.[139] Hoover assured oil officials that he had established contact with the antitrust chief, Stanley Barnes, at the Department of Justice, and with Herbert Brownell, the US attorney general. He requested that American oil majors explore a cooperative agreement among themselves and with British companies so long as no actual agreement was signed. In response to similar inquiries by the oil companies, Secretary of State John Foster Dulles explained that the State Department for its part had "no objection" to the oil companies attending the meeting set up by Fraser, chair of AIOC, at the behest of Hoover.[140] Dulles assured American oil company representatives that the matter had been "cleared by the Department of Justice." The attorney general confirmed this, recommending that a representative of the US government such as Hoover attend the meetings.[141]

There were uncertainties involved, however, in excusing the major oil companies from antitrust proceedings for the sake of restoring Iranian oil to world markets. The attorney general explained that approval by the Council of NSC 175 (US policy toward Iran) amounted to adopting a policy in the interest of national security, which was contrary to the antitrust laws of the United States.[142] As with the boycott, however, the provisions of the Defense Production Act of 1950 could be deployed to safeguard "those involved" from charges of violating antitrust laws in the first phase of consultation.[143] As for future phases, "implementation of plans would certainly involve violation." It would be necessary to go to Congress "for relief," argued the attorney general, and decisions might specifically be directed toward the problem of oil companies and Iran or all US companies doing business abroad. The national defense framing enabled the recasting of American oil interests as necessary for securing "strategic concerns" and the safe flow of cheap oil from the Middle East, without which "the allies' economic plans for the postwar reconstruction of Europe would have been all but impossible."[144] The engineering of these arrangements out of flows of Iranian oil supported American oil policy in the Middle East and its expansion.

Explanations put forward by the US State Department and oil majors about blocking communism and securing the continued supply of Middle Eastern oil have been reproduced in the scholarship on oil in the Middle East.[145] I am arguing that the problem was not about restoring the supply of Iranian oil but impeding and slowing it down to place constrictions on the flow of oil and income through strategies of delay. The mechanism of the consortium arrangement was precisely the means through which the control of Iranian oil production levels could be placed in the hands of the oil majors and block alternative political arrangements. Such an arrangement had to be successfully negotiated with the newly installed Zahedi government.

American and British government and oil company interests appeared to be in alignment on key issues essential to managing Iran's oil nationalization in a shifting international order. The question of compensation was transformed into a government-negotiated issue and the consortium delegation negotiated all other company arrangements and daily activities. This negotiating arrangement ensured that control of the industry was not relinquished to militant nationalism. National and local political actors—the Majlis, national public opinion, and oil workers—needed to be convinced that this particular arrangement of the energy system was the solution to resolving the crisis.

British government officials agreed with their American counterparts that stabilizing Iran's political situation and protecting the country from communist infiltration were more important than compensating AIOC.[146] The United States hoped to promote political and economic stability by providing military and economic aid to expedite the settlement of the Anglo-Iranian dispute on terms that "assuaged Iranian nationalism," protected the British position, and "squared with the oil agreements that US companies were negotiating with governments elsewhere."[147] Zahedi received emergency aid from the United States worth approximately $45 million but still required revenues from the sale of Iranian oil on the international market, which "depended on a successful resolution."[148] The Zahedi government resumed diplomatic relations with Britain in December 1953. As a "gesture to Iranian nationalism," Zahedi would not allow AIOC to return as the sole operator of the Iranian oil industry, but only as a minority member in an international consortium that included the major American oil companies.[149]

To reach a solution, British and US oil companies coordinated the exchange of information and policy, as well as reintroducing Iranian oil to world markets, which risked disrupting other energy flows and triggering a cutback in production in other Middle East countries.[150] Coordination and consultation between the US and British governments were essential to managing transit problems in Syria and Lebanon and imminent price talks between Saudi Arabia and Aramco as well as between Iraq and IPC. These larger connections to flows of oil elsewhere needed to be factored into any consortium arrangement with Iran. AIOC and British government officials discussed their concerns about the Eastern Hemisphere's dependence on Middle East oil in light of the new arrangements worked out in various Middle Eastern countries. The Middle East was forecasted to supply 90 percent of the Eastern Hemisphere's oil requirements by 1958 and 70 percent of Europe's total requirements. However, there was a lot of uncertainty concerning the pricing of oil in Iraq and Saudi Arabia as these countries were "ganging up" on the oil companies with the aim of "extorting every cent they can."[151] The "Arabs are as greedy as the Persians," claimed British officials, "and have not learnt that extortion does not pay." Pipeline countries such as Syria and Lebanon were also demanding some form of 50–50 profit sharing for transit rights.[152] Aramco must not yield to Saudi pressure, which would "prejudice the position of other British and US Companies in the Middle East."[153]

Backed by the Anglo-American installed Zahedi government, Iranian negotiators attempted to build connections between flows of oil, income, and national control differently, but within the framework of a new consortium. Abdullah Entezam, the Iranian foreign minister, maintained that the return of the former oil company in any future settlement was "impossible."[154] From his government's perspective, negotiations looking toward the sale, transportation, and distribution of oil should be initiated with representatives of "a group of large international companies." This consortium would purchase oil from the NIOC, Iran's national oil company, and undertake to handle its transportation and distribution. If other British firms wished to join the group, "their total shares must not exceed fifty per cent." In helping the Iranian oil industry recover, the Iranian government expected "considerable financial aid" from the oil companies, which would be reimbursed from revenues accruing in future years. A settlement must be arranged between the consortium and the former oil company with "no claim for loss of profits ... taken into consideration." Finally, "whenever the price of oil increases," Iran's government should "benefit from such increases and its income should at no time be less than the maximum which accrues to others."[155] Iran had learned from past disputes over the calculation of royalties and production rates to propose formulations that attempted to deconstruct the formulations of the new consortium members, but they were in a weaker bargaining position.

Questions about percentage shares, compensation, management domicile, operational control, production levels, world energy supplies, and Iranian public opinion were central to the working out of a consortium arrangement in Iran and reaching a final resolution to the nationalization crisis. Different sets of interests were at stake for different groups, and there was no clear divide between the so-called private interests of the oil companies and the public interests of their respective governments.

With American oil companies on board, US antitrust laws out of the way, and the consent of both the British government and AIOC to

American participation, Hoover moved to the complicated task of working out a consortium arrangement that Anglo-American parties could present to the Iranian government. The American government made sure that percentage shares of participation were settled before definitive negotiations commenced with the Iranian government.[156] The State Department and British Foreign Office finally agreed that AIOC along with the five US oil companies should each hold a 40 percent share in the consortium, AIOC with 40 percent and the five others splitting 40 percent equally. The remaining 20 percent would be divided between Royal Dutch Shell and the Compagnie Française des Pétroles (CFP).[157] The US position regarding AIOC's percentage participation was tied to its particular oil interests and its goal of bringing stability to Iran, which US officials continued to argue was crucial in view of the neighboring Soviet threat.[158]

The final agreement saw CFP settling for a 6 percent share, 14 percent to Shell, and 40 percent each to AIOC and the group of five US oil majors. The British Chancellor of the Exchequer expressed his government's unwillingness to see the American share rise above 40 percent as any greater amount would have a damaging effect on the British balance of payments.[159] Despite the substantial British shareholding in Royal Dutch Shell, Heiss says its 14 percent share did not disturb the Americans as much as the 8 percent proposed for CFP, "perhaps because the Dutch owned 60% of the company," or because they saw in Shell an "independent partner," which would help appease the Iranian side. Thus, the British share in the consortium came to less than 50 percent, which AIOC finally agreed to in March 1954.[160]

The British hoped that by conceding the size of AIOC's share in the consortium, it might earn a "US quid pro quo" on the question of compensation to be paid to AIOC for the loss of its dominant position in Iran. This was a key British demand and major sticking point in negotiations with Mosaddiq. On the other hand, the British fully understood the danger they faced from an independent US policy in Iran and the Middle East.[161] The American and British governments agreed they should decide between themselves how much compensation should be paid to AIOC but disagreed on the precise amount Iran should pay.[162] To prevent another breakdown of Anglo-American cooperation and due to their weaker bargaining position, British officials lessened their demands, arguing that if a satisfactory profit-sharing arrangement could be worked out with the Iranians, the British would settle for compensation somewhere between

$280 million, finally accepted by AIOC, and the mere $5 million proposed by Hoover.

An assemblage of Anglo-American government and corporate interests with differing degrees of strength was taking shape as a set of connections within a larger political project to develop the post–World War II international petroleum order on the basis of a particular arrangement of energy flows, finance, labor politics, international regulatory regimes, and US dominance. "Dressed up" as nationalization, particularly to the Iranian public, the new arrangement was merely a shift from the concession to the consortium format, in which the worldwide production, pricing, and distribution of oil continued to follow monopoly arrangements set up by the largest transnational oil corporations, which NIOC was not one of.

"APQ" and the Infamous Law of Supply and Demand

Consortium members aimed to settle the dispute with the Iranian government according to their own plans for compensation, management domicile (nationality), and operational control, production levels, and pricing. The working out of these details would enable the largest oil companies to reengineer the postwar oil system into one based on placing limits on supplies.[163] One weapon to accomplish this involved a formula allowing the consortium alone to decide how much oil Iran would produce after the third year of the contract, having finally set the first three years of production at fifteen, twenty-five, and thirty million tons respectively.[164] Known as the "Aggregate Programmed Quantity" or APQ, this calculating technology had been secretly agreed on by the eight participating companies in order to implement a secret arrangement to maintain total control over oil production levels.

In 1954, Herbert Hoover Jr., U.S.-government-appointed spokesperson and negotiator at the consortium talks, met with representatives of Standard Oil of New Jersey, Socony-Vacuum, the Gulf Oil Company, and the Texas Oil Company. They discussed the "probable terms" of a future agreement between the American oil industry, AIOC, and other foreign oil companies in which the role of APQ was revealed. First, the consortium members "would determine the total production and each member would take its proportion (based on capital contributions) of the oil and products." Second, each consortium member was free to sell oil to any party and "group members will adjust over and under liftings among themselves by special agreement."[165] Thus, the oil companies used the calculating technology of APQ to ensure that the Iranian settlement would not cause trouble with production levels in Iraq, Kuwait, Saudi Arabia, and other oil-producing countries.

The secret role of APQ in working out the consortium arrangement was finally made public in 1974.[166] It was the price that American oil majors extracted for agreeing to join the consortium. Combined with government quotas and price controls in the United States, and cartel arrangements to govern the worldwide distribution and marketing of oil, the consortium arrangement helped restrict the development of new oil discoveries in the Middle East.[167] Along with the secret mechanism of the APQ, this set of techniques and controls shaped the development of the transnational oil corporation, which emerged as a long-distance machinery for maintaining limits on the supply of oil in a period of abundance.

Consortium members—British Petroleum, Shell, Mobil, CFP, Esso, Texaco, Gulf, Socal, and Iricon[168]—voted on production levels, which were factored into a "rigid formula agreed to by Consortium participants in 1954."[169] To calculate the APQ, nominations by each company on production levels were translated into a production total required to provide each company with its desired volume based on its equity share in the consortium. Then, the "APQ volume" for the year was decided. Nominations were listed in descending order of magnitude until a cumulative 70 percent equity level was reached. Fesharaki explains that the "crude short companies" such as Mobil, Shell, CFP, and Iricon consistently voted for the highest production levels. On the other hand, Socal, Exxon, Texaco, or Gulf, with large production in Saudi Arabia and Kuwait, usually voted lower (table 6.1).

US Senate hearings on the international oil cartel and its antitrust activities were first conducted in 1974. Jerome Levinson, chief counsel for the Senate committee, observed that "the historical APQ tables in the Iranian Consortium show that when you go to 70 percent to arrive at APQ, in virtually every year we find Exxon, Texaco, Socal, and Gulf on the low side."[170] Such calculations, explains Fesharaki, brought to light "some unexpected peculiarities indicating very close collaboration between the participants."[171] British Petroleum was the most important force in determining production volumes, and together with any group of small equity holders, it could "impose its will on the others."

Senator Frank Church, chair of the Senate hearings, admitted that the formula worked by connecting itself to other kinds of political actors and interests. Formulas were designed, through the Iranian consortium

Table 6.1
Historical APQ tablings in Iranian consortium (in thousands of barrels per day). Transcribed from Subcommittee on Multinational Corporations, *Multinational Oil Corporations and US Foreign Policy* (Washington, DC: US Government Printing Office, 1975), part 7, 255.

1957		1958		1959		1960		1961		1962	
Co.	MBD	Co.	MBD	Co.	MBD	Co.	MBD	Co.	MBD	Co.	MBD
BP	750	CFP	827	Iricon	912	Mobil	1111	Mobil	1213	Iricon	1370
CFP	750	Iricon	826	Mobil	890	Iricon	1023	Iricon	1201	BP	1370
Shell	715	Shell	822	BP	850	BP	984	BP	1192	Shell	1370
Iricon	700	Mobil	814	Shell	850	Shell	960	Shell	1192	CFP	1370
Exxon	690	BP	787	Socal	850	CFP	911	CFP	1192	Mobil	1342
Socal	603	Exxon	759	Exxon	841	Exxon	900	Exxon	980	Texaco	1195
Gulf	574	Socal	724	CFP	822	Gulf	880	Texaco	980	Socal	1170
Texaco	528	Texaco	700	Texaco	805	Texaco	875	Socal	973	Exxon	1115
Mobil	528	Gulf	655	Gulf	680	Socal	874	Gulf	940	Gulf	1000

1963		1964		1965		1966		1967		1968	
Co.	MBD	Co.	MBD	Co.	MBD	Co.	MBD	Co.	MBD	Co.	MBD
BP	1575	Iricon	1740	CFP	1863	Iricon	2030	CFP	2274	Iricon	2698
CFP	1575	Mobil	1721	Iricon	1860	BP	2027	Shell	2233	Shell	2678
Iricon	1535	Shell	1708	BP	1836	Shell	2027	BP	2219	Gulf	2571
Shell	1534	BP	1680	Shell	1836	Mobil	1964	Mobil	2178	BP	2568
Mobil	1521	CFP	1680	Mobil	1836	CFP	1945	Iricon	2178	CFP	2568

Table 6.1 (continued)

1963		1964		1965		1966		1967		1968	
Co.	MBD	Co.	MBD	Co.	MBD	Co.	MBD	Co.	MBD	Co.	MBD
Socal	1360	Texaco	1470	Gulf	1686	Exxon	1890	Exxon	2160	Mobil	2568
Texaco	1299	Exxon	1400	Texaco	1589	Texaco	1712	Texaco	2005	Socal	2391
Exxon	1200	Gulf	1314	Exxon	1575	Gulf	1700	Socal	1973	Exxon	2363
Gulf	1068	Socal	1214	Socal	1370	Socal	1644	Gulf	1871	Texaco	2363

1969		1970		1971		1972		1973	
Co.	MBD	CO.	MBD	Co.	MBD	Co.	MBD	Co.	MBD
Iricon	3062	Iricon	3547	Iricon	4054	Iricon	4678	Gulf	5428
BP	3014	Shell	3346	Mobil	3986	CFP	4645	Iricon	5342
Shell	3014	Mobil	3342	BP	3973	Exxon	4638	CFP	5297
Mobil	3014	BP	3329	Shell	3973	Shell	4590	Exxon	5250
CFP	3014	CFP	3329	CFP	3973	Mobil	4516	BP	5137
Gulf	2900	Exxon	3250	Exxon	3973	Socal	4500	Mobil	5068
Exxon	2849	Gulf	3200	Socal	3767	BP	4372	Shell	5055
Socal	2795	Socal	3164	Texaco	3562	Gulf	4257	Texaco	5043
Texaco	2679	Texaco	3134	Gulf	3286	Texaco	4200	Socal	4857

Company whose tabling sets APQ

Companies tabling below APQ

arrangement, "to counteract governmental pressures" and give companies a means for "controlling the total lift, oil, crude lift each year." The purpose of the consortium, like other arrangements in the Middle East, was to enable companies to "work through formulas mutually agreed upon." More equitable possibilities for oil production were blocked precisely through this technical work of calculating how to close out situations in an open market in which "elaborate sharing arrangements and production formulas are not involved and the oil is bid for by the various companies in a completely free, open and competitive way."[172]

Contrary to the laws of supply and demand, the formula's work demonstrated that certain parties among the oil majors with interests in Saudi Arabia, Kuwait, and Bahrain sought to minimize the amount of the APQ and thus produce the minimum amount of Iranian oil to keep prices high and avoid conflict with their other oil interests in the Middle East. In this instance, the Iranian government did not have a choice in the definition of the formula and its world. Rather, the construction of the APQ worked in secret to manage oil supplies and reengineer the terms of national control. First, it intervened to manage negotiations among the largest Anglo-American oil companies backed by the British and American governments. Second, having secured the incalculability of alternative arrangements to the energy system, the formula opened the space for negotiating a compromise with the newly installed, pro-Western Iranian government.

As in earlier crises, Iranian government negotiators such as Entezam, the minister of foreign affairs, and Ali Amini, the minister of finance, continuously expressed their dissatisfaction with the proposed volumes of production. They demanded that a rate be reached within four years that totaled one-third of total Middle East oil production.[173] Unaware of the APQ arrangement, Entezam argued that the production quantities were disappointingly low. In the absence of any further guarantees, he feared that the situation would open itself to misrepresentation, implying that in view of the consortium's interests in other Middle East oil-producing countries, it was intending to differentiate against Iran. Entezam remarked on the very rapid increase in production in Kuwait, which seemed wrong considering that Iran's needs were much greater than those of this small state. Mohammad Reza Shah, the ruler of Iran and son of Reza Shah, who had abdicated in 1941, eventually came to learn of the APQ when the French company, Compagnie Française des Pétroles, expressed its dissatisfaction with the consortium's levels of production and gave the shah a copy of the APQ arrangements.[174]

The reengineering of national control in the format of an Iranian oil consortium was, in part, a battle over the dynamics of a formula as in previous concession disputes. What the Iranians did not know, however, was that they had also agreed to the secret workings of the APQ formula that sought to protect consortium members by managing their collaborative efforts to control world oil supplies and keep prices high. Thus, there is an obvious asymmetry here. AIOC appears to have designed formulas in the earlier periods to legitimize to the Iranian government the arrangements favored by the oil company. But when the formula becomes secret, as with the APQ, it no longer serves this function of legitimization. Rising American political and economic dominance meant that AIOC was now in a position of weakness. The British oil company was unable to resume its operations as before and was forced to negotiate the terms of Iran's production levels with the largest transnational oil corporations to protect American participation in any future compromise with the Iranian government. This was also a different historical moment, which, by the end of World War II, marked the creation of a nation-state system and the spread of anticolonial nationalist movements that prevented the continued imposition of sovereign power over other countries by European states and colonizing corporations. The role of formulas in contractual arrangements could not be made public because they would be seen as acting against the national interests of oil-producing countries. Consequently, transnational oil corporations of the mid-twentieth century did not have the same privileges as the older colonizing corporations from which they originated, yet they continued to use secret formulas to redefine the post–World War II petroleum order according to new monopoly arrangements dominated by the largest American oil companies.

During the course of each dispute, the Iranian government had contested, at multiple sites domestically and internationally, the decision made by the oil companies to limit production. But it was bargaining from a position of weakness, confronting the long-distance machinery of global monopoly arrangements that had, through a series of private agreements among the largest Anglo-American oil companies, concession agreements with other oil-producing countries, and collaborations with national host governments, managed to secure total control of oil production, distribution, and transportation networks. Following the construction of the APQ has exposed the formula's connection to politics—that is, the technologies, procedures, and tools enabling the collaborative efforts of consortium members to block rivals within a set of constraining relationships and to manage Iran's oil supplies precisely through the nationalization of its oil industry.

"Right Wording" for Nationalization

Iranian ministers assured consortium members that they did not intend to intervene in the details of management control, but sought promises that proper regulatory functions on their behalf and that of NIOC could be effectively implemented. The wording of the consortium agreement must be made "palatable to the Iranian Majlis and public." Consortium members proposed that considerations of the right wording could be found, defining the Iranian government as a "regulator" or "inspector" to avoid the suggestion of direct management.[175] Iranian ministers were now working to enroll the Iranian government in restoring foreign control of its oilfields. Amini assured American officials that while the reaction of the parliament was certain to be explosive, Iranian deputies provided every indication of "aggressively supporting" the results of consortium negotiations before the Majlis and the public, even though it contained a number of "distasteful features."[176] A satisfactory solution would be found shortly, explained Loy Henderson, American ambassador in Tehran, once the Iranians frankly admit that it was "primarily [a] problem [of] finding words not offensive to Majlis and public."[177]

As in earlier disputes, the Iranian state was being molded in important ways by the struggle over the control and distribution of oil. Stabilizing the mechanism of the consortium arrangement in relation to the oilfields of southwest Iran was precisely where this power was worked out, but it threatened to fall apart in the face of mounting criticism from the Iranian public. The Iranian government was "determined to do whatever was necessary to outmaneuver any opposition to ratification in the Majlis."[178] For a brief moment, the Majlis threatened to disrupt the consortium arrangement on the question of production levels. Iran's Mixed Oil Committee argued that Article 20 of the proposed plan placed Iran at the "mercy of the consortium," leaving it free to "produce and sell just as much Iranian oil as it pleased—3 or 50 million tons—without violating contract."[179] Amini proposed that the consortium write a letter clarifying or interpreting the article by changing the wording to say that "after [the] third year there would be adjustments in production of Middle East oil so that Iran would be able to sell its fair share." The committee was more likely to ratify the agreement if such clarification was made. It would demonstrate to the Iranian public that the committee was not a "mere rubber stamp." Howard Page, representing Standard Oil of New Jersey, drafted a letter modifying the wording of Article 20. Statements about plans for decreased production and consumption in future years were reworded to say that the consortium would treat the production and export of Iranian oil "equitably" and in relation to other Persian Gulf sources of supply.[180] No mention of the APQ arrangement was made.

Finding the right wording for nationalization helped transform the Majlis into a rubber stamp precisely as the shah, in his ratification of an oil agreement, effectively overturned any real nationalization of the oil industry. Thus, the Majlis had moved from a relative position of strength in 1946–1951, banning contracts with foreign concerns, fighting to improve the treatment of oil workers, and guaranteeing Iranian control over its oil industry through a series of nationalization laws, to a position of weakness. This was made possible, in part, by the Anglo-American overthrow of the Mosaddegh government and reinstallation of the shah and General Zahedi, both of whom supported the return of an international oil consortium. The Majlis approved the consortium arrangement by a vote of 113 to 5, with one abstention, on October 21, 1954.

The final outcome saw a nominally nationalized group of Western firms with full rights to manage oil output and prices,[181] leaving Iran with a formal title to all its oil and a 50–50 division of net profits of production, but no share of marketing or distribution operations. Consortium members approved a second revised version in 1955, which allocated 5 percent of American participation to independent oil companies.[182] It was the price of blocking any future threat of competition and uncertainty.[183]

The consortium agreement did not come into force until it received approval from Brownell, the US attorney general, the eight oil majors, and the Iranian Majlis and Senate.[184] Following his earlier support in blocking US antitrust proceedings, Brownell approved the planned consortium even though the secret production agreement (APQ) constituted a cartel arrangement. Thus, the APQ did the work of protecting the monopoly arrangement as before, but under American tutelage instead of British.

Alongside the secret workings of the APQ, mechanisms of debt and Iran's desire for military and economic aid helped weaken its bargaining position. Iran was entering a new relationship among flows of energy, finance, and violence as a new client state of the United States and according to a new oil-dollars-weapons arrangement that would help organize the international political economy of oil and "reconfigure the intersecting elements of carbon democracy" in the second half of the twentieth century.[185] To help defend the mid-twentieth-century oil order against nationalist and popular pressures in the Middle East, the possibilities for a more democratic politics of oil production in Iran were closed off, excluding the oil workers who had struggled to build an alternative and more just arrangement.

New International Petroleum Order

The Mosaddiq government's decision to nationalize AIOC in 1951 had triggered a boycott of Iranian oil production by the dominant Anglo-American oil companies. This led to a restabilizing of world oil supplies by increasing production in neighboring Middle East countries. The shift in production to neighboring countries such as Saudi Arabia benefited the American oil industry. Thus, American participation in the consortium agreement meant that production levels could not be shifted back to favor Iranian production.

The failure of oil nationalization in Iran entailed the restoration of a resoundingly successful Anglo-American consortium arrangement set up in terms of 50–50 profit sharing and production quotas built into the managing technology of the APQ.[186] Nationalization helped solidify the international oil cartel's control, also known as the "Seven Sisters," over the oil reserves of the entire Middle East.[187] It was, therefore, an overwhelming success for the international oil industry and the American and British governments, who managed to enroll the Iranian government in helping to restore foreign control of the seven largest oil corporations over its oil industry for at least another forty years.

The moment also marked the closure of alternative political possibilities for running Iran's oil operations, as formerly demanded by the more militant alliance of oil workers, the communist Tudeh Party, and national control. As this discussion and the previous chapter revealed, a set of technologies for the management and control of oil—sanctions and a boycott, 50–50 profit sharing, the APQ, the racial-technical organization of labor, and legal-economic metrology—intervened in the dispute to weaken Iran's bargaining position drastically while enabling Anglo-American interests to claim to have stepped outside of local constraints. The nationalization dispute unraveled at a critical moment of political, legal, and financial rearrangement in the post–World War II international order. The oil infrastructure of southwest Iran constituted one of the sites in which this complex rearrangement was worked out.

Iran's position suddenly shifted from a position of strength to one of weakness as Anglo-American oil companies and governments reengineered mechanisms for protecting the oil-rich and strategically important Middle East in terms of a Cold War battle against Soviet expansion, militant nationalism, and populist politics.

Studies of Iran's oil nationalization crisis and its resolution often assume a predetermined set of historical actors that overlook the pivotal activities of nonhuman agents such as laws, mechanisms of international governance, formulas, and the US national security apparatus. These technologies of management and control ensured that the central issues—production levels, pricing, and percentage participation—were rendered nonnegotiable and settled beforehand among the American oil majors and then with the British. One effect was to undermine Iran's national sovereignty over its natural resource. As a collective agency, oil workers, the Tudeh Party, Iranian technologists, and public opinion demanded alternative, more just arrangements of oil, but their bargaining positions were gradually weakened and redirected toward other arrangements not necessarily within their control. These efforts shaped the Iranian state in important ways, enrolling the Majlis initially as an opponent and subsequently as a supporter of foreign, international control over its oil industry. The Iranian government's decision to enter a consortium arrangement plunged the country into a different kind of international oil order from before, one that was beginning to rely on a peculiar relationship between oil, dollars, and the US arms industry, promising further closure of political possibilities and demanding new kinds of equipment with which to address oil crises to come.

Conclusion: Petroleum Collectives—
Translating the Power of Oil into Politics

The infrastructure of the global oil industry as it was constructed in southwest Iran remains active today. The oilfields, pipeline, refinery, staff bungalows, and leisure clubs are intact, albeit in need of extensive refurbishment. The old town of Abadan and the bazaar are bustling places where Arabic- and Persian-speaking Iranians continue to thrive. Many of them are employees of the National Iranian Oil Company (NIOC) and have memories of the company town, passed down from their grandparents, former employees who still refer to its residential roads as "lanes," a term first used by AIOC's British imperial architects. The history of the collective life that has emerged in Khuzistan is a history of infrastructural transformation that rather than eliminating technical uncertainties, caused sociotechnical controversies to proliferate in the pursuit of alternative, petroleum-based political collectives.

The previous chapters inquired into the procedural techniques, points of vulnerability, opportunity, and control that made the first foreign-controlled oil concession in the Middle East operational. *Machineries of Oil* has discussed the organizational work involved in transforming the world into a place in which the Anglo-Iranian oil industry would survive. The set of remarkable actors that emerged, namely, the concession contract, formulas, international law, technoscience, and strategic ignorance, helped make AIOC a success in the first fifty years of its existence, only to end abruptly with its departure in 1954. By contrast, the company's representation of its expert knowledge greatly underplayed the role of its organizational work locally. The uncertainty of the energy system built into the heart of each oil dispute enabled certain machineries of control to equip actors with the power to reassemble political possibilities in unpredictable ways. Starting with a rhetoric that represented local actors in terms of the economic rationality they lacked, these devices were black-boxed over time. The execution of concession terms was all that mattered, not the arguments and equipment that went into making them operational.

Reassembling the history of BP in Iran has exposed previously unstudied petroleum-based assemblages that managed political outcomes in the twentieth century. There are more powerful forces at play that models of the rentier state do not take into account when explaining the failure of democracy as symptomatic of oil-producing countries. Rather than narrowing the focus either to the social construction of the technological or to the technological construction of the social, *Machineries of Oil* has placed the study of oil infrastructure at the center of the analysis, yet without limiting it to this alone or to what some would call its economic and political impact. In doing so, the book has identified a *sociotechnical* process of simplification by which relations between politics and markets in oil are made durable. In Iran, oil infrastructure was the site of power struggles and the constructing and deconstructing of boundaries made between technical and economic issues inside oil operations and the politics of property claimants, national sovereignty, and striking oil workers outside. These distinctions between what was political and what was technical were *the outcome* of the sociotechnical process of manufacture under investigation and not the input. Thus, the oil infrastructure in the Middle East, as constructed in Iran, was constituted precisely by building alliances that did not respect apparent divides between material and abstract, technical and social, ignorance and knowledge, representation and violence. Each chapter can be seen as working through this proliferation of a set of distinctions that at first glance appear fixed and natural, but in practice, are highly vulnerable and shape political outcomes. The kinds of problems that emerged in this history have to do with the specificity of Anglo-Iranian oil's sociotechnical properties.

Undemocratic machineries of oil emerged in the shape of formulas, concession terms, and international law, which were among the central actors entangled in the history of Anglo-Iranian oil and the Iranian state in the first half of the twentieth century. Only by intertwining histories of the Middle East with business history and science and technology studies has the history recounted in this book been able to open up the dynamics of diverse sociotechnical devices and calculating technologies in an interconnected process of formulating similar but moving problems within a local history. As a business strategy, firms such as AIOC made use of this equipment as an entry point or field of negotiation that could be used advantageously over other devices. Why were these devices so effective?

Depending on markets or industrial relations, the relation between politics and economy took on different forms with different strategic variables, making it increasingly difficult to assemble democratic politics out of the production of oil. For example, the issue of volume of production was present on multiple occasions regarding profits, the company's recruitment policy, and production rates. The importance of the strategic character of formulas turned around one strategic variable, volume of production, and the subordination of other parameters to this one, making the market in oil predictable. Thus, in each of the disputes, formulas and other sociotechnical devices organized relations between markets and politics, although this could vary from one dispute to another. Different formulas for calculating profits, Iranian labor recruitment, and the APQ, in particular, were all built around figures indicating the notion of an extension of petroleum formulas in the first half of the twentieth century.

Quantitative formulas have played a decisive role in the long history of Anglo-Iranian oil but were also used in connection with other types of formulas such as international law, which, while claiming universal applicability among a community of states, was devised to deal with a specific type of economic development agreement (e.g., concessions) to which new states of the Global South were the primary parties. Formulas were decisive even in the organization of the international scientific community—volume of production was important for scientists in the formulation of reserve estimates and other measures. Thus, it is important to note that the circulation of formulas did occur outside relations between governments and firms, such as in the scientific community and in international legal arenas of security and finance. Formulas in general offered the possibility of clear negotiations and revisions. Formulas and other kinds of sociotechnical devices have been so important in this history because they equipped actors with varying degrees of agency to frame issues advantageously and legitimize their respective negotiating positions in each oil dispute.

The machineries of Anglo-Iranian oil also served as useful devices to control the sensitive question of drawing boundaries between private/public and secrecy/transparency. Within the variables of mathematical formulas and legal and scientific formulation, the interests and organizational forms of public (governmental) and private (nongovernmental, corporate) entities appeared intertwined from the start. Thus, the history of these devices can also be seen as a history of maintaining a distinction between secret and private or open and public boundaries. Built into the notion of formulation is the idea of what is said and not said. The circulation of a formula such as the APQ (discussed in chapter 6) could be secret for a period of time (in this instance, ten years or so). In pursuing compromises, they enabled the calculation of interests—political, cultural, technical, and economic—and were therefore helpful in the management and calculation of heterogeneous interests, particularly the threat of militant nationalism allied with striking oil workers and communism.

Petroleum devices and formulas were bound up with and shaped by competing political projects within the history of Anglo-Iranian oil. These specific technologies required certain kinds of social worlds to survive. AIOC's conceptions about local political forces in terms of the Western civilization/expertise they lacked informed the company's technological choices and were inscribed in its conceptions of legality, valuation, and the hierarchical classification of English law terms as superior to those in Farsi. Likewise, racial-cultural assumptions about Iranian oil workers as incapable of learning petroleum technology were inscribed in the company's economic decisions about profits and recruitment. In practice, these modern techniques of power and domination produced concerned groups, such as oil workers and the Communist Party, who battled to reconstruct the techno-econo-social collectives at the heart of each dispute in advantageous ways to block rivals and represent the many silent actors of the worlds they mobilized.

AIOC mobilized petroleum devices to limit political possibilities in favor of the cartel and monopoly arrangements of the major Anglo-American oil corporations. More generally, these devices did not describe a preexisting world, they helped format a world of which they were a truer reflection.[1] Throughout the history of oil in Iran, these calculative technologies equipped AIOC officials with the capacity to redirect political outcomes against competing sociotechnical programs put forward by Iranian oil workers, the Communist Party, reformist nationalism, and public opinion demanding more equitable forms of oil production. Thus, formulas and the work of formulation (legal, economic, scientific, administrative) can be central actors in the establishment, negotiation, and strengthening of sociotechnical assemblages, thus equipping people and things with new forms of agency.

Reconfiguring historical analysis as a sociotechnical process such as this respects the connections, translations, and networking performed by the diverse human and nonhuman actors, which led to the constitution of the shifting arrangements. In practice, the building of the global oil industry occurred through the formation of a series of technological zones.

These zones of qualification, measurement, and regulation involved a multitude of international and national bodies, along with experts such as private consultants, scientists, and other professionals employed by multinationals and institutions of global economic governance. The effect was to establish a border over the course of the first half of the twentieth century, one whose limits did not correspond to or get determined by the existing border of the nation-state, but to a shifting sphere of operations of companies and of international institutions such as the League of Nations, the International Court of Justice, and the World Bank. With such borders established, the oil corporation was allowed to step outside of local constraints and make universal claims to international guidelines and standards.

This study has intervened to consider the role of sociotechnical devices in the reconfiguration of politics.[2] It has done so by engaging in an analysis of the historical construction of particular political and economic spaces and the specificities of the materials, practices, and locations that technological zones within the global oil industry in the Middle East transform, connect, exclude, and silence. A focus on the technological zones of Anglo-Iranian oil has made visible the ways in which the formation of these zones has become critical to the constitution of a distinction between global/Western political and economic forms and their non-Western others.[3] Southwest Iran served as a laboratory for producing knowledge about the social and technical worlds of oil, and for hybridizing them. Laboratory work was the infrastructural work of holding formulas and oil workers together so that alternative forms of control, such as nationalization, would not happen.

By recovering the process through which sociotechnical devices were constructed and deconstructed in Iran's history, *Machineries of Oil* has followed their remarkable role in shaping the powers of the oil-producing state and the global oil industry. But the goal of this study has not been to unveil the underlying mechanisms of sociotechnical devices in already well-known events and policies, for such nonhuman actors have always been upfront and on the surface. The world of these devices became actual not through the work of numeric, legal, administrative, and scientific formulation alone. Rather, they combined with other forces that helped assemble the machinery of oil in standardized terms to redirect political outcomes and agencies in favor of the multinational oil corporation. At the same time, engineering political relations out of flows of oil made certain actors in the history of Anglo-Iranian oil incompatible with these devices. This incompatibility created inherent vulnerabilities within the petroleum-based energy system, allowing the possibility for alternative political arrangements, events, and worlds to strike back in future oil crises.

The Iranian oil consortium succeeded in reintegrating Iranian oil production into the postwar petroleum order, but the arrangement was not long lasting. In the second half of the twentieth century, rival petroleum collectives emerged to counter the powers of the Seven Sisters, all of which participated in the consortium. Venezuela and the main oil-producing countries of the Middle East, including Iran, Saudi Arabia, and Iraq, coordinated in 1960 to establish a rival cartel, the Organization of the Petroleum Exporting Countries (OPEC), based on the universal right of all countries to exercise permanent sovereignty over their natural resources. Their initial goal was to coordinate the control of oil production from their fields and deconstruct the "posted price" of oil, or the unilateral fixing of prices controlled by the seven oil corporations to keep prices artificially high.[4] In southwest Iran, protesting Iranian oil workers used their specialist knowledge to disrupt oil operations on multiple occasions with the most transformative action occurring during the 1978–1979 Islamic Revolution. The oil workers successfully shut down specific nodes in the energy system to paralyze the Pahlavi shah's regime. The workers linked their action in solidarity with striking teachers and called for the release of political prisoners.[5] They again connected calls for the Iranianization of the oil industry to improvements in housing and work, but their demands were directed at a different political actor, NIOC, first established during the oil nationalization crisis of 1953. The revolution marked NIOC's official declaration of the end of Iran's relations with consortium member companies. All exploration and production contracts with foreign companies before the revolution were declared null and void. The Islamic Republic took control of the oil production variable in the pursuit of more democratic forms of control, but as discussed in the previous chapters, tools were readily available within the energy system to adjust to such disruptions in supply by relying on alternative sources and apparatuses of control to protect international oil markets.

Led by Saudi Arabia, OPEC continues to wield its power in the twenty-first century as it battles on multiple fronts to shut out independent shale oil competitors in the United States and to readjust to the return of Iranian oil production to international markets in the aftermath of the July 2015 nuclear deal and the lifting of economic sanctions. The problem is that many of the giant oilfields of Khuzistan, discussed in the previous chapters, are already depleted or nearing depletion, and the massive Abadan refinery is in need of renovation. With the second largest natural gas reserves, Iran now hopes to lure back many of the international oil companies that it fought hard to evict from its oil industry in the last century. But the negative impact of fossil fuel consumption on global climate change has triggered an energy shift toward the development of renewable sources of energy. A sociotechnical analysis such as this offers better tools to address these uncertainties in the pursuit of alternative energy systems and collective life.

Abbreviations

BP	British Petroleum Archives, University of Warwick, Coventry
INA	Asnad-e Melli Iran [Iranian National Archives], Tehran
IOR	India Office Records, British Library, London
FO	Foreign Office, National Archives, Kew Gardens, London
LAB	Ministry of Labor, National Archives, Kew Gardens, London
CAB	Cabinet Papers, National Archives, Kew Gardens, London
T	Treasury, National Archives, Kew Gardens, London
PREM	Prime Minister's Office, Kew Gardens, London
NARA	National Archives and Records Administration, College Park, Maryland
AHC	American Heritage Center, Laramie, Wyoming

Notes

Introduction

1. Studies in political science assume an inherent tendency among countries with large oil revenues to pursue excessive degrees of centralization and authoritarian rule with little prospects for democracy. For example, see Michael Ross, "Does Oil Hinder Democracy?," *World Politics* 53, no. 3 (2001): 325–361; Ellis Goldberg, Erik Wibbels, and Eric Mvukiyehe, "Lessons from Strange Cases: Democracy, Development, and the Resource Curse in the U.S. States," *Comparative Political Studies* 41, nos. 4–5 (2008): 477–514. In economics, the "resource curse" or "Dutch disease model" describes how an abundance of natural resources, such as oil, creates obstacles to growth (e.g., productivity, trade, and consumption) rather than democracy. See Jeffrey D. Sachs and Andrew M. Warner, "Natural Resource Abundance and Economic Growth," *National Bureau of Economic Research Working Paper Series*, no. 5398 (1995): 1–47.

2. Michael Watts made an important contribution to this scholarship by factoring in the material properties of oil in reconfiguring "governable spaces" in local and national Nigerian politics. (Michael Watts, "Resource Curse? Governmentality, Oil, and Power in the Niger Delta, Nigeria," *Geopolitics* 9 (2004): 50–80. Another exception that takes seriously the materiality of oil in terms of its transformation into property and rents is Fernando Coronil's *The Magical State: Nature, Money, and Modernity in Venezuela* (Chicago: University of Chicago Press, 1997).

3. A. S. Alsharhan and A. E. M. Nairn, *Sedimentary Basins and Petroleum Geology of the Middle East* (New York: Elsevier, 2003), 693. In geological terms, the combination of carbon-rich source rocks, porous and permeable reservoir rocks, and effective caprocks characterizes the Middle East (e.g., Iran, Iraq, Syria, Kuwait, Saudi Arabia, Bahrain, Qatar, the United Arab Emirates, Oman, and Yemen) to a high degree, producing high-quality oil.

4. At the onset of the First World War, virtually all the leading navies—those of Britain, Germany, Italy, Russia, and the United States—were in the process of converting to oil. See Bruce Podobnik, *Global Energy Shifts: Fostering Sustainability in a Turbulent Age* (Philadelphia: Temple University Press, 2006), 65–67.

5. Ibid., 5.

6. Daniel Yergin, *The Prize* (New York: Free Press, 1992). Also, see the first two volumes of the three-volume official history of AIOC, which laud the company's pioneering impact and consistent showing of ingenuity in confronting setbacks posed by oil operations and political groups in Iran: R. W. Ferrier, *The History of the British Petroleum Company,* Volume 1: *The Developing Years, 1901–1932* (Cambridge: Cambridge University Press, 1982); James H. Bamberg, *The History of the British Petroleum Company,* Volume 2: *The Anglo-Iranian Years, 1928–1954* (Cambridge: Cambridge University Press, 1994); James H. Bamberg, *British Petroleum and Global Oil, 1950–1975: The Challenge of Nationalism* (vol. 3 of *The History of the British Petroleum Company*) (Cambridge: Cambridge University Press, 2000).

7. My goal is not to study the impact of oil, intact as a separate sphere, on society and the economy because that is what political economy does. I am arguing that the identity of oil is not well bounded but is intrinsically entangled in all sorts of environments, regulatory regimes, information, and machineries. Concession terms, property relations, cartel arrangements, and the law were internal to the technical definition and production of the oil "itself," so to speak. The historical identity of oil is not possible without all the representational and physical work that must be done to produce and sell it. See Marianne De Laet and Annemarie Mol, "The Zimbabwe Bush Pump: Mechanics of a Fluid Technology," *Social Studies of Science* 30, no. 2 (2000): 225–263.

8. Within the field of STS, there has been no investigation showing how human and technological forces clash and get rearranged in quite the same way, and with as broad consequences, as occurred in the building of Iran's British-controlled oil industry. Geoffrey Bowker's investigation of the oilfield services company, Schlumberger, is one of the few studies of the oil industry drawing on methods from STS to think about the organizational work of building an oil industry. But it focuses on the building of Schlumberger's laboratory and its world, rather than drawing out the further connections for our understanding of politics. See Geoffrey Bowker, *Science on the Run: Information Management and Industrial Geophysics at Schlumberger, 1920–1940* (Cambridge, MA: MIT Press, 1994).

9. Ross, "Does Oil Hinder Democracy?"; Alan H. Gelb, *Oil Windfalls: Blessing or Curse?* (New York: Oxford University Press, 1988).

10. On the "oil curse" and the importance of looking at both domestic social forces and external geopolitics, also see Andrew Rosser, "Escaping the Resource Curse: The Case of Indonesia," *Journal of Contemporary Asia* 37, no. 1 (2007): 38–58. Political economy improves on the political science approach by taking social classes and the "rentier state" into account. See, for example, Hazem Beblawi and Giacomo Luciani, eds., *The Rentier State, Nation, State, and Integration in the Arab World,* vol. 2 (New York: Croom Helm, 1987); Terry Lynn Karl, *The Paradox of Plenty: Oil Booms and Petro-States* (Berkeley: University of California Press, 1997).

11. The concept of the "rentier state" serves as the dominant paradigm for studying the impact of oil on state formation. Oil is explored only in terms of

its economic properties as a rent, overlooking its physical or material properties as well as the role of oil workers and oil company operations in shaping the powers of the state. See, for example, Hossein Mahdavy, "The Patterns and Problems of Economic Development in Rentier States: The Case of Iran," in *Studies in the Economic History of the Middle East: From the Rise of Islam to the Present Day*, ed. M. A. Cook, 428–467 (London: Oxford University Press, 1970).

12. Donald MacKenzie, "An Equation and Its Worlds: Bricolage, Exemplars, Disunity and Peformativity in Financial Economics," *Social Studies of Science* 33, no. 6 (2003): 831–868.

13. Thomas Hughes, *Networks of Power: Electrification in Western Society, 1880–1930* (Baltimore: Johns Hopkins University Press, 1983).

14. Ferrier, *History of British Petroleum*; Bamberg, *History of British Petroleum*.

15. For STS studies of calculation in a non-Western setting, see for example, Helen Verran, *Science and an African Logic* (Chicago: University of Chicago Press, 2001).

16. On the company's relation to the Iranian and British governments, many studies have framed the issue as a conflict between private, commercial versus public, government interests, rather than leaving the distinction an open issue. See, for example, Gregory P. Nowell, *Mercantile States and the World Oil Cartel, 1900–1939* (Ithaca, NY: Cornell University Press, 1994).

17. Koray Caliskan and Michel Callon, "Economization, Part 1: Shifting Attention from the Economy towards Processes of Economization," *Economy and Society* 38, no. 3 (2009): 368–398; Michel Callon, Yuval Millo, and Fabian Muniesa, eds., *Market Devices* (Malden, MA: Blackwell, 2007).

18. Michel Callon, "What Does It Mean to Say That Economics Is Performative?," in *Do Economists Make Markets? On the Performativity of Economics*, ed. Donald A. MacKenzie, Fabian Muniesa, and Lucia Siu, 311–357 (Princeton, NJ: Princeton University Press, 2007); Bruno Latour, *Reassembling the Social: An Introduction to Actor-Network Theory* (Oxford: Oxford University Press, 2005).

19. To avoid burdening readers, I refer to STS and "sociotechnical" in broad terms but am drawing on specific scholarship from Actor Network Theory (ANT), performativity of economics, social shaping of technology, and infrastructure studies. ANT is exemplified in Michel Callon, "Some Elements of a Sociology of Translation: Domestication of Scallops and the Fishermen of St. Brieuc Bay," in *The Science Studies Reader*, ed. Mario Biagioli, 67–94 (New York: Routledge, 1999), esp. 68.

20. Callon, "What Does It Mean to Say That Economics Is Performative?," 348; Caliskan and Callon, "Economization, Part 1," 12–13.

21. Bruno Latour, *Politics of Nature: How to Bring the Sciences into Democracy*, trans. C. Porter (Cambridge, MA: Harvard University Press, 2004), 12.

22. Historians and sociologists of technology have been studying infrastructures (e.g., large-scale technical systems, informational infrastructure) seriously and as an analytical category since the 1980s and 1990s. Exemplary works include Paul N. Edwards, *A Vast Machine: Computer Models, Climate Change, and the Politics of Global Warming* (Cambridge, MA: MIT Press, 2010); Ashley Carse, *Beyond the Big Ditch: Politics, Ecology, and Infrastructure at the Panama Canal* (Cambridge, MA: MIT Press, 2014); Geoffrey Bowker and Susan Leigh Star, *Sorting Things Out: Classification and Its Consequences* (Cambridge, MA: MIT Press, 2001).

23. Vincent Lepinay, "Parasitic Formulae: The Case of Capital Guarantee Products," in *Market Devices*, ed. Michel Callon, Yuval Millo, and Fabian Muniesa, 261–283 (Malden, MA: Blackwell, 2007); MacKenzie, "An Equation and Its Worlds."

24. For an important analysis that highlights the limits of traditional democracies in managing sociotechnical controversies by drawing on tools from STS to propose "technical democracy in action," see Michel Callon, Pierre Lascoumes, and Yannick Barthe, *Acting in an Uncertain World: An Essay on Technical Democracy*, trans. Graham Burchell (Cambridge, MA: MIT Press, 2009).

25. See, for example, Michel Callon, "Society in the Making: The Study of Technology as a Tool for Sociological Analysis," in *The Social Construction of Technological Systems: New Directions in the Sociology and History of Technology*, ed. Wiebe E. Bijker, Thomas P. Hughes, and Trevor J. Pinch, 83–103 (Cambridge, MA: MIT Press, 1987) on electric cars, and Bruno Latour, *Aramis or the Love of Technology*, trans. Catherine Porter (Cambridge, MA: Harvard University Press, 1996) on the building of a new rail system in France. In recent years, scholarship has started to address this bias, such as in Edwards, *A Vast Machine*, on the work of modeling in relation to the climate, and Carse, *Beyond the Big Ditch*, on the building of the Panama Canal.

26. Gabrielle Hecht, *The Radiance of France: Nuclear Power and National Identity after World War I* (Cambridge, MA: MIT Press, 1998). Hecht's study of French nuclear power uses the concept of "technopolitical regime" to nicely map some of the larger sociotechnical processes at work in the building of the nuclear industry and national state. Also see Samuel Lussac, "The State as a (Oil) Company? The Political Economy of Azerbaijan," *GARNET Working Paper* 74/10 (Warwick, UK: Center for the Study of Globalisation and Regionalisation, University of Warwick, 2010). Lussac's study of the building of the Azerbaijani oil network is another example of the extent to which energy and politics intertwine when one scrutinizes the connections between the so-called state and nonstate actors (human and nonhuman) involved.

27. My discussion of Khuzistan as a laboratory draws on Bowker, *Science on the Run*. For a discussion of STS scholarship on real-world experiments, see Matthias Gross, "Give Me an Experiment and I will Raise a Laboratory," *Science, Technology & Human Values* 41, no. 3 (2015): 1–22. On the relation between imperialism and scientific expertise in transforming Africa into a laboratory, see Helen Tilley, *Africa as a Living Laboratory: Empire, Development,*

and the Problem of Scientific Knowledge, 1870–1950 (Chicago: University of Chicago Press, 2011).

28. Andrew Barry, "Technological Zones," *European Journal of Social Theory* 9, no. 2 (2006): 239–253. Scholarship that draws on STS has identified strategies of unknowns, notably those of uncertainty, ignorance, and risk, but also ambiguity and forms of temporization practiced by international economic organizations such as the World Bank and the IMF. See the special issue on "Strategic Unknowns: Towards a Sociology of Ignorance" edited by Linsey McGoey in *Economy and Society* 41, no. 1 (2012).

29. More and more, historians are moving away from cultural history toward histories of the sociotechnical, influenced by STS. See Michael T. Allen and Gabrielle Hecht, eds., *Technologies of Power: Essays in Honor of Thomas Parke Hughes and Agatha Chipley Hughes* (Cambridge, MA: MIT Press, 2001); Tony Bennett and Patrick Joyce, eds., *Material Powers: Cultural Studies, History and the Material Turn* (New York: Routledge, 2010); Sara B. Pritchard, *Confluence: The Nature of Technology and the Remaking of the Rhone* (Cambridge, MA: Harvard University Press, 2011). Environmental historians, such as Richard White, are also paying more attention to the materiality of commodities and writing sociotechnical histories, in the way that William Cronon has been doing for some time. See William Cronon, *Nature's Metropolis: Chicago and the Great West* (New York: Norton, 1991), and Richard White, *Railroaded: The Transcontinentals and the Making of Modern America* (New York: Norton, 2011). The history of the corporation figures in work like that of Philip Stern, *The Company State: Corporate Sovereignty and the Early Modern Foundations of the British Empire in India* (Oxford: Oxford University Press, 2011).

30. Timothy Mitchell, *Carbon Democracy: Political Power in the Age of Oil* (London: Verso, 2011). Toby Jones has also conceptualized the activities and meanings of scientific and technical knowledge in the building of oil and water infrastructure as internal to the work of building the national, Saudi Arabian state. The analysis departs significantly from mine, however, in treating science and technology as mere "instruments" of state power. See Toby Jones, *Desert Kingdom: How Oil and Water Forged Modern Saudi Arabia* (Cambridge, MA: Harvard University Press, 2010). On the importance of materials in politics, also see Andrew Barry, *Material Politics: Disputes along the Pipeline* (Malden, MA: Wiley, 2013).

31. Donald MacKenzie, "Nuclear Missile Testing and the Social Construction of Accuracy," in *The Science Studies Reader*, ed. Mario Biagioli, 343–357 (New York: Routledge, 1999). Regarding the testing of nuclear missiles, decision-making processes involve a kind of heterogeneous engineering in which the technical characteristics of missiles, such as their accuracy, are determined based on design criteria. But at the same time, the decisions are organizational because the well-being of particular corporations, project offices, and whole branches of the armed services can depend on them, as can a multiplicity of careers. Thus, interests clash.

32. Jacqueline Best, "Bureaucratic Ambiguity," *Economy and Society* 41, no. 1 (2012): 86–106, esp. 101.

33. According to the dominant narrative, British imperial interests in naval expansion and in blocking Russian imperial encroachment from the north of Iran triggered Winston Churchill's decision, as First Lord of the British Admiralty, to get the British government involved in D'Arcy's oil venture by switching the Royal Navy's source of fuel from coal to an almost total dependency on oil.

34. Kenneth Pomeranz, *The Great Divergence: China, Europe, and the Making of the Modern World* (Princeton, NJ: Princeton University Press, 2000), 23. The energy shift to fossil fuels, particularly from wood to coal, produced a dramatic easing of northwest Europe's land constraints during industrialization.

35. Ibid., 276, 280.

36. Steven G. Galpern, *Money, Oil, and Empire in the Middle East: Sterling and Postwar Imperialism 1944–1971* (Cambridge: Cambridge University Press, 2009), 1–2.

37. Timothy Mitchell, "McJihad," *Social Text* 73 (2002): 1–18, quote on 7. In his discussion of the global economy of oil, Mitchell locates a shift in the form of empire and its method of creating political control particular to the emergence of the oil corporation in the interwar period.

38. See Stern, *The Company State*, which studies the English East India Company on its own terms as the product of overlapping sovereignties, legal structures, and boundary making shaping the corporation's formation over an extended period, with important consequences for our understanding of the powers of the British Empire.

39. An account of Iran's historical experience with state building usually sees oil as central to understanding the patterns and problems of the state's social and economic development over the course of the twentieth century. The moment of despotic rule under the Pahlavi monarch, Reza Shah (1925–1941), is the usual point of departure for political and economic accounts that insist on the failure of Third World "pseudomodernist" leaders. These leaders are described as alienated from the culture and history of their own society yet as rarely demonstrating any "real understanding of European ideas, values, and techniques" in their problematic projects of rapid industrialization. See, for example, Homa Katouzian, *The Political Economy of Modern Iran: Despotism and Pseudo-Modernism, 1926–1979* (New York: New York University Press, 1981), 102-103.

40. For a critique of the boundary problem in theories of the state, and in relation to society, see Timothy Mitchell, "Limits of the State: Beyond Statist Approaches and Their Critics," *American Political Science Review* 85 (1991): 77–96. Mitchell argues that technical innovations in the organization and control of modern society are better understood as techniques of power and domination internal to social processes rather than occurring in terms of an external relationship between separate spheres.

41. Large-scale infrastructure projects were projects of state building, but they did not work simply as practices of reduction and simplification to render,

as James Scott argues, the holder of authority's "field of vision more legible and hence more susceptible to careful measurement and calculation." Scott's argument highlights the failures of methods of abstraction used to organize and control populations and industry, yet at the same time it assumes that there are more "real" or correctly scientific ways for a state authority to represent the objectives of its farmers, for example. See James C. Scott, *Seeing Like a State: How Certain Schemes to Improve the Human Condition Have Failed* (New Haven, CT: Yale University Press, 1998), 11.

42. The best discussion of how Western "democracies" relate to fossil fuels is provided by Mitchell's *Carbon Democracy*, which I draw on in this study to identify particular infrastructural configurations as incompatible with democracy. While democracies appear to thrive in Western contexts, they rely on the pursuit of undemocratic arrangements in other parts of the world, often in terms of the control, distribution, and sale of Middle East oil.

43. Antony Anghie, *Imperialism, Sovereignty, and the Making of International Law* (Cambridge: Cambridge University Press, 2007), 212.

44. Podobnik, *Global Energy Shifts*, 72–73. According to Podobnik, the history of these "complex commercial agreements" that were "designed and repeatedly revised, as broader geopolitical relationships ... underwent major shifts" begins after World War I with the San Remo Agreement of 1920, which granted British and French companies rights to the oilfields of Mesopotamia. Dutch interests controlled oil production in the East Indies and "Britain controlled the Indian and Burmese oil sectors."

45. Anthony Sampson, *The Seven Sisters: The Great Oil Companies and the World They Shaped* (New York: Viking Press, 1975), 6.

46. Thomas Brockway, "Britain and the Persian Bubble, 1888–1892," *Journal of Modern History* 13, no. 1 (1941): 36–47.

47. Ranin Kazemi, "The Tobacco Protest in Nineteenth-Century Iran: The View from a Provincial Town," *Journal of Persianate Studies* 7 (2014): 251–295.

48. The term *Seven Sisters* became widely used for the largest oil corporations in the 1950s when Enrico Matteo, head of the Italian state oil company, ENI, first popularized it. The term referred to Exxon, Shell, BP, Gulf, Texaco, Mobil, and Socal (or Chevron).

49. Barry, "Technological Zones," 246. Other zones include social and environmental matters affected by oil industry projects, corruption, and violence.

Chapter 1

1. Ozokerite is a paraffin-like member of the bitumina family. See Alison Frank, *Oil Empire: Visions of Prosperity in Austrian Galicia* (Cambridge, MA: Harvard University Press, 2007), 54.

2. D'Arcy's legal counsel signed the actual text on his behalf. See Ferrier, *History of British Petroleum*, 643.

3. The British company was known as the Anglo-Persian Oil Company until 1935, when it was renamed AIOC. For consistency, I refer to the company as AIOC and the country as Iran, rather than Persia, except in direct quotations.

4. See "Concessions," http://www.iranicaonline.org/articles/concessions.

5. A number of studies in the field of environmental history have examined the social and technical aspects of the global oil industry's development within their respective geographic locations by considering the question of property development in relation to mineral rights. For a discussion of the history of property rights and the building of the oil market in California, see Paul Sabin, *Crude Politics: The California Oil Market, 1900–1940* (Berkeley: University of California Press, 2005); for the building of the oil market in Austrian Galicia, see Frank, *Oil Empire*.

6. *Ferae naturae* means "nature [wild] animals" or any animals that are not designated domesticated animals by law. As Peter Shulman explains, "In the late nineteenth century, the Pennsylvania Supreme Court in the United States ruled that on account of the migratory nature of oil and gas (it flowed underground), ownership of the resource belonged to the individual upon whose property the fuel was first brought to the surface" (Peter A. Shulman, "'Science Can Never Demobilize': The United States Navy and Petroleum Geology, 1898–1924," *History and Technology* 19, no. 4 (2003): 365–395, quote on 378).

7. Ricardo Hausmann, "State Landed Property, Oil Rent and Accumulation in Venezuela: An Analysis in Terms of Social Relations," doctoral dissertation, Cornell University, 1981, 64–65.

8. Ibid. Fernando Coronil also adds that oil production was significantly cheaper in all other oil-producing nations than in the United States, "in part because they had richer oilfields but also because oil extraction was more rationally organized than in the United States. Since the subsoil in these countries is public property, oil fields are not anarchically subdivided into scattered small plots, as in the United States. ... Given uniform market prices determined by high-cost US oil, this cost difference was the source of surplus profits for the oil companies that operated overseas" (Coronil, *The Magical State*, 53). Coronil sees state ownership of the subsoil as a more "rational form" of management control, but this chapter argues somewhat differently that forms of property developed due to a combination of the oil's physical properties, and battles with local forms of property control that generated alternative political arrangements, rather than any (ir)rational forms of property management.

9. Michel Callon, "What Does It Mean to Say That Economics Is Performative?"

10. For a brief history of concessions (*emtiazat*) in Iran, see http://www.iranicaonline.org/articles/concessions. For the Middle East, see Henry Catton, *The Evolution of Oil Concessions in the Middle East and North Africa* (Dobbs Ferry, NY: Oceana Publications, 1967).

11. Studies that consider the history of Anglo-Iranian oil, partially or in its entirety, focus mainly on concession disputes over royalty terms and postwar oil crises concerning nationalization of the oil industry. See, for example, Ervand

Abrahamian, *The Coup: 1953, the CIA, and the Roots of Modern U.S.-Iranian Relations* (New York: The New Press, 2013); Massoud Karshenas, *Oil, State, and Industrialization in Iran* (Cambridge: Cambridge University Press, 1990); Nikkie R. Keddie, *Modern Iran: Roots and Results of Revolution* (New Haven, CT: Yale University Press, 2003).

12. Wolff was previously responsible for having secured the famous British tobacco concession, which ultimately led to the first organized political struggle in Iran, uniting religious clerics, students, and merchants against foreign concessionary control. As a result, the ruling monarch had no choice but to revoke the concession in 1892. Wolff was also a leading figure behind the opening of the Karun River to international traffic. See Brockway, "Britain and the Persian Bubble, 1888–1892," 37–38.

13. Ferrier, *History of British Petroleum*, 5. Reuter's agent, Edouard Cotte, discussed de Morgan's reports with Kitabji at the Paris Exhibition in 1900. See Arash Khazeni, *Tribes and Empire on the Margins of Nineteenth-Century Iran* (Seattle: University of Washington Press, 2009), 117.

14. See Article 6 of the 1901 concession in Ferrier, *History of British Petroleum*, 641. On the connection to Russian oil, see Robert W. Tolf, *The Russian Rockefellers: The Saga of the Nobel Family and the Russian Oil Industry* (Stanford, CA: Hoover Institution Press, 1976); Nowell, *Mercantile States*, 53.

15. Nowell, *Mercantile States*, 50–51. On the role of the Swedish Nobel family and French Rothschild family in building the Russian oil industry, see Tolf, *The Russian Rockefellers*, 156.

16. Nowell, *Mercantile States*, 53.

17. Keddie, *Modern Iran*, 61–62, 69–71. The Anglo-Russian agreement was signed in 1907. According to Brockway, Russia strengthened its position in Iran from the 1880s on by "organizing under tsarist officers the only effective military force in the country" (Brockway, "Britain and the Persian Bubble," 37).

18. Iran's government actively sought loans to solve its chief financial problems and to reestablish control of the provinces. An alternative Baku-Batum pipeline route ultimately prevailed due to the lower cost it incurred for exports to the Far East via the Suez Canal as well as exports to Europe from the Black Sea. The route also avoided antagonizing the British. The construction of the pipeline was delayed through 1901, however, because the Russian minister of transportation continued to back the Persian route (Nowell, *Mercantile States*, 51).

19. This argument draws on Timothy Mitchell's critical account of the beginnings of the oil industry in the Middle East in "The Prize from Fairyland," *Carbon Democracy*, chap. 2. Nowell also questions the imperialist framing (Nowell, *Mercantile States*, 51). The BP company history takes on the imperialist narrative in Ferrier, *History of British Petroleum*.

20. Thus, Nowell explains that as long as D'Arcy met his contractual obligations to explore for oil, "the Russian pipeline project was stymied" (Nowell, *Mercantile States*, 53).

21. Mitchell says that after securing the 1901 concession in Iran, D'Arcy started to compete with the Deutsche Bank to win rights in Mesopotamia from Turkey (Mitchell, *Carbon Democracy*, 50). His attempts to obtain a concession received approval from the British Foreign Office in 1904 and diplomatic support starting in 1908. See Marian Jack (Kent), "The Purchase of British Government's Shares in the British Petroleum Company, 1912–1914," *Past & Present* 39, no. 1 (April 1968): 139–168, esp. 141.

22. Ferrier, *History of British Petroleum*, 70; Nowell, *Mercantile States*, 53.

23. Nowell, *Mercantile States*, 52–54.

24. Mitchell, *Carbon Democracy*, 50. After further exploration failures, the firm cut back its commitments in Iran: "It was primarily the protection of its Indian investments that took Burmah Oil Co into Persia" (in Ferrier, *History of British Petroleum*, 69, 92). Burmah Oil took a year to incorporate AIOC as there was little interest in developing a new source of supply at Masjid Suleiman. The company took another three years to lay an 8-inch pipeline to carry oil 140 miles to Abadan and build a set of machinery with which to refine it (Mitchell, *Carbon Democracy*, 54).

25. Nowell, *Mercantile States*, 52–53. Royal Dutch and Shell Trading and Transport entered the oil business in the Far East in the 1890s. Competition with Standard Oil forced these two companies to cooperate on Russian production as a supplement to their holdings in the Far East—this was the Asiatic Petroleum Company. Shell was strongest in transport, Royal Dutch in marketing, and Rothschild's company, Bnito, in production.

26. As Mitchell explains, one alliance formed to manage Asian trade and the other for Europe. As mentioned above, in 1905, the Shell group forced Burmah Oil to agree to a division of Asian sales, an agreement that Standard Oil also followed. On the European side, Rothschild joined the other large Caspian producer, Nobel, in 1906 and with Deutsche Bank, its partner in Romania, formed the European Petroleum Union (EPU) to manage the western European kerosene and fuel oil markets. The cartel agreed with Standard Oil to divide up European sales. The EPU worked in tandem with the 1905 cartel agreement, Asiatic Petroleum. See Mitchell, *Carbon Democracy*, 46–47; Nowell, *Mercantile States*, 59.

27. Pennsylvania (North America), Baku, Burma, Sumatra, Austrian Galicia, and Romania constituted the six main oil-producing regions of the world at the time. Oil companies knew of several sites in the Middle East where rival companies might develop large new sources of petroleum, "threatening the world with additional supplies" (Mitchell, *Carbon Democracy*, 46–47).

28. The Burmah Oil Company was threatened on two fronts—by the Shell group's (Asiatic Petroleum) oil exports from Baku eroding its sales in India, and by the potential development of oil in neighboring Mesopotamia. On the oil and railway scheme in Mesopotamia that threatened to circumvent AIOC's control of Iran's oilfields, see Mitchell, *Carbon Democracy*, 57–59.

29. Ferrier, *History of British Petroleum*, 24, 27–28. According to Ferrier, the first "serious geological reference" comes from W. K. Loftus's study, "On the

Geology of Portions of the Turco-Persian Frontier and the Districts Adjoining," submitted in August 1855 to the Geological Society in London. Prior to going to Iran, de Morgan had worked as an Egyptologist. He published his findings on oil deposits in Iran in a multivolume, encyclopedic work, *Mission Scientifique en Perse*. See Khazeni, *Tribes and Empire*, 116.

30. Geoffrey Jones, *The State and the Emergence of the British Oil Industry* (London: Macmillan, 1981), 132.

31. Ferrier, *History of British Petroleum*, 55. The British Admiralty formed a very positive opinion of the "value of the Concession," having had the advantage of reading geologists' reports, "all pronounc[ing] favourably as to the petroliferous possibilities of the territory" ("Admiralty Commission on the Persian Oilfields," April 6, 1914, no. 23–47, L/P&S/10/410, India Office Records (IOR), London).

32. Khazeni, *Tribes and Empire*, 116.

33. The region was transferred to Iraq after 1914. As Mitchell points out, it was also located near the Tehran-Baghdad road that passed through the town of Khanaqin, a location that might have been pursued by the Baku oil firms to bypass D'Arcy's monopoly over pipeline construction to the Gulf (Mitchell, *Carbon Democracy*, 50).

34. Ferrier, *History of British Petroleum*, 66–67, 86–87.

35. Mitchell, *Carbon Democracy*, 59.

36. Mitchell, *Carbon Democracy*, 61. Mitchell adds that on the domestic front, the British government's rescue of AIOC "provided a means to evade public scrutiny of the cost of a switch from coal to oil, but did nothing to address causes of those expenses—monopolistic control of oil prices." The bailout also avoided the "democratic obstacle" of a rising labor force, seen in the movement by Welch coal miners. Also see Jack (Kent), "The Purchase of British Government's Shares in the British Petroleum Company"; Nowell, *Mercantile States*, 54–57.

37. See A. A. Fursenko, *The Battle for Oil: The Politics and Economics of International Corporate Conflict over Petroleum, 1860–1930* (Greenwich, CT: JAI Press, 1990), 141. TPC was renamed the Iraq Petroleum Company (IPC) in 1928. The agreement excluded Egypt and Kuwait, which together with southern Iran were already under British control. See chapter 3.

38. Ferrier underplays the company's relation to the British Admiralty and the role of the British government. It takes for granted the nature of relations between transnational corporations and governments. See Ferrier, *History of British Petroleum*, 10, 202–204, 260–261.

39. Keddie, *Modern Iran*, 54. On the Reuter concession see "Iran Under the Later Qajars, 1848-1922," in Peter Avery, G. R. G. Hambly, and C. P. Melville, eds., *The Cambridge History of Iran*, Volume 7: *From Nadir Shah to the Islamic Republic* (Cambridge: Cambridge University Press, 1991), 174-212 esp. 187–191 and "The Iranian Oil Industry" in Peter Avery, G. R. G. Hambly, and C. P. Melville, eds., *The Cambridge History of Iran*, Volume 7: *From Nadir Shah to*

the Islamic Republic (Cambridge: Cambridge University Press, 1991), 639-705 esp. 640. Other renting precedents involving foreign concerns included European trading companies (Keddie, *Modern Iran*, 132). Ferrier explains that the first 1872 Reuter concession specifically granted rights in the extraction of coal, iron, copper, lead, and petroleum (Ferrier, *History of British Petroleum*, 24).

40. Brockway, "Britain and the Persian Bubble, 1888–1892," 38.

41. Ibid., 39.

42. Ibid., 41. As Ferrier notes, the ruling Qajar government had granted a few smaller petroleum concessions in the late nineteenth century but the operations were unsuccessful (Ferrier, *History of British Petroleum*, 24–27).

43. Ibid., 42. Also see George N. Curzon, *Persia and the Persian Question*, vol. 2 (London: Longmans, Green, 1892); Mitchell, *Carbon Democracy*, 52. He also signed an agreement with the ruler of Kuwait to prevent the building of a pipeline.

44. Mitchell, *Carbon Democracy*, 53; Ferrier, *History of British Petroleum*, 43–44, 69; Nowell, *Mercantile States*, 53. Mitchell explains that this group of India-based British imperialists was attempting to expand the reach of Britain's Indian empire by establishing greater control of the Persian Gulf and encouraging local British monopolies in trade, steam navigation, and other enterprises.

45. Ferrier, *History of British Petroleum*, 33.

46. On the history of the doctrine of ownership in Iran see Ann K. S. Lambton, *Landlord and Peasant in Persia: A Study of Land Tenure and Land Revenue Administration* (London: Oxford University Press, 1953; reprint New York: I. B. Tauris, 1991), v (page citations are to the reprint edition); Afsaneh Najmabadi, *Land Reform and Social Change in Iran* (Salt Lake City: University of Utah Press, 1987); Eric J. Hooglund, *Land and Revolution in Iran, 1960–1980* (Austin: University of Texas Press, 1982). For the development of this argument in the Ottoman context, see Huri Islamoglu, *State and Peasant in the Ottoman Empire* (Leiden: Brill, 1994); Martha Mundy and Richard Saumarez Smith, *Governing Property, Making the Modern State: Law, Administration and Production in Ottoman Syria* (New York: I. B. Tauris, 2007). For the case of Egypt, which builds on Islamoglu's argument, see Timothy Mitchell, *Rule of Experts: Egypt, Techno-Politics, Modernity* (Berkeley: University of California Press, 2002), 57. On the transformation of local land tenure arrangements in relation to Mexican oil, see Myrna I. Santiago, *The Ecology of Oil: Environment, Labor, and the Mexican Revolution, 1900–1938* (Cambridge: Cambridge University Press, 2009).

47. Gene R. Garthwaite, *Khans and Shahs: A Documentary Analysis of the Bakhtiyari in Iran* (Cambridge: Cambridge University Press, 1983), 19. Also see Khazeni, *Tribes and Empire*, chap. 4, passim.

48. Abadan is located approximately 30 miles east of Basra.

49. Ahmad Kasravi, *Tarikh-i Pansad Sal-i Khuzistan (Five Hundred Year History of Khuzistan)* (Tehran: Payam Press, 1950), 208.

50. Ibid. Ferrier also explains that the oil deposits of Iran have been utilized for centuries for "caulking boats, bonding bricks, setting jewelry, and in flaming missiles." See L. Lockhart, "Histoire du Pétrole en Perse jusqu'au Début du XX Siècle," *La Revue Pétrolifère* (Paris, 1938), cited in Ferrier, *History of British Petroleum*, 24.

51. According to Lambton, the official unit of currency in Iran was the *rial* (rs) composed of 1,000 *dinars*. "*Tuman*, while not in official use," she explains, was "commonly employed to designate 10 rs." The official exchange rate at the time of Lambton's research in the 1940s and early 1950s was 89.40 *rial* to the British pound sterling (Lambton, *Landlord and Peasant*, 409–410).

52. Membership in these communities is usually understood in terms of kinship relations, so outsiders, especially European powers such as the British, called them "tribes"—a vague term for people the state does not control.

53. Kasravi, *Tarikh-i Pansad Sal-i Khuzistan*, 211. Also see William Theodore Strunck, "The Reign of Shaykh Khaz'al Ibn Jabir and the Suppression of the Principality of 'Arabistan: A Study in British Imperialism in Southwestern Iran, 1897–1925," doctoral dissertation, Indiana University, 1977, 349, 389.

54. In the region around Masjid Suleiman where the Bakhtiyari winter pastures were located, over 250 plows were known to exist. Crops such as wheat, barley, and small amounts of rice were under cultivation at the time of oil discovery. See Arash Khazeni, "Opening the Land: Tribes, State, and Ethnicity in Qajar Iran, 1800–1911," doctoral dissertation, Yale University, 2005, 226.

55. Khazeni, "Opening the Land," 74, 78, 86, 98. Khazeni explains that as leader (*ilkhan*) of the Bakhtiyari confederation from the 1870s onward, Husayn Quli Khan led a migratory administration, taxing on behalf of the state both the pastoral and agrarian wealth of the "tribes" through the *maliyat shakhi* ("horn" or flock tax) and the *khish* (plow tax). As they became increasingly sedentary, the Bakhtiyari became easier targets for state taxation and control.

56. Lambton, *Landlord and Peasant*, 283.

57. Ibid., liv–lv.

58. Najmabadi, *Land Reform and Social Change*, 44–45.

59. Also see Lambton's discussion of the multiple forms of *tuyūl* in the Qajar period (p. 139). She explains that "as the control of the [Qajar] government weakened so the tendency grew to convert *tuyūls* into *de facto* private property, inheritable and alienable by sale."

60. Najmabadi refers to Lambton's history as a "classic reference" for the study of property in Iran. See Najmabadi, *Land Reform and Social Change*, 43:214n1. For her treatment of *milk* (private property), see Lambton, *Landlord and Peasant*, lii, 51, 456. At one point, Lambton makes the translation while quoting a primary Farsi source dating from the fourteenth century that is discussing property arrangements in the tenth century.

61. Lambton, *Landlord and Peasant*, lv.

62. For example, Lambton explains that "all *divani* land, was ... potentially *khasseh*, since the *shah* could declare it so whenever he wished" (ibid., lv, 108).

63. See the four "older" aspects characterizing an imperialist empire's method of achieving political control in a foreign land in the context of the shift to a new kind of colonizing corporation, the transnational oil corporation (Mitchell, "McJihad," 7).

64. Ferrier, *History of British Petroleum*, 53. D'Arcy worked hard to relieve himself of the financial responsibilities of maintaining the concession, drilling for oil, and setting up a company by attempting to organize a syndicate prior to the creation of a company. In a violation of the concession's terms, he failed to do so, but he finally succeeded in setting up an exploitation company rather than a concessionary company as he was in no position to do so without oil or financial support. With the significant financial involvement of Burmah Oil in 1905, a new Concessions Syndicate was subsequently formed.

65. *Farman* refers to a "decree, command, order, judgement," often denoting a royal or government decree, "that is, a public and legislative document promulgated in the name of the ruler or another person holding elements of sovereignty." See http://www.iranicaonline.org/articles/farman.

66. Grey to Barclay, January 4, 1911, no. 266, L/P&S/10/144, IOR.

67. Young to Lamb, March 22, 1912, 70335, British Petroleum Archives (BP), Coventry, UK.

68. Cox to Lorimer, March 9, 1909, 71194, BP.

69. "Lorimer to Ahwaz British Consulate containing extract of letter from Wilson to Reynolds," March 16, 1909, 71194, BP.

70. Messieurs Strick, Scott and Co. to Wilson, September 30, 1910, no. 255, L/P&S/10/144, IOR.

71. Lambton, *Landlord and Peasant*, 238–239. For this period, Lambton no longer refers to *divani*.

72. Ibid., 253.

73. Reynolds to Messrs. Concessions Syndicate, April 17, 1909, 71194, BP.

74. Ibid.

75. "Memorandum by Mr. Kitabji on the Affairs of the D'Arcy Oil Company in Persia," Spring-Rice to Grey, enclosure in no. 49, March 21, 1907, 416/32, British National Archive, Records of the Foreign Office, Kew, UK (FO).

76. Jenkins to D'Arcy, March 27, 1902, 69403, BP.

77. Ibid.

78. Wallace to "Winchester House, London," April 28, 1909, 71194, BP.

79. Ibid.

80. Agent for APOC to Minister of Mines and Public Works, June 5, 1911, 240–014783/18/(0–217), Iranian National Archive (INA), Tehran.

81. In *Governing Property*, Mundy and Saumarez Smith make a similar argument for the history of property and land registration in Ottoman Syria.

82. Mitchell, "McJihad," 7.

83. Ferrier, *History of British Petroleum*, 48.

84. Ibid.

85. Ali Reza Abtahi, *Naft va Bakhtiyariha (Oil and the Bakhtiari)* (Tehran: Iranian Institute for Contemporary Historical Studies, 2005). Abtahi is citing Reynolds to Ali Sultaneh, November 16, 1904 (1322 Safar 12th), 42–43, Iranian Foreign Ministry Archive, Tehran. Also see Ferrier citing Reynolds to Jenkins, March 31, 1904, H12/27, BP, that the Bakhtiari were not interested in the distinction between surface and subsurface rights.

86. See Ali Morteza Samsam Bakhtiari, *The Last of the Khans* (New York: iUniverse, 2006), 90.

87. Stephanie Cronin, "The Politics of Debt: The Anglo-Persian Oil Company and the Bakhtiyari Khans," *Middle Eastern Studies* 40, no. 4 (2004): 1–34, esp. 4.

88. Cronin, "Politics of Debt," 4.

89. Laurence Paul Elwell-Sutton, *Persian Oil: A Study in Power Politics*, History and Politics of Oil Series (Westport, CT: Hyperion Press, 1976), 19.

90. Cronin, "Politics of Debt," 4.

91. Ibid., 6.

92. On the concept of technical device in relation to property and fish, see Petter Holm and Kare Nolde Nielsen, "Framing Fish, Making Markets: The Construction of Individual Transferable Quotas (ITQs)," in *Market Devices*, ed. Michel Callon, Yuval Millo, and Fabian Muniesa, 173–195 (Malden, MA: Blackwell, 2007).

93. Abtahi, *Naft va Bakhtiyariha*, 60.

94. "Memorandum by Mr. Kitabji on the Affairs of the D'Arcy Oil Company in Persia," Spring-Rice to Grey, March 21, 1907, 416/32, FO.

95. Cronin, "Politics of Debt," 5.

96. Mostafa Elm, *Oil, Power, and Principle: Iran's Oil Nationalization and Its Aftermath* (Syracuse, NY: Syracuse University Press, 1992), 14.

97. Abtahi, *Naft va Bakhtiyariha*, 144. Extract from "Memorandum on the Governorship of Bakhtiaristan," by M. Young, n.d., no. 386, 416/54, FO.

98. Elwell-Sutton, *Persian Oil*, 21.

99. Quoted in Elwell-Sutton, *Persian Oil*, 21.

100. On the first group of Bakhtiyari pastoral nomads and sedentary villagers recruited to join the oil industry as laborers working at the wellhead or refinery or as security guards, see Touraj Atabaki, "From 'Amaleh (Labor) to Kargar (Worker): Recruitment, Work Discipline and Making of the Working Class in the Persian/Iranian Oil Industry," *International Labor and Working-Class History* 84 (Fall 2013): 159–175.

101. Jones, *The State and the Emergence of the British Oil Industry*, 95–96. Charles Greenway managed Burmah Oil's campaign to win and maintain the Indian kerosene market and as its managing director, directed AIOC's

negotiations with the British government after 1912. He was a senior managing partner in Shaw, Wallace, and Company, which worked with Burmah and whose senior partner, Charles Wallace, joined Burmah Oil's board in 1902. Greenway discredited Shell in discussions with the government of India, suggesting a link to Standard Oil.

102. Greenway to Lloyd Scott & Co., April 8, 1910, 70275, BP.

103. Reynolds to Crow, February 25, 1909, 71194, BP.

104. Bakhtiari, *Last of the Khans*, 62–63. Bakhtiari cites Layard's work as *Early Adventures in Persia, Susiana, and Babylonia* (London: John Murray, 1894). Also see Khazeni, *Tribes and Empire*, 82. Khazeni explains that a Farsi translation of parts of Layard's travelogue was included in *Tarikh-i Bakhtiyari*.

105. Lloyd to Reynolds, April 10, 1910, 71194, BP.

106. Lambton, *Landlord and Peasant*, 282.

107. Garthwaite, *Khans and Shahs*, 18–19.

108. Lambton, *Landlord and Peasant*, 285.

109. Garthwaite, *Khans and Shahs*, 24–26. Khazeni also discusses this in *Tribes and Empire*.

110. "Bakhtiyari Affairs: The Land Problem," Dr. Young, Maydan-i Naftun, January 12, 1911, 70335, BP.

111. Young to ?, January 12, 1911, 71691, BP. See Ferrier, *History of British Petroleum*, 76.

112. Ibid.

113. Ferrier, *History of British Petroleum*, 146. The pipeline extended 130 miles from Masjid Suleiman to the Abadan refinery. The company had acquired 9.58 miles at Masjid Suleiman and a strip of land 17 miles long and 1 foot wide for a section of the pipeline extending from Masjid Suleiman to the limits of the Bakhtiyar Mountains.

114. Young to Lamb, April 18, 1911, 317–319, 70335, BP.

115. Ranking to Barclay, May 7, 1911, no. 112–113, L/P&S/10/144, IOR.

116. Young to Lamb, April 18, 1911, 220-221, 71691, BP. In March 1911, the khans employed a man from Isfahan, Farajullah, to "walk about with a stick measuring" the flatlands and the hills. This covered an area that G. B. Scott and "four natives" surveyed and mapped of the oilfields in preparation for the company's negotiations with the khans. For a detailed discussion of these land negotiations and surveying procedures, see Khazeni, *Tribes and Empire*, 154–155.

117. Khazeni, "Opening the Land," 303–304.

118. According to Lloyd to Reynolds, April 7, 1910, 70275, BP, AIOC offered 15 *tuman* per *jarib* for cultivated land and 3 *tuman* for grazing ground with an advance of £2500. Also see Lambton, *Landlord and Peasant*, 407.

119. Lambton, *Landlord and Peasant*, 406–407.

120. Ibid.

121. Ibid.

122. Timothy Mitchell, *Colonising Egypt* (Berkeley: University of California Press, 1991), 33, 44.

123. See *acre* in the *Oxford English Dictionary* (2nd ed., 1989).

124. Young to Lamb, April 25, 1911, 70335, BP. Also see Khazeni, *Tribes and Empire*, 155–156.

125. Ibid. The Farsi word to which they refer is not indicated in the documents.

126. Young to ?, 1911, 194–200, 71691, BP.

127. Major Noel to ?, October 10, 1921, 71691, BP.

128. Cronin, "Politics of Debt," 7.

129. The agreement was known as the Bakhtiyari Land Agreement of 1911. In total, the khans had sold almost 7,000 acres of winter pastureland, which included customary grazing fields and arable lands. See Khazeni, *Tribes and Empire*, 125.

130. See "Bakhtiyari Affairs: The Land Problem," Dr. Young, January 12, 1911, 70335, BP; cited in Khazeni, *Tribes and Empire*, 152–153.

131. See Young to Lamb, July 17, 1911, 70335, BP.

132. For a definition of translation as it works in STS, see Bruno Latour, *Pandora's Hope: Essays on the Reality of Science Studies* (Cambridge, MA: Harvard University Press, 1999), 311.

133. Reynolds to Concessions Syndicate (Glasgow), February 16, 1909, no. 42, 71994, BP.

134. "Iraqi-Persian Relations and the Situation in Khuzistan," February 27, 1928, 68419, BP. Known as Arvand Rud in Farsi, the Shatt al-Arab waterway is a river formed by the confluence of the Tigris and Euphrates rivers. The southern end of the river forms the border between Iraq and Iran down to the mouth of the river, which empties into the Persian Gulf. The Karun River, Iran's only navigable river, also flows into the Arvand Rud. The major cities along the river are Abadan (site of the oil refinery) and Khorramshahr on Iran's side, and the city of Basra in Iraq. The region remains one of the largest sources of dates in the world.

135. Cox to Lorimer, March 9, 1909, 71994, BP.

136. Sheikh Khazzal to APOC (translation), May 15, 1909, 71194, BP.

137. Strunck, "The Reign of Shaykh Khaz'al Ibn Jabir and the Suppression of the Principality of 'Arabistan," 396.

138. Cox to Sheikh of Mohammerah, May 15, 1909, L/P&S/10/132, IOR.

139. July 1, 1909, Enclosure 1, no. 350, L/P&S/10/144, IOR.

140. Cox to Wilson, November 6, 1910, no. 256, L/P&S/10/144, IOR.

141. Ibid.

142. Ibid.

143. Greenway to Mallet, December 5, 1910, 71573, BP.

144. Keddie, *Modern Iran*, 69–70, 75–76.

145. Cox to Butler, 1908, no. 148, L/P&S/10/132, IOR.

146. Strunck, "The Reign of Shaykh Khaz'al Ibn Jabir and the Suppression of the Principality of 'Arabistan," 396–397.

147. Ervand Abrahamian, *Iran between Two Revolutions* (Princeton, NJ: Princeton University Press, 1982), 120.

148. Bakhtiari, *Last of the Khans*, 157–160.

149. Ibid.

150. Ibid.

151. Ibid., 180. See also Abrahamian, *Iran between Two Revolutions*, 150; Strunck, "The Reign of Shaykh Khaz'al Ibn Jabir and the Suppression of the Principality of 'Arabistan," chap. 7, passim.

152. "Revolt in Persia," *The Times*, n.d., 68419, BP.

153. Ibid.

154. Ibid.

155. Chamberlain to Clive, November 4, 1927, no. 133, 416/81, FO.

156. Cronin, "Politics of Debt," 9.

157. Ibid., 11, 17.

158. Ibid., 14, 19.

159. Lambton, *Landlord and Peasant*, 19.

160. Najmabadi, *Land Reform and Social Change in Iran*, 45.

161. See Lambton, *Landlord and Peasant*, 183.

162. "Diary of HBM's Consul for Khuzistan, Ahwaz, no. 2, for month of February 1929," signed by R. G. Moneypenny (HBM's Consul for Khuzistan), L/P&S/11/290, IOR.

163. ?, 1935, no. 496, 482, L/P&S/12/3400, IOR. For example, a representative of Nizam-us-Saltaneh arrived to claim many properties in the bazaar quarter of Ahwaz, in Khuzistan Province, but was ordered by government authorities of the local Finance Department to leave.

164. See Najmabadi, *Land Reform and Social Change in Iran*, 45.

165. Peter J. Beck, "The Anglo-Persian Oil Dispute 1932–33," *Journal of Contemporary History* 9, no. 4 (1974): 123–151.

166. "Persian Concession: 1932–1993," n.d., 1–32, 70223, BP. See Article 4 in particular. Specific clauses within Article 4 also defined the procedure by which land would be acquired by the company. Lands belonging to the government would undergo valuation at prices "not to exceed the current price of lands of the same kind and utilized in the same manner in the district." Lands not belonging to the government "shall be acquired by the Company by agreement with the parties interested, and through the medium of the Government." The

government would not allow owners of such lands to "demand a price higher than the prices commonly current for neighbouring lands of the same nature" and with no regard to "the use to which the Company may wish to put them."

167. Amin Banani, *The Modernization of Iran, 1921–1941* (Stanford, CA: Stanford University Press, 1961), 143. A new national mining law was finally passed in 1953, after the nationalization of the oil industry. See "Persian Mining Law 1953," n.d., 71888, BP.

168. Democratic forms of control would have to involve local actors, claimants, and public opinion in the decision-making process to devise novel scenarios for the organization of oil operations, negotiating contracts, setting production rates, and pricing. As Mitchell says, processes of democratization have generally depended on "engineering such forms of vulnerability ... that render the technical processes of producing concentrations of wealth dependent on the well-being of large numbers of people" (Mitchell, *Carbon Democracy*, 146). On "technical democracy," also see Callon, Lascoumes, and Barthe, *Acting in an Uncertain World*, 228.

169. One newspaper reported that AIOC was busy demolishing "2000 houses of the poor people in order to have a road," questioning why 20,000 people should be made homeless for a road. In another instance, Braim village, designated as the site for the construction of residential areas for European workers, was represented by the company as the site of "extreme congestion and highly insanitary conditions," a danger to public health and "in consequence, to the operations of the Refinery proper." See Jacks to Iranian Department of Industry, Ministry of National Economy, November 19, 1931, pp. 64–66, 240–005548/102/36/(0–97), INA. The company proposed to the Iranian government that property owners in locations such as Braim be compensated. This was to occur "on the basis of a valuation to be fixed by the Governor-General of Khuzistan, whereafter they should be allotted fresh sites in a new area to be laid out." The company agreed with local Iranian government authorities that after having compensated individual property owners such as Mullah Salman and Mullah Ghanem, the village should be "evacuated and the local residents transferred to Abadan Town and not to a new site adjacent to Braim village." See Iranian Interior Ministry (Khuzistan) to Head of Finance Cabinet, 1932, pp. 2–3, 240–00548/102/36/(0–97), INA. The inhabitants had no choice but to transfer their possessions outside the company compound. The company proposed that dwellings at Braim village be demolished and the owners allotted sites in Ahmadabad to construct new dwellings at their own expense. There were 210 complaints by displaced and resettled people regarding the valuation of their homes by the Department of Industry. See "Ministry of Finance (at Khuzistan) Report on Braim Lands" by Yussef Mofakham Sane'aee (Head of Khuzistan Province Finance/Fiscal Affairs), August 23, 1934, pp. 22–29, 240–018256/329/37/(0–67), INA. The local government at Khuzistan argued that the right of ownership to the land belonged to the natives of the area, but at the time of the transaction, these rights were not recognized and the residents still had not been justly compensated to rebuild homes elsewhere. See Rouhani to Ministry of Finance (Department of Concessions, Petroleum, & Revenue from Mines), April 22,

1942, pp. 47–48, 240–003836/71/16/(0–50), INA. Also see Rouhani to Ministry of Finance, January 26, 1942, pp. 22–23, 240–004029/71/17/(0–56), INA; Davari to Department General of Concessions and Petroleum and Revenues from Mines, May 28, 1950, pp. 119–123, 240–003102/53/35/(0–176), INA; Reza Deeba to Ministry of Finance, 1328/11/7, 112, 240–02367/26/31/(0–132), INA; Davari to Controller of Finance (Ahvaz), January 28, 1950, pp. 127–128, 240–003102/53/35/(0–176), INA.

170. For claims made by the descendants of Khaz'al, see Jakins to Rupert Hay, March 24, 1951, no. 7, 371/91265, FO. For those of the Bakhtiyari khans, see for example Sultaneh Morad Montazem ad-Dowleh to Ministry of Finance (commission concerning Bakhtiyari affairs), 1941–1942, pp. 12–13, 230–001924/(0–93), INA.

171. Bayat (NIOC) to General Bureau of Petroleum, Concessions, and Fisheries, 1954, p. 10, 240–001708/21/55/(0–132), INA.

172. Timothy Mitchell, "The Properties of Markets," in *Do Economists Make Markets? On the Performativity of Economics*, ed. Donald MacKenzie, Fabian Muniesa, and Lucia Siu, 244–275 (Princeton, NJ: Princeton University Press, 2007), esp. 267.

173. Bruno Latour, *Science in Action* (Cambridge, MA: Harvard University Press, 1987), 2, 131–132.

Chapter 2: Petroleum Knowledge

1. Podobnik, *Global Energy Shifts*, 2, 67. As mentioned in the introduction, the invention of the internal combustion engine in the late nineteenth century had first brought oil into direct competition with coal, rather than serving as a complementary source of energy. Podobnik explains that the oil industry initially provided products to support the preexisting coal system. Kerosene lamps were used in coal mining and oil-based lubricants were consumed in coal-powered rail locomotives and ships (p. 49). Colonial campaigns in Africa and Asia commenced the deployment of oil-powered vehicles and diesel-powered ships for war. This global energy shift, with countries such as the United States emerging as pioneers in petroleum development, posed a threat to British imperial power, and it triggered British state intervention in potential oil sectors such as aviation as well as naval and automotive transport (p. 65). Also see R. J. Forbes and D. R. O'Beirne, *The Technical Development of the Royal Dutch / Shell 1890–1940* (Leiden: Brill, 1957), which usefully parallels various technological developments discussed here within a different context.

2. Paul Lucier, *Scientists and Swindlers: Consulting on Coal and Oil in America, 1820–1890* (Baltimore: John Hopkins University Press, 2008), 268.

3. Podobnik, *Global Energy Shifts*, 72. Forbes and O'Beirne explain that to deal with the sulfur problem, AIOC adopted the hypochlorite process on a large scale (Forbes and O'Beirne, *Technical Development of the Royal Dutch / Shell*, 362).

4. As discussed in the previous chapter, imperial interest in fuel oil, among other factors, eventually encouraged the British Admiralty to step in with the essential commercial backing to transform AIOC into a new government-corporate entity, providing capital, a guaranteed demand, and political protection from regional unrest in Khuzistan. In exchange, the British Admiralty won an unlimited supply of fuel oil far below market prices and a controlling share in the company of 51 percent. Forbes and O'Beirne also explain that the British Navy, which was used to the Penna quality of turbine oil, only became interested in giving AIOC's oil a trial when the Amsterdam Laboratory succeeded in making good Mexican and Venezuelan substitutes (Forbes and O'Beirne, *Technical Development of the Royal Dutch / Shell*, 425).

5. Ferrier, *History of British Petroleum*, 14. For the company history of Anglo-Iranian oil framed as a business history, see Ferrier, *History of British Petroleum*; Bamberg, *History of British Petroleum*.

6. According to a petroleum engineering book published in the 1930s, petroleum expertise is the product of a universal history encompassing theories of oil accumulation that could be applied to any local setting. Building on evidence from American journals, company data, oil industry experts, and his professional experience, Dorsey Hager's *Fundamentals of the Petroleum Industry* (New York: McGraw-Hill, 1939) describes the history of oil in terms of the history of his own discipline—petroleum engineering. For a similar treatment, see Forbes and O'Beirne, *Technical Development of the Royal Dutch / Shell*.

7. Mitchell, "Carbon Democracy," 31–32.

8. For example, the 1928 oil cartel, as Nowell has shown, was actually a broader hydrocarbon cartel because it involved an agreement not just to restrict the production of oil but to prevent access to patents that would allow the coal industry to move into the production of synthetic oils using hydrogenation technology. See Nowell, *Mercantile States*, and my discussion of the 1928 cartel in chapter 3.

9. Instead, most of the scholarship situates the history of the first three decades of the concession's existence within the larger geopolitical and socioeconomic context. This context includes the formation of the Iranian state and the ruling monarch's campaign to modernize it by directing oil revenues toward large-scale military and industrialization projects, campaigns against foreign influence, and the suppression of semiautonomous nomadic groups that often served as intermediaries, provided security, and were a source of labor in the oilfields.

10. Bowker, *Science on the Run*.

11. See Callon, "Some Elements of a Sociology of Translation," 67–94; Latour, *Reassembling the Social*.

12. John Enos, *Petroleum Progress and Profits: A History of Process Innovation* (Cambridge, MA: MIT Press, 1962); Harold Williamson, *The American Petroleum Industry: The Age of Energy, 1899–1959*, vol. 2 (Evanston, IL: Northwestern University Press, 1963). These studies portray the history of oil in terms of a kind of technological determinism of innovation and progress propelling the expansion of an inevitably capitalist system.

13. Located in the southwestern province of Khuzistan bordering the Persian Gulf, 60 percent of the Bakhtiyari district—defined as an administrative unit by Iran's Qajar government—was mountainous. Given the relative inaccessibility of the region, it was usually beyond the direct control of the government. See Garthwaite, *Khans and Shahs*, 19.

14. Other geologists built on his finding, but the various theories on the nature and origins of the rock formations across the Zagros mountain range remained controversial. See, for example, Ernest Raymond Lilley, *The Geology of Petroleum and Natural Gas* (New York: Van Nostrand, 1928), 155; R. K. Richardson, "The Geology and Oil Measures of South-West Persia," *Journal of the Institute of Petroleum Technology* 10 (1924): 1–3; Edgar Wesley Owen, *Trek of the Oil Finders: A History of the Exploration for Petroleum* (Tulsa, OK: American Association of Petroleum Geologists, 1975), 1257, 1262–1263.

15. Ferrier, *History of British Petroleum*, 272.

16. Ibid., 191.

17. J. W. Williamson, *In a Persian Oil Field* (London: Ernest Benn, 1927), 63.

18. Ferrier, *History of British Petroleum*, 272.

19. Ibid., 135. One early problem in refining was that the kerosene continued to have a yellow tinge and the benzene fraction had an unpleasant smell. See Ferrier, *History of British Petroleum*, 152.

20. Ibid., 272.

21. Ibid., 455.

22. Bowker, *Science on the Run*, 2, 110.

23. The viscosity of fuel oil plays a role in the quality of combustion inside a diesel engine and must be regulated to achieve adequate engine efficiency (Hager, *Fundamentals of the Petroleum Industry*, 342). Viscosity is the internal friction between molecules of oil measured by resistance to flow. On the API (American Petroleum Institute) scale, so-called light oils are indicated by the higher degrees and the heavy oils, such as Iran's oil, by the lower degrees. Among the numerous uses of the various fractions of oil, the gas fraction (natural gas) was used for heating, gasoline for automobile engines, fuel and gas oil for ocean transport, kerosene for lighting lamps, lubricating oil for petrolatum jelly, and asphalt for tractors and industry.

24. Bamberg, *History of British Petroleum*, 189.

25. Ferrier, *History of British Petroleum*, 148.

26. Ibid.

27. Cadman to Lloyd, January 17, 1923, 71537, BP.

28. Ibid.

29. F. E. Smith, "Obituary Notices of Fellows of the Royal Society: John Cadman, Baron Cadman, 1877–1941," *Royal Society* 3, no. 10 (1941): 915–928, esp. 917 On John Cadman's early role in shaping what would eventually become the Department of Chemical Engineering at the University of Birmingham,

see A. J. Biddleston and J. Bridgewater, "From Mining to Chemical Engineering at the University of Birmingham," in *One Hundred Years of Chemical Engineering*, ed. N. A. Peppas, 237–244 (Dordrecht, The Netherlands: Kluwer Academic Publishers, 1989).

30. "Birmingham University's New Oil Building," *NAFT* II, no. 4 (July 1926): 24–30, cited in Michael E. Dobe, "A Long Slow Tutelage in Western Ways of Work: Industrial Education and the Containment of Nationalism in Anglo-Iranian and Aramco, 1923–1963," doctoral dissertation, Rutgers University, 2008, 47.

31. Ferrier, *History of British Petroleum*, 689. Cadman (1877–1941) had acquired early work experience in England's collieries. He was a leading advocate of the application of science in all operations tied to the oil industry. See Smith, "Obituary," 1941.

32. Donald MacKenzie, "Missile Accuracy: A Case Study in the Social Processes of Technological Change," in *The Social Construction of Technological Systems: New Directions in the Sociology and History of Technology*, ed. Wiebe E. Bijker, Thomas P. Hughes, and Trevor J. Pinch, 195–222 (Cambridge, MA: MIT Press), esp. 219.

33. Keddie, *Modern Iran*, 75.

34. Ferrier, *History of British Petroleum*, 27.

35. Owen, *Trek of the Oil Finders*, 1263.

36. Jones, *The State and the Emergence of the British Oil Industry*, 158n51. AIOC built its first chemistry laboratory in Abadan, Iran, in 1935.

37. Owen, *Trek of the Oil Finders*, 1257. For a history of consulting geologists in North America, see Lucier, *Scientists and Swindlers*.

38. AIOC's geologists, engineers, and chemists published mainly in the *Journal of the Institute of Petroleum* (in the early years, from 1918; a British journal), the *World Petroleum Congress Proceedings* (in 1933, the first issue), and in older American journals such as the *Bulletin of the American Association of Petroleum Geologists* (AIOC published here in the 1930s) and the *American Mining Engineering Bulletin*.

39. Owen, *Trek of the Oil Finders*, 1254. On Redwood's work in Nigeria, see Phia Steyn, "Oil Exploration in Colonial Nigeria, c. 1903–58," *Journal of Imperial and Commonwealth History* 37, no. 2 (2009): 249–274. Also see Ferrier, *History of British Petroleum*, 32; Jones, *The State and the Emergence of the British Oil Industry*, 95.

40. Jones, *The State and the Emergence of the British Oil Industry*, 97. As Jones explains, Redwood was a chemist by training who became involved in the petroleum industry after 1870 and made important technical innovations and inventions. He also published what was, as Jones explains, the "standard textbook on petroleum matters" until after World War I.

41. Owen, *Trek of the Oil Finders*, 1259; Jones, *The State and the Emergence of the British Oil Industry*, 98.

42. Owen, *Trek of the Oil Finders*, 1256. Ferrier says that Redwood was on the British government's fuel oil committee and most likely informed them of D'Arcy's difficulties in funding oil exploration (Ferrier, *History of British Petroleum*, 61).

43. Jones, *The State and the Emergence of the British Oil Industry*, 99. His advice, suggests Jones, is what convinced Pretyman, president of the Admiralty Oil Committee of 1903–1906, to "oppose the entry of 'foreign trusts' into the oilfields of the Empire, and to support smaller British companies such as Burmah Oil."

44. Ibid., 104.

45. Ibid., 134.

46. Ibid.

47. Through their contacts with G. & J. Weir & Co., Burmah Oil also handled most of AIOC's first pipeline construction. Shaw Wallace & Co., Burmah's managing agents in India, provided a variety of services for AIOC in the early years and hired Strick Scott & Co., also former managing agents for Burmah Oil. Ibid., 135.

48. Owen, *Trek of the Oil Finders*, 1260.

49. Ferrier, *History of British Petroleum*, 691. Garrow explained that "we are not dealing with a field similar to Burmah or other fields like it which draw their production from sand." In fact, "our conditions here ... resemble more those of Mexico" (D. Garrow to Cadman, "Persian Notes," October 1, 1924, 1–12, 72018, BP).

50. Ibid. On Baku, see Ferrier, *History of British Petroleum*, 8.

51. Cadman to Sir Francis Ogilvie, November 6, 1922, 71537, BP.

52. "Confidential File on Considerations for the Establishment of a Formal Geological Survey Department with Permanent Staff," 1922–23?, 1–15, 71537, BP. Most geologists were in their late twenties with degrees in mining, oil geology, petroleum technology, or chemistry from Cambridge, Oxford, Edinburgh, or Birmingham as well as oil experience in Venezuela, Burma, Mexico, Africa, and Mesopotamia. See "Particulars of Geological Staff," February 26, 1923, 71537, BP.

53. Owen, *Trek of the Oil Finders*, 1261.

54. Bowker, *Science on the Run*, 21–22. As Bowker explains, geological considerations provided some help because a geologist could to a certain extent say whether "it was possible for oil to be found in such and such an area, though they could not determine with certainty whether it could be." The range of geophysical surface methods yielded additional data, and sometimes this data could be correlated with underground structures and sometimes those structures could be correlated with oil, but "no link in this chain was firm."

55. A geological fold occurs when a flat surface or stack of surfaces, such as sedimentary strata, are bent or curved as a result of some sort of stress underground, due to a change in pressure or temperature, for example.

56. Hager, *Fundamentals of the Petroleum Industry*, 138–139.

57. Owen, *Trek of the Oil Finders*, 1261.

58. Ibid., 1266. The concept of a hydrostatic unit coincided with the notion of "unit control" of production, which encouraged single ownership of an oilfield to eliminate competitors in adjacent blocks.

59. Reynolds to Williams, May 30, 1909, 70275, BP.

60. W. H. Dalton to Boverton Redwood, July 12, 1904, 70275, BP.

61. Owen, *Trek of the Oil Finders*, 1261.

62. Ibid.

63. Ibid.

64. Bowker, *Science on the Run*, 83.

65. H. G. Busk and H.T. Mayo, "Some Notes on the Geology of the Persian Oilfields," *Journal of the Institute of Petroleum* 5 (1918–1919): 3–26, esp. 7.

66. Busk and Mayo, "Some Notes on the Geology of the Persian Oilfields," 3.

67. Ibid.

68. Ibid., 5, 7.

69. Ibid., 6. Each "series" consisted of a combination of clays, shales, and sandstone, while some also included limestone with the main oil zones.

70. Ibid.

71. Latour, *Science in Action*, 16.

72. Lucier, *Scientists and Swindlers*, 244.

73. Gabrielle Hecht, "Rupture-Talk in the Nuclear Age: Conjugating Colonial Power in Africa," *Social Studies of Science* 32, nos. 5–6 (December 2002): 691–727, esp. 699. The inclusion of details of ethnic composition and relations in AIOC's reports followed a long British colonial tradition, particularly prevalent in India, in which scientists recorded ethnographic, geological, archeological, and other information outside their own domains of interest.

74. Busk and Mayo, "Some Notes on the Geology of the Persian Oilfields," 4.

75. Ibid.

76. Busk and Mayo, "Some Notes on the Geology of the Persian Oilfields," 14.

77. See Cronin, "Politics of Debt."

78. "APOC Concession in Persia: Policy of Geological Exploration in Persia," Lister James to Cadman, February 23, 1923, 71537, BP.

79. Strunck, "The Reign of Shaykh Khaz'al Ibn Jabir and the Suppression of the Principality of 'Arabistan," 396.

80. Asef Bayat, "With or without Workers in Reza Shah's Iran: Abadan, May 1929," in *The State and the Subaltern: Modernization, Society and the State in Turkey and Iran*, ed. Touraj Atabaki, 111–122 (London: I. B. Tauris, 2007).

81. Bowker, *Science on the Run*, 10.

82. Michel Callon, "Society in the Making," 86. See note 38 for a list of international journals that company geologists, engineers, and chemists published their findings in.

83. "Admiralty Commission on the Persian Oilfields," April 6, 1914, no. 23–47, L/P&S/10/410, IOR.

84. Ibid.

85. Cadman on "Gas Pressure Problems on Maidan Naftun Field and Conservation of Resources," July 26, 1923, 70503, BP.

86. Ibid.

87. Ibid.

88. Ibid.

89. For example, the gradual fall in flowhead pressures, the relation of flowhead pressures to rock pressure, and the deductions to be drawn from pressure observations as to locations best calculated to give prolonged production.

90. The history of categories constructed for the estimation of oil reserves and lifespan is unclear in AIOC's case. On similar cases in the United States, see Shulman, "Science Can Never Demobilize," 376. Also see Gary Bowden, "The social construction of validity in estimates of US Crude Oil Reserves," *Social Studies of Science* 15, no. 2 (1985): 207–240.

91. Cadman on "Gas Pressure Problems on Maidan Naftun Field and Conservation of Resources," July 26, 1923, 70503, BP.

92. Terms included *rock pressure, flowhead pressure, datum pressure, normal oil, transitional oil, gas, transitional zone, rock elevation*, etc. Formulas included one to determine "Zero Level" at any given time; the "Normal Oil Level" of any proved gas dome; the difference in "Normal Oil Level" between any two gas domes; the relation between "Gas Reservoir Pressures" and "Gas Flowhead Pressures," and so on.

93. Bowker, *Science on the Run*, 33.

94. For a similar argument, see De Laet and Mol, "The Zimbabwe Bush Pump."

95. De Böckh was a Hungarian and former Under Secretary of State for Mines who was retained by AIOC to conduct surveys that were the beginning of a systematic program designed to map, on a one-fourth-inch to a mile scale, unexplored parts of the concession area. He recommended six anticlines for testing in Iran, four of which were proved giant fields! See Owen, *Trek of the Oil Finders*, 1262–1263.

96. "Preliminary Report on the Principal Results of My Journey to Persia," by Dr. H. de Böckh, May 15, 1924, 1–93, 70501, BP.

97. Ibid.

98. Ibid.

99. D. Garrow to Fraser, March 21, 1929, 72437, BP.

100. Ibid.

101. Barry, "Technological Zones," 249.

102. D. Garrow to Fraser, March 21, 1929, 72437, BP.

103. "M.I.S. Crude Oil Reserves," Garrow to Cadman, October 25, 1929, 72437, BP.

104. Comins to Jameson and Cadman, November 7, 1929, 72437, BP. Like "proved reserves," "unproved reserves" change over time. Unproved reserves refer to oil and gas underground that cannot be readily accessed because the infrastructure is not in place, or it does not make financial sense, or the presence of the oil or gas is assumed but not confirmed.

105. As Latour says, while a piece of technoscience is being developed, what counts as "nature" to be included in a theory and what counts as "background conditions" to be factored out are very much up for grabs. Once a new theory or technology is accepted, then nature itself is often reconfigured. "Technoscien-, tists" will now claim that their theory had always described the way nature was. There was no way of saying a priori whether or not the bases for calculating reserve estimates were a better method. See Latour, *Science in Action*, 49.

106. Comins to Jameson and Cadman, November 7, 1929, 72437, BP.

107. Ibid.

108. Barry, "Technological Zones," 246.

109. "Report on Visit to Middle East Oilfields" by Dr. Nuttal (technical adviser to Petroleum Division), June 13, 1947, L/P&S/12/1195, IOR.

110. C. A. P. Southwell to Kemp, September 28, 1933, 8713, BP.

111. Other geological studies included were G. M. Lees, "Reservoir Rocks of Persian Oil Fields" and "Saltzgletscher in Persien"; J. V. Harrison, "The Geology of Some Salt-Plugs in Laristan (Southern Persia)"; J. A. Douglas, "A Marine Triassic Fauna from Eastern Persia" and "Contributions to Persian Paleontology."

112. Prior to the formation of this department as the central authority, the Ministry of Agriculture, Commerce, and Public Works (renamed the Ministry of Mines and Public Works after oil was first discovered in 1908) managed most problems organized around oil (e.g., property, exploration, geology surveys, the building and expansion oil facilities). See Agent for APOC to Minister of Mines and Public Works, June 5, 1911 (1329 lunar), 128, 240–014783/18/(0–217), INA; also see Agent for APOC to Ministry of Agriculture, Commerce and Public Works, April 3, 1924, 174, 240–014783/18/(0–217), INA. The organization of interests and concerns about oil at the state level occurred after Reza Shah came to power in 1926 and established the Bureau of Concessions and Oil within the Ministry of Finance. In the years immediately after the signing of the 1901 D'Arcy Concession until AIOC's establishment in 1909, most correspondence occurred directly between the First Exploitation Company and local political communities (e.g., Bakhtiyari khans, Shaykh Khaz'al), governors representing the central Qajar government, or ministers in Tehran.

113. "Memorandum on Jehangir, Petroleum Director," Elkington to London, October 25, 1933, 8713, BP.

114. Mayhew to Lefroy, October 11, 1933, 8713, BP.

115. Ibid.

116. Greenhouse to Fraser, January 27, 1934, 8713, BP.

117. "Information for the Persian Government: Article 13," Mayhew to ?,
March 22, 1934, 8713, BP.

118. Ibid.

119. "Director Production Department's Conference: Exploration and Exploitation—Iran," July 1949, 44627, BP.

120. Ibid.

121. Bowker, *Science on the Run*, 36. The company pursued a similar strategy
in organizing its labor regime along racial-technical lines. This would help limit
the possibility of oil workers using their technical knowledge to disrupt the flow
of oil from the wells, through the main pipeline, and out of the refinery.

122. Ibid., 46.

123. Ibid., 38.

124. A. E. Dunstan and George Sell, eds., *World Petroleum Congress Proceedings* 1 (1933): xix.

125. Ibid.

126. Bowker, *Science on the Run*, 13.

127. Ibid., 12.

128. John Cadman, "Science in the Petroleum Industry," *World Petroleum
Congress Proceedings* 1 (1933): 563–570, esp. 563. Michael Faraday isolated
"benzole (benzene)" in the course of studying gas from cracked oil (the breaking
up of hydrocarbon molecules by heat application to produce lighter fractions).
This technical invention was followed by James Dewar's work, which made
possible the commercial extraction of helium and low-temperature rectification
of liquefied gases.

129. Established in 1799 by British scientists, the Royal Institution in London
was devoted to scientific education and research.

130. In chemical and petroleum engineering, "unit operation" refers to one
converter or distillation column rather than to a transformation taking place in
a single factory or facility.

131. Cadman, "Science in the Petroleum Industry," 564.

132. Ibid., 566.

133. Ibid., 569.

134. Ibid.

135. Bowker, *Science on the Run*, 14.

136. For details on AIOC's in-house circulation of production reports in Iran
and London, see "Director Production Department's Conference: Exploration
and Exploitation—Iran," July 1949, 44627, BP.

137. On the "anatomy of the claim to universality," see G. Bowker, "How to Be Universal: Some Cybernetic Strategies, 1943–70," *Social Studies of Science* 23 (February 1993): 107–127.

138. Mitchell, *Rule of Experts*, chap. 1, passim. Mitchell has demonstrated this for the case of Egypt as technical experts tried to learn from the failures of dam building and the construction of fertilizer plants.

Chapter 3: Calculating Technologies in Crisis

1. Davar, the finance minister, made the announcement with the approval of the cabinet.

2. Ferrier, *History of British Petroleum*, 607.

3. Ibid.

4. Clive (British Legation, Tehran) to A. Henderson (FO), April 30, 1931, no. 4–16, enclosure no. 6–8 (translation), *Shafagh-i-Surkh*, April 24, 1931, L/P&S/12/3453, IOR.

5. Ferrier, *History of British Petroleum*, 642. Article 12 of the concession stipulated that "the workmen employed in the service of the Company shall be subjects of … the Shah, except the technical staff such as the managers, engineers, borers and foremen."

6. Ferrier, *History of British Petroleum*, 294.

7. Keddie, *Modern Iran*, 101.

8. Article 3 of the 1901 concession: "The above deduction (i.e. the refining and distributing allowances) shall be made from the total net profits of any Company … and if such deductions more than absorb the whole of the profit then any deficiency so caused shall not be carried forward and … shall not be set against the net profit in the case of any other Company."

9. Other companies included the First Exploitation Company, Scottish Oils Ltd., Société Générale des Huiles de Pétrole, L'Alliance S.A. Belge, Commonwealth Oil Refineries Ltd. (Australia), and Det Forenede Oile Komaniet A.S. (Denmark).

10. Beck, "The Anglo-Persian Oil Dispute," 125.

11. Ferrier, *History of British Petroleum*, 605.

12. Ibid., 601.

13. Ibid., 602.

14. The 1928 figure was based on 1 shilling (s.) 10 pence (d.) royalty per ton, whereas the 1927 figure was 5s. 10d. Until 1971, the old British currency system used pounds (£), shillings (s.), and pence (d.). There were 12d. to the shilling and 20s. or 240d. to the pound. The adoption of decimal currency in February 1971 replaced the old system such that 100p = £1 or 100p = 240d.

15. "Royalty," J. B. Lloyd, August 21, 1928, 1–4, 72176, BP.

16. Feroughi to Iranian minister, December 12, 1932, 88373, BP.

17. ?, 1928, 71074, BP.

18. Yergin, *The Prize*, 261.

19. Ibid. A price war was launched by William Deterding of Royal Dutch / Shell against Standard Oil of New York in retaliation for its purchases of Russian oil. AIOC also complained about the "phenomenally low freight rates and short routes" as well as the "most violent competition" between all the distributing companies for European markets, which made their position highly unstable. See Cadman to Teymourtache, August 7, 1931, in "Concession Revision and Royalty Negotiations, April 1929–May 1932," 7, 88373, BP.

20. Yergin, *The Prize*, 263–264.

21. See Nowell, *Mercantile States*, 237–239. The I. G. Farben chemical industry made a collaborative deal with Standard Oil concurrently with the "As Is" oil negotiations. Fears over oil shortages before the mid-1920s were linked with attempts to convert coal/lignite to oil, using the Bergius process as licensed by BASF in Germany to American and British corporations.

22. The system relied on "antimarket" mechanisms to secure profits and maintain prices at "the relatively high price at which oil was produced and sold in Texas" (Mitchell, "McJihad," 6, 8). To keep oil prices high, a state body known as the Texas Railroad Commission enforced a "system of rationing, by which the demand in any one month was shared among oil producers" (Sampson, *The Seven Sisters*, 76).

23. Yergin, *The Prize*, 204, 263–264. Sampson further explains that "the basis of the cartel was to be the maintenance of American prices." World oil prices were set by the "Gulf Plus System," which protected American oil by fixing the price of oil from "anywhere else … at the price in the Gulf of Mexico," from where most US oil was shipped abroad, "*plus* the standard freight charges for shipping the oil from the Gulf to its market." The system was "a blatant device to keep up prices," because American oil was threatened by cheaper oil from Iran and Venezuela. If AIOC sold cheap oil from Iran to a European country, for example, "the oil would be charged as if it came from the Gulf of Mexico … and the saving would make a large profit for the company" (Sampson, *The Seven Sisters*, 73–74).

24. See chapter 1 for a discussion of this point.

25. Ferrier, *History of British Petroleum*, 582. See also Edward Peter Fitzgerald, "Business Diplomacy: Walter Teagle, Jersey Standard, and the Anglo-French Pipeline Conflict in the Middle East, 1930–1931," *Business History Review* 67, no. 2 (1993): 207–245, esp. 207.

26. The new Iraqi government had first validated the oil concession claimed by TPC under the British mandate in March 1925.

27. Ferrier, *History of British Petroleum*, 585; Elwell-Sutton, *Persian Oil*, 46. Only after oil started to flow from southwest Iran did rivals such as Deutsche Bank pursue oil development in Mesopotamia (Mitchell, *Carbon Democracy*, 56).

28. In a joint venture with Gulf Oil, signed in 1934, AIOC gained a 50 percent share in the Kuwait Oil Company.

29. Ferrier, *History of British Petroleum*, 583. Consortium participants agreed that the British company should receive a 10 percent overriding oil royalty as compensation.

30. Iraqi oil was not exported until 1934. In March 1931, IPC renegotiated the 1925 agreement, granting itself an exclusive concession over the entire northeast of Iraq and tax exemptions in exchange for annual payments in gold until exports began. Oil revenues constituted 20 percent of government revenues on the signing of this agreement, compared to zero the previous year. See Charles Tripp, *A History of Iraq*, 2nd ed. (Cambridge: Cambridge University Press, 2002), 71.

31. On the role of technical devices, see Callon, "Some Elements of a Sociology of Translation," 67–94; also see Holm and Nielsen, "Framing Fish, Making Markets."

32. Beck, "The Anglo-Persian Oil Dispute," 126–127.

33. Ibid., 133.

34. Ibid., 147. The British government first became concerned with Anglo-Iranian relations in October 1932 in part because of its position in the Persian Gulf and the sanctity of international agreements, which, according to Beck ("The Anglo-Persian Oil Dispute," 147), "assumed far greater significance than oil interests." The shah was impatient with Britain's failure to acquiesce to his demands and began to search for a means of striking at British interests. AIOC interests offered an obvious target since the company constituted a British interest and the British government was the majority shareholder in the company. At the time, the shah was working to rid the Persian Gulf of British influence and this caused "friction with Britain" over Bahrain and Henjam. A series of "Persian notes" refusing recognition of the British protectorate over Bahrain and demanding British evacuation of Henjam resulted in "the Cabinet displaying an interest in Anglo-Persian relations." On November 2, 1932, prior to the cancellation decree, the British cabinet made the decision to adopt a firm line against Iran's demands on Bahrain and Henjam, while the company attempted to keep the two issues separate. On this latter point, see Beck, "The Anglo-Persian Oil Dispute," 127–128, 130. Also see no. 34128, "Copy in Translation. From the Persian Minister for Foreign Affairs Addressed to His Majesty's Minister, dated 21st Azer 1311 (12th December 1932)," 88373, BP.

35. Beck, "The Anglo-Persian Oil Dispute," 130–131.

36. Ibid., 145.

37. Ibid., 131–132.

38. Anghie, *Imperialism, Sovereignty, and the Making of International Law*, 209.

39. Beck, "The Anglo-Persian Oil Dispute," 133.

40. Under Optional Clause, Article 41 of the Statute. League of Nations Official Journal, *Annex 1419c: Memorandum by His Majesty's Government in the United Kingdom*, December 1932, 2299.

41. The Court was replaced by the International Court of Justice (ICJ) in 1946 when the United Nations was established.

42. Beck, "The Anglo-Persian Oil Dispute," 136.

43. Ibid., 137.

44. Ibid., 135–136.

45. League of Nations Official Journal, *Annex 1419c: Memorandum by His Majesty's Government in the United Kingdom*, December 1932, 2299–2308. Also, see "Notes on Points Raised by Persian Government—12.12.1932," December 1932, 88373, BP. AIOC was asked by the Foreign Office to address various points concerning Iranian government grievances as well as the potential for disturbances to arise to company property and personnel in the oilfields during the interim period of litigation between the two governments, with the notes intended for the use of British government representatives in Geneva.

46. Ibid.

47. Ibid. According to Ford, "it is an elementary principle of international law that a state has the right to protect its nationals when they have been injured by the internationally illegal conduct of another state." Further, "a state can interpose on behalf of a corporation incorporated under its own laws, the nationality of the corporation being derived from the place of incorporation." See Alan W. Ford, *The Anglo-Iranian Oil Dispute of 1951–1952: A Study of the Role of Law in the Relations of States* (Berkeley: University of California Press, 1954), 181.

48. League of Nations Official Journal, *Optional Clause Recognizing the Court's Jurisdiction as Described in Article 36 of the Statute*, January 1933, 17.

49. Ibid. The resolution referred to the *Optional Clause Recognizing the Court's Jurisdiction, as Described in Article 36 of the Statute*.

50. League of Nations Official Journal, *Annex 1422b: Memorandum from the Persian Government to the Secretary General of the League of Nations Regarding the Dispute between the United Kingdom and Persia in Connection with the Concession Held by APOC*, no. 2, January 1933, 288–302.

51. Ibid.

52. Ibid. In a letter to Teymourtash on royalty questions, Cadman wrote that "like you, I am anxious to avoid arbitration and very much prefer to settle all questions by mutual agreement." See Cadman to Teymourtache, August 7, 1931, in "Concession Revision and Royalty Negotiations, April 1929—May 1932," 9, 88373, BP.

53. League of Nations Official Journal, *Annex 1422b*, 295.

54. In this early period, the infrastructure, rules, and precedents of international law regarding concession contracts and the sovereignty of states were emerging

as a kind of metrological order, "a gigantic enterprise to make of the outside a world inside which facts and machines can survive" (Latour, *Science in Action*, 250–251).

55. Anghie, *Imperialism, Sovereignty, and the Making of International Law*, 204, 213. Anghie explains that "many of the controversies regarding the impact of new states on the rules of international law emerged in pointed form in disputes generated by the doctrine of state responsibility as related to the protection of foreign investment" (p. 209). Third World countries such as Iran argued that "nationalization was to be determined according to national rather than international standards, thus attacking ... the rules of state responsibility relating to foreign investment" (p. 213).

56. Beck, "The Anglo-Persian Oil Dispute."

57. For a discussion of Latour's approach to law as law by association, see Ron Levi and Mariana Valverde, "Studying Law by Association: Bruno Latour Goes to the Conseil d'État," *Law & Social Inquiry* 33, no. 3 (2008): 805–825, esp. 806.

58. Latour, *Science in Action*, 254.

59. League of Nations Official Journal, No. *3215: Dispute between the United Kingdom and in Regard to the Concession Held by the Anglo-Persian Oil Company*, February 1933, 198–199.

60. Ibid.

61. Ibid.

62. Ibid.

63. Ibid. For example, the Iranians had not taken any economic risks in exploiting the concession and it was a baseless accusation that the company was seeking to limit profitable production of oil in Persia, etc.

64. League of Nations Official Journal, No. *3215: Dispute between the United Kingdom and in Regard to the Concession Held by the Anglo-Persian Oil Company*, February 1933, 201.

65. Robert Vitalis, *America's Kingdom: Myth-Making on the Saudi Oil Frontier* (Stanford, CA: Stanford University Press, 2006), 19.

66. League of Nations Official Journal, No. *3215: Dispute between the United Kingdom and in Regard to the Concession Held by the Anglo-Persian Oil Company*, February 1933, 205.

67. Ibid., 207.

68. Ibid., 208.

69. League of Nations Official Journal, No. *3206: Dispute between the United Kingdom and Persia in Regard to the Concession Held by the Anglo-Persian Oil Company: Appointment of a Rapporteur*, no. 2, January 1933, 192.

70. League of Nations Official Journal, No. *3237: Dispute between the United Kingdom and Persia in Regard to the Concession of the Anglo-Persian Oil Company (continuation)*, February 1933, 252–253.

71. Ibid.

72. Ibid., 253.

73. League of Nations Official Journal, No. 3284: Dispute between the United Kingdom and Persia in Regard to the Concession of the Anglo-Persian Oil Company, and Letter from the Persian Government to the Secretary General of the League of Nations, July 1933, 827–828, 996.

74. Anghie, Imperialism, Sovereignty, and the Making of International Law, 224.

75. Ibid., 209.

76. Beck, "The Anglo-Persian Oil Dispute," 142.

77. It should be noted here that events on the ground in the oilfields were constantly changing in spite of the apparent stability of legal proceedings in the international arena. The company was increasingly anxious about the land around the Abadan refinery, as it feared a military response from the government. This in turn would provoke general disturbances on the island and endanger refining operations. See Elkington to London, January 27, 1928, 68419, BP. The recent completion of construction work at Abadan had also created the problem of unemployed surplus labor, leading to "mob action" and crowds gathering outside AIOC's labor office with a "hostile attitude" and stoning the office. See Jacks to Teymourtache, March 14, 1928, 68419, BP. There were rumors in March 1928 about the company's intention to fire 10,000 Iranians and replace them with Indian and Iraqi laborers. As a result, more than 2,000 Iranian workers gathered in front of the company's labor office in Abadan and threw stones at it. See Bayat, "With or without Workers in Reza Shah's Iran," 117. Also, disturbances among Arabic-speaking nomadic groups and local farmers, who inhabited and worked the land that was suddenly transformed into the oil regions, emerged in their refusal to pay taxes to the state on their date crops. See "Revolt in Persia," The Times, January 19, 1928, 68419, BP.

78. "The Oil Prices in Persia," Mostafah Fateh to APOC, 1932–33, enclosure A, 88373, BP.

79. Ibid.

80. "A Basis of Royalty Payments," 1933, Appendix 3, 70223, BP.

81. Bamberg, History of British Petroleum, 47.

82. "A Basis of Royalty Payments," 1933, Appendix 3, 70223, BP.

83. Ibid.

84. For illustrations of the other two schemes, see "A Basis of Royalty Payments," 1933, Appendix 3, 70223, BP.

85. See "Scheme 1. Draft Clause," "Scheme 2. Draft Clause," and "Investigation of the Problem in General Terms," n.d., 70223, BP.

86. Ibid.

87. Bamberg, History of British Petroleum, 45–47.

88. Ibid., 50.

89. By 1925, the new shah implemented his tribal policy to ethnically homogenize the state through a string of military campaigns against semiautonomous groups threatening his power, particularly in the oil regions of Khuzistan. He ousted Shaykh Khaz'al, confiscated oil shares controlled by the rival Bakhtiyari confederation, and forced its chiefs to sell all their land and disarm. See Abrahamian, *Iran between Two Revolutions*, 120; Cronin, "Politics of Debt."

90. Latour, *Reassembling the Social*.

91. I am building on Michel Callon's concept of translation (see Callon, "Some Elements of a Sociology of Translation," 68).

92. Accounts of the concession crisis of 1932–1933 in Iran mention the dispute only in passing. See Katouzian, *The Political Economy of Modern Iran*; Keddie, *Modern Iran*; Abrahamian, *Iran between Two Revolutions*; Peter Avery, G. R. G. Hambly, and C. P. Melville, eds., *The Cambridge History of Iran*, Volume 7: *From Nadir Shah to the Islamic Republic* (Cambridge: Cambridge University Press, 1991), 643–648.

93. Hughes, *Networks of Power*, 33–34.

94. "Persian Concession, 1933–1993," 10, 70223, BP.

95. The Iranian parliament voted to reopen negotiations concerning Article 16 on Persianization, along with the rest of the terms of the 1933 concession, in 1947.

96. *Shafagh-i-Surkh*, April 22, 1931, and April 23, 1931, no. 12–16, L/P&S/12/3453, IOR.

97. "Persian Concession. Article 16—Personnel in Persia," November 16, 1933, 1–5, 52889, BP.

98. Ibid.

99. "Details Estimated Costs and Notes on the Company's General Plan," 1934, 52890, BP.

100. Or at least the recognition of common terms like *steam oil*, and the use of various measuring devices such as thermometers, hydrometers, gauges, and the operation of boilers and pumps. See "Persian Concession. Article 16—Personnel in Persia," November 16, 1933, 1–5, 52889, BP.

101. "D—Organization of the Oil Co. Apprentice Trainees" in "List of Proclamations and Pamphlets Published during the Recent Strike and Disturbances," March/April1951, 68908, BP.

102. "Report on Matters Relevant to Article 16 Covering Period May 1933 to December 1934" by Elkington, 1934, 52885, BP.

103. Elkington to Gass, August 20, 1935, 52885, BP.

104. Ibid.

105. Ibid.

106. On the history of AIOC's racial organization of living and working conditions in the oil regions of southwest Iran, see chapter 4.

107. "General Plan (2 April 1936)," July 16, 1936, 126413, BP.

108. See graph, "Percentage of Foreign to Total Personnel," in "General Plan," July 16, 1936, 126413, BP. The graph shows an overall decline in the proportion of foreign personnel to total personnel (including unskilled labor) from 14 percent in January 1932 to about 8.5 percent in August 1935. Not including unskilled labor, the value shifts from approximately 27 to 17.5 percent for the same period.

109. Neveen Abdelrehim, Josephine Maltby, and Steven Toms, "Imperialism, Employment, and Racial Discrimination: The Anglo-Iranian Oil Company, 1933–51," paper presented at the Business History Conference, University of Georgia, Athens, March 2010, http://www.thebehc.org/annmeet/program10.html#E5, 3, 7.

110. See "Report on Visit to Tehran 31st August to 26th October 1948," 36, 126407, BP, cited in Abdelrehim, Maltby, and Toms, "Imperialism, Employment, and Racial Discrimination," 30. The latter series of negotiations pressing for Persianization had occurred in the context of AIOC's attempts to renegotiate its oil drilling rights with the ruling government.

111. As noted earlier, Reza Shah changed the country's name from Persia to Iran in 1935. The terminology used in British archival documents after 1935 notes this shift by referring to "Persianization" as "Iranianization." For consistency, I have kept the term as "Persianization."

112. Abdelrehim, Maltby, and Toms, "Imperialism, Employment, and Racial Discrimination," 31–32.

113. Ibid., 33–34.

114. Timothy Mitchell, "Culture and Economy," in *The Sage Handbook of Cultural Analysis*, ed. Tony Bennett and John Frow, 447–466 (London: Sage, 2008), esp. 458–459.

115. Social histories of Iran's oil workers for this period focus on the question of unionization, political ideology, and degree of organization without considering how oil's technical properties and the terms of the concession contract concerning Persianization shaped disputes. See for example, Bayat, "With or without Workers in Reza Shah's Iran"; Fred Halliday, "Iran: Trade Unions and Working Class Opposition," *MERIP Reports*, no. 71 (1978): 7–13. A conference paper has investigated more carefully the terms of Article 16 of the 1933 concession, but still reads the problem as a mere reflection of the "negative attitude" and "anti-Iranian" sentiment of AIOC managers (Abdelrehim, Maltby, and Toms, "Imperialism, Employment, and Racial Discrimination"). On the specificity of the oil worker as compared with coal workers and the consequences for democratic forms of politics, see Mitchell, "Carbon Democracy."

116. On the work of technical devices see Michel Callon, Yuval Millo, and Fabian Muniesa, eds., *Market Devices* (Malden, MA: Blackwell, 2007). Also see Callon's introduction in Callon, *Market Devices*. He discusses how economics helps format the world it claims to describe.

117. Mitchell, *Rule of Experts*, chap. 1, passim.

118. Ibid., 34.

119. Callon, "Some Elements of a Sociology of Translation," 71, 74–75.

Chapter 4: What Kind of Worker Does an Oil Industry Require to Survive?

1. The majority of unskilled and skilled workers (excluding managerial and engineering staff who were British) were recruited locally from Arab nomadic groups in Khuzistan and from Bakhtiyari nomads of the surrounding mountains, or in the case of skilled workers, from Isfahan and Tehran. Technical and clerical workers were brought from India, especially in the early years. According to Halliday, who is citing Charles Issawi, oil production created the "first substantial section of the Iranian proletariat" from Iranian migrant workers in the oilfields and towns of the Caucasus in southern Russia at first, and subsequently in the oilfields of Iran. In 1915, there were 13,500 Iranians in the oil workforce of Russian Azerbaijan, and by 1920, 20,000 Iranians were employed by the oil company in southwest Iran (Halliday, "Iran," 9).

2. For national and local histories of Iran's oil workers in AIOC, see Danesh Abbasi-Shahni, *Tarikh-i Masjid Suleiman (History of Masjid Suleiman)* (Tehran: Hirmand Press, 1995); Touraj Atabaki, "Far from Home, but at Home: Indian Migrant Workers in the Iranian Oil Industry," *Studies in History* 31, no. 1 (2015): 85–114; Atabaki, "From 'Amaleh (Labor) to *Kargar* (Worker)"; Kaveh Bayat, "With or without Workers in Reza Shah's Iran"; Patrick Clawson, "Capital Accumulation in Iran," in *Oil and Class Struggle*, ed. Petter Nore and Terisa Turner, 143–171 (London: Zed Press, 1980); Kasravi, *Tarikh-i Pansad Sal-i Khuzistan*; Habib Ladjevardi, *Labor Unions and Autocracy in Iran* (Syracuse, NY: Syracuse University Press, 1985). For scholarship on oil workers in the Middle East, see Joe Stork, "Oil and the Penetration of Capitalism in Iraq," in *Oil and Class Struggle*, ed. Petter Nore and Terisa Turner, 172–198 (London: Zed Press, 1980); Vitalis, *America's Kingdom*.

3. Here, I build on Vitalis's definition of paternalism (Vitalis, *America's Kingdom*, 19–21).

4. On labor and race in the operations of firms, see Dobe, *A Long Slow Tutelage in Western Ways of Work*; Philip J. Mellinger, *Race and Labor in Western Copper: The Fight for Equality, 1896–1918* (Tucson: University of Arizona Press, 1995); Carlos A. Schwantes, *Vision & Enterprise: Exploring the History of Phelps Dodge Corporation* (Tucson: University of Arizona Press, Phelps Dodge Corporation, 2000); White, *Railroaded*.

5. T. A. B. Corley, *A History of the Burmah Oil Company 1886–1924* (London: Heinemann, 1983), 146–164.

6. Mitchell, "Carbon Democracy," 413. Mitchell is discussing the first set of connections between fossil fuels and democracy, in which the mobilization of new political forces—mass political movements—depended in part on the organization of the flow of unprecedented amounts of nonrenewable fossil fuels such

as coal and oil. To suggest that more democratic forms of energy production are possible, I am arguing that oil workers, oil-producing states, and public opinion, for example, would have a say in the decision-making process of designing the industry and its markets. Thus, the enormous amounts of wealth generated from technical operations of oil production, distribution, and marketing would be tied to the well-being of large numbers of people. See Mitchell, *Carbon Democracy*, 146.

7. Mitchell, "Carbon Democracy," 407–413.

8. Podobnik, *Global Energy Shifts*, 47–49.

9. Ibid., 47.

10. Vitalis, *America's Kingdom*, 20.

11. Mark Crinson, "Abadan: Planning and Architecture under the Anglo-Iranian Oil Company," *Planning Perspectives* 12, no. 3 (1997): 341–359, esp. 342–343. The new neighborhood was built in 1912 and first named "Coolie Lane," then renamed "Sikh Lane," and finally changed to "Indian Lane." See Atabaki, "Far from Home," 102.

12. Williamson, *In a Persian Oil Field*, 14.

13. Crinson, "Abadan," 346.

14. Ibid., 342.

15. Ibid., 345–346.

16. Crinson, "Abadan," 342. Also see Mark Crinson, *Modern Architecture and the End of Empire* (Burlington, VT: Ashgate, 2003).

17. Cracking oil refers to a refining process that involves the physical and chemical breakdown of heavy carbon chains of crude oil into lighter fractions to produce multiple petroleum byproducts such as gasoline.

18. Crinson, "Abadan," 346.

19. Mitchell, *Colonising Egypt*, 163–164. As Mitchell has argued with regard to the colonial ordering of Cairo, both "economically and in a larger sense, the colonial order depended upon at once creating and excluding its own opposite."

20. "Economic Report No. 8: Persian Employees of the Anglo-Persian Oil Company" by B. Temple (Major, Commercial Secretary, HBM Legation, Tehran), March 11, 1922, L/P&S/11/224, IOR. See Trevor to Des Bray, October 18, 1922, ibid.

21. Atabaki, "Far from Home," 88–89. Shaw Wallace & Co. Ltd. was the agent recruiting labor in India for AIOC operations in Iran.

22. Secretary of Government of India (Foreign and Political Department) to Chairman of the Board of AIOC, November 25, 1920, 68731, BP. On the suspension and reinstatement of the Emigration Act, see Atabaki, "Far from Home," 92–93, 97–98.

23. Ibid.

24. Ibid.

25. "Appendix to Correspondence: Abadan Refinery Strike Commencing 14 March Terminating 22 March, 1922," n.d., ibid. Complaints of "ill-treatment" of Indian labor against the company triggered the Emigration Act's subsequent reinstatement. Garrow (AIOC) to Secretary of India, December 21, 1920, 68731, BP. The escalation of labor protests by Indian workers in 1922 triggered the decision to amend the Emigration Act "to end the practice of indentured labor, extensively practiced during the War." AIOC could only recruit migrant labor for a maximum period of one year, and this created labor shortages and costs (Atabaki, "Far from Home," 97).

26. See Dr. Ghore, "Indian Workers in Persia (Miserable Conditions)," *Bombay Chronicle*, January 10, 1922, in Trevor to Des Bray, March 15, 1922, enclosure 3, L/P&S/11/213, IOR.

27. General Committee of the Workmen (Indians) to L. F. Bayne (Joint Works Manager, AIOC-Abadan), March 10, 1922, L/P&S/11/213, IOR.

28. "Report by HBM's Consul H. G. B. Peel, Ahwaz" in Trevor to Des Bray, March 15, 1922, enclosure 5, L/P&S/11/213, IOR.

29. Ibid.

30. Ibid.

31. Cox to Lorimer, March 9, 1909, 71994, BP.

32. Kasravi, *Tarikh-i Pansad Sal-i Khuzistan*, 211.

33. Trevor to Des Bray, March 28, 1922, L/P&S/11/224, IOR.

34. The documents do not explain where the Chittagonian laborers came from, but the Chittagonian language is spoken today by people in southeastern Bangladesh. Perhaps it is safe to assume that they arrived in Iran from the northeast region of India known as Bengal. Touraj Atabaki explains that the majority of migrant workers recruited to the British-controlled oil industry in Iran from Burma were "Indians" employed by the Burmah Oil Company. These workers were "mainly Chittagonian Sunni Muslims" who had gained experience working in Burma's oil industry since the 1890s. Other workers included Punjabi Sikhs. See Atabaki, "Far from Home," 90, 96.

35. Cox to Lorimer, March 9, 1909, 71994, BP.

36. Trevor to Des Bray, March 28, 1922, L/P&S/11/224, IOR.

37. Nichols (Managing Director, AIOC) to General Managers at Strick, Scott, and Co., Ltd., March 28, 1922, 54496, BP.

38. Viceroy to Secretary of State, March 27, 1922, BP 54496.

39. Political Resident (Persian Gulf, Bushire) to Foreign Secretary of Government of India in Foreign and Political Department, March 27, 1922, no. 14, 54496, BP.

40. Ibid.

41. J. Rieu (Commissioner in Sind) to Secretary of Government of Bombay in Home Department, April 13, 1922, 54496, BP.

42. Bayat, "With or without Workers," 112.

43. I am drawing on Barry's concept of "technological zone" in his study of the global oil industry (see Barry, "Technological Zones," 239–253).

44. Dobe, *A Long Slow Tutelage in Western Ways of Work*, 36.

45. Young to Finlay, August 12, 1911, 159–161, 70335, BP.

46. H. E. Nichols to R. I. Watson, May 9, 1921, 54506, BP, cited in Atabaki, "Far from Home," 97.

47. Nichols to General Managers at Strick, Scott, and Co., Ltd., March 28, 1922, 54496, BP. AIOC officials prevented Indian laborers from getting rehired by other oil companies, such as the Burmah Oil Company in Rangoon.

48. "Appendix to Correspondence: Abadan Refinery Strike Commencing 14 March Terminating 24 March 1922," n.d., L/P&S/11/224, IOR.

49. Bayat, "With or without Workers," 112.

50. Crinson, "Abadan," 346.

51. Ibid., 348.

52. Crinson, *Modern Architecture and the End of Empire*, 62–63.

53. The decision was made at the party's second congress, held in December 1927 (Bayat, "With or without Workers," 114). It must be noted that the Iranian government, led by Reza Shah, banned all trade unions starting in 1927. Many communist and socialist labor organizers were arrested in cities across Iran such as Abadan. See Abrahamian, *Iran between Two Revolutions*, 139.

54. Bayat, "With or without Workers," 115. Through the collaboration of the communist and socialist parties, the Central Council of Federated Trade Unions (CCFTU) was established in 1921 and successfully organized over thirty unions throughout Iran (Abrahamian, *Iran between Two Revolutions*, 129).

55. Ibid., 115, 118.

56. Prior to Reza Khan's rise to power in the 1920s, it had been advantageous for AIOC to strengthen semiautonomous political communities in southwest Iran in the event of the ruling Qajar government's collapse. See Cox to Butler, 1908, no. 148, L/P&S/10/132, IOR. The British government dramatically shifted its strategy after 1921, engineering the installation of Reza Khan and a strong centralized government to suppress any further threats from Iran's provinces, especially Khuzistan. Formerly a Cossack Brigade member, Reza Khan changed his name to Reza Pahlavi and formalized the execution of the British-supported coup by ordering the constituent assembly to depose the Qajar government in 1925.

57. "'Note' Chairman's Folder on Labor Agitations," and "Telegrams Re Labour Agitations in Persia," May 28, 1929, 1–70, 68899, BP. Atabaki says there were approximately 9,000 workers (Atabaki, "Far from Home," 109).

58. Elkington to Greenhouse, May 4, 1929, 59010, BP.

59. Abadan to Chairman, May 4, 1929, 59010, BP.

60. Ibid.

61. "Extract from 'Hablul Matin' Paper No. 20 & 21: 'Strike in Abadan,'" June 4, 1929, 59010, BP. Also see Atabaki, "Far from Home," 109.

62. Abadan to Cadman, May 6, 1929, 59010, BP.

63. Abadan to Cadman, May 7, 1929, 59010, BP.

64. Jacks to Cadman, December 23, 1929, 59011, BP.

65. *Shafagh-i Sorkh*, Mordad 1309 (August 1929), Library of the Majlis (Iranian parliament).

66. "Report by Mohammad Hassan Bade'a to the Foreign Ministry on the Abadan Oil Workers' Agitation due to Discrimination amongst Them and the Indian Workers by way of the Anglo-Persian Oil Company," no. 36, 24 Ordibihisht 1308 (May 14, 1929), cited in *Naft Dar Durah-i Reza Shah* (Tehran: Sazman-i Chap va Intisharat-i Vizarat-i Farhang va Irshad-i Islami, 1378), 107–111.

67. "A Report by the Head of Khoramshahr's Post and Telegraph, Office on the Abadan Disturbances," no. 1026, 16 Ordibihisht 1308 (May 6, 1929), *Naft Dar Durah-i Reza Shah*, 99–100.

68. "Extract from 'Hablul Matin' Paper No. 20 & 21: Strike in Abadan," June 4, 1929, 59010, BP.

69. Ibid.

70. Ibid. Also see Cadman to Nichols, March 25, 1926, 72344, BP.

71. Kaveh Ehsani, "Social Engineering and the Contradictions of Modernization in Khuzestan: A Look at Abadan and Masjed-Soleyman," *International Journal of Social History* 48, no. 3 (2003): 361–399, esp. 385. Thus, as more Iranian labor was accommodated "inside" operations, there was continuity from the 1920s in terms of the racially segregated layout of facilities but also and simultaneously along technical (occupational) lines. The professionally planned layout of residential estates was introduced in the 1930s. See Crinson, "Abadan," 345.

72. Ehsani, "Social Engineering," 362.

73. Prior to the 1929 strike, AIOC's "Central Official Relations Department" served as the "final authority" on labor matters, managing disputes among employees and petitions and complaints to local authorities. The department ensured that "blacklisted men" were not reengaged in other oil operations abroad. A "Finger Print Bureau" worked to deter the employment of "undesirables," to identify and provide information on laborers previously engaged by the company at other centers, and to "bring deviant acts to the attention of local authorities" by monitoring labor movements, propaganda, and Soviet activities. The "Labor Department" was responsible for discharging labor, training artisans, testing workers into various grades, ensuring standard wages and accommodations for each class and grade of labor, and receiving petitions, complaints, and applications for work. See "Report on Relations between Company and Its Labour and Suggestions for Improvement of Existing System," May 1, 1929, 59011, BP.

74. Vitalis, *America's Kingdom*, 23. On petroleum advertising in the interwar years, see R. W. Ferrier, "Petrol Advertising in the Twenties and Thirties," *Journal of Advertising History* 9, no. 1 (1986): 29–51. On the first company film advertising AIOC's achievements in Iran and distributed to audiences worldwide beginning in the 1950s, see Mona Damluji, "The Oil City in Focus: The Cinematic Spaces of Abadan in the Anglo-Iranian Oil Company's Persian Story," *Comparative Studies of South Asia, Africa and the Middle East* 33, no. 1 (2013): 75–88.

75. Crinson, "Abadan," 347.

76. Ibid., 346.

77. Ibid., 347.

78. Crinson, "Abadan," 351. The architectural design of Bawarda was inspired by Lutyens's remodeling of the garden city in New Dehli (Crinson, *Modern Architecture and the End of Empire*, 66).

79. Ibid.

80. Ibid.

81. Crinson, *Modern Architecture and the End of Empire*, 67.

82. Crinson, "Abadan," 357.

83. Ibid. It is interesting to note that even after the establishment of an international oil consortium in 1954, British architectural work resumed, and in 1959, Fry, Drew, and Partners designed a new town at Gach Saran for the consortium. Wilson also returned to work for the consortium, but for political reasons, formed a joint practice with an Iranian architect, Aziz Farmanfarmanian, an aristocrat whose brother, Manouchehr, had worked as an engineer at AIOC (Crinson, *Modern Architecture and the End of Empire*, 71).

84. Ibid., 349. Crinson says that the plan for these estates originated in the 1920s, when APOC addressed the threat of "social disorder posed by the overcrowded 'town.'" This "nuclei of small towns," rather than large townships, would help reduce political activism by creating distance between them and the center of town (Crinson, *Modern Architecture and the End of Empire*, 64–65).

85. "Visit to the Operations of the Anglo-Iranian Oil Company in Iran. Spring 1945," by A. Hudson Davies, July 4, 1945, LAB 13/519, BNA. The report was circulated to the Ministries of Labor and Fuel and Power.

86. Ibid.

87. Ibid.

88. Ibid.

89. Ibid.

90. See "Social and Municipal Development in Abadan and the South Persian Oilfields," Fraser to Chancellor Dalton, August 28, 1946, 43762, BP.

91. Dobe, *A Long Slow Tutelage in Western Ways of Work*, 63, 73. Over the course of nine years, enrollment in the degree program increased from 50 to 150 students.

92. Fraser to Jacks (extract), December 7, 1933, Article 16—Allocation of the Grant (Iran), 52887, BP, cited in ibid., 66. Workers were not given "first class status" on their return from abroad and even the granting of "second class status" was considered dangerous.

93. "Visit to the Operations of the Anglo-Iranian Oil Company in Iran. Spring 1945," by A. Hudson Davies, July 4, 1945, 13/519, LAB. The report was circulated to the Ministries of Labor and Fuel and Power.

94. Ibid. The British workers had their own groups that also expressed their grievances about lack of accommodations for spouses: the Abadan Shift Society, which was an association of men who worked on shift at Abadan, mainly in the Process Department. Another group was the Association of Scientific Workers, which included technical recruits from the United Kingdom, especially chemists (see "Notes on Other Abadan Organizations for Employees" from the Davies report "Visit to the Operations of the AIOC in Iran, February–May 1945," no. 69–111, 371/45483, FO).

95. Halliday, "Iran," 8. The initials in English were CCUTU but known in Farsi as *Shorayet Motahedi Markazi*. There were 400,000 members by 1945 with 186 unions affiliated with it.

96. Tehran to FO (Cabinet Distribution), May 7, 1946, 371/52713, FO.

97. Halliday, "Iran," 8. On clashes with the British-backed Arab Union during the strike, see Ladjevardi, *Labor Unions and Autocracy in Iran*, 133–134.

98. Tehran to FO (Cabinet Distribution), May 7, 1946, no. 69–70, 371/52713, FO. Ladjevardi also says this event marked the first meeting between AIOC and the Khuzistan Workers and Toilers Union when Mehdi Hashemi-Najafi, correspondent for *Zafar* and *Rahbar* (Tudeh-backed newspapers), demanded a tour of operations and residential areas. When the company refused to negotiate with Hashemi-Najafi, the union decided to demonstrate on May Day. The strike ended May 6 when the company agreed to reinstate locomotive strikers and provide full pay for strike days. The union in turn agreed not to dismiss the British superintendent. See Ladjevardi, *Labor Unions and Autocracy in Iran*, 124–126.

99. "General Strike—Effect on Production," 1946, 130264, BP.

100. See the document signed by Abdul Hussein Mukhless in "Miscellaneous Documents Relating to Strike, June 1946," 130263, BP. Also see Ladjevardi, *Labor Unions and Autocracy in Iran*, 133.

101. "Miscellaneous Documents Relating to Strike, June 1946," n.d., 130263, BP. (1) The Workers' Commission should select a representative for each oil region by the oil workers of AIOC and not the company to deal only with the complaints and petitions of workers in a specific work region. (2) The AIOC had raised the wages from 14 to 35 rials without negotiating with the representative of the Workers' Commission, and the ration coupons (daily value of 6 rials) were no longer provided. The council was against the pay raise because it did not benefit the wages of unskilled workers in real terms, and had to

establish a minimum wage. Another strike was therefore necessary to achieve this. (3) The Workers' Commission should supervise the food distribution center. (4) AIOC had no right to discharge, fine, or deport a worker without the approval of the commission. (5) Commission meetings should occur twice per week. (6) A map for laying out a railway to transport the workers from their homes to the oil regions would also be submitted to the commission. (7) Funds were necessary to organize the new contract laborers hired by the company. (8) The labor representative's request had led to the approval by the Workers' Commission of a 30 percent pay raise for contract workers until the Labor and Insurance Bill was implemented.

102. "Memorandum on Tudeh Party Activities amongst AIOC Labour March/May 1946," by Elkington, n.d., no. 79–81, 371/52713, FO.

103. Le Rougetel to Foreign Office (Cabinet distribution), May 13, 1946, no. 73–74, 371/52713, FO.

104. Ibid. The party made the following demands on their behalf: (1) a 100 percent increase in wages, (2) double pay for overtime, (3) an annual one-month holiday with pay, and (4) a Friday holiday with pay. Also see Ladjevardi, *Labor Unions and Autocracy in Iran*, 131.

105. "Memorandum on Tudeh Party Activities amongst AIOC Labour March/May 1946," by Elkington, n.d., no. 79–81, 371/52713, FO.

106. "Report on Delegation to Persia" by J. Jones, June 5, 1946, 16256, BP. Ladjevardi also explains that Jones, though unsympathetic toward Iranian labor, shared the minority view with the oil refinery manager, and the labor manager that the trade union movement emerged because of a lack of improvement in working conditions and was not inspired by outside (Soviet) influences, as the British government and company liked to claim (Ladjevardi, *Labor Unions and Autocracy in Iran*, 137).

107. Bevin to Sargent, June 23, 1946, no. 80–81, 371/52715, FO. It is interesting to note that Bevin was himself a socialist, one of the architects of the British welfare state, former head of the Transport and General Workers' Union, and the son of a mine worker. This may have shaped his somewhat critical view of AIOC, although did not do too much to improve his broader understanding of the problems of colonialism, as Louis points out (see William Roger Louis, *The British Empire in the Middle East, 1945–1951* (Oxford: Clarendon Press, 1984), 4–5, 48–50, 633–634).

108. "Report on Delegation to Persia" by J. Jones, June 5, 1946, 16256, BP.

109. Ibid.

110. British Members of Parliament to Bevin (Secretary of State for Foreign Affairs), June 19, 1946, 43762, BP. Also see Ladjevardi, *Labor Unions and Autocracy in Iran*, 137–138. Ladjevardi argues that AIOC officials were aware that they would receive little support from a Labor government for their repression of the oil workers' union "if they admitted that the union's demands were non-political." Thus, framing the workers' union as Soviet inspired was a calculated tactic deployed by AIOC managers. Ladjevardi also notes that the

Foreign Office based its analysis of the situation in the oilfields on misinformed reports spread by Colonel Underwood's informers at the British consul in Khorramshahr.

111. The law recognized the bargaining power of trade unions, but as Ladje-vardi explains, the provisions of the law provided the government with the legal means to intervene in industrial relations, "reducing the influence of organized labor, and preventing it from representing the workers before the employers and the state." With the national government as spokesperson, this would eliminate the need for a militant trade union. See Ladjevardi, *Labor Unions and Autocracy in Iran*, 62.

112. "Labour Bill" enclosure, Northcroft to Jones, May 21, 1946, 53012, BP.

113. It also listed holidays and dealt with working conditions for children, contracts, hygiene, loss of employment, unions, wages, dispute settlement, and the appointment of a "high labor council" to supervise and draft labor laws. Clause 6 of the law stipulated that "if the employer does not pay his workmen wages for Fridays, he is bound to fix their weekly wages in such a manner that the wages paid them for six days shall be equivalent to wages for seven days." Clause 24 stipulated that "the minimum wages of a workman shall be determined in accordance with conditions in each different part of the country, in such manner as to meet the expenses of living of himself and his family. ... The minimum rate of wages in each part of the country shall be fixed by the Board mentioned in Article 34 at the beginning of each year for a period of one year, and shall be enforced with the approval of the High Labour Council." See "B. Diary of Events (14/18 July 1946)," July 1946, 130264, BP.

114. Cooke to Elkington, June 14, 1946, 16256, BP.

115. Further negotiations took place between Britain's visiting labor attaché and the Labor Department in Tehran, including CCUTU leaders and representatives. Ladjevardi says it is certain that the agreement reached on setting a minimum wage was reached without the participation and consent of the oil workers' union in Abadan (Ladjevardi, *Labor Unions and Autocracy in Iran*, 132).

116. Enclosure to "Aide Memoire: Tudeh Party Activities amongst AIOC Labor" by Elkington, June 11, 1946, 16256, BP.

117. "Note to Chairman " July 25, 1946, 16256, BP.

118. See "Report on Delegation to Persia" by J. Jones, June 1946, 1–14, 16256, BP.

119. The delegates visited the mud huts of "3rd class employees" in Ahmad-abad as well. They spoke to an oil worker with thirty-seven years of experience. He had four children but was still living in a place that was "good only for dogs." Workers approached the delegates as they inspected the schools created for the workers' children. One worker asked, "Have you seen such a stable-like school in England?" The delegates then visited the home of Qarbali Ibrahim in Bahmashir, who was an older man of sixty years with thirty-five years of experience with the company. He lived in just one small room with his family. Next

door, another worker introduced his nine children and observed that he and his family thought their house looked "rather like a prison than a dwelling place." The worker stated that he was paid only 65 rials per day. Another worker commented that not even "German prisoners of war were being kept in such dirty places." See "Report on MP's (J. Jones) Visit to Persia," June 1946, 16256, BP.

120. Cooke (Ministry of Fuel and Power) to Pattinson, June 29, 1946, 16256, BP.

121. Ibid.

122. Ibid.

123. "Report on Delegation to Persia" by J. Jones, June 1946, 1–14, 16256, BP.

124. "Pay-E Kalam Delneshine Sagha," *Peyk-e Naft* (Bahman 1367 (1989)): 52–53. In this interview, Gholi-Ghanavati also recalls the years of nationalization of the oil industry and the arrival of American supervisors in the oilfields thereafter. He describes in detail the main forms of violence and humiliation suffered by Iranian workers at the hands of their American supervisors. Young men, viewed as potential employees and so as a threat to British and American domination of the workforce, were sometimes lured or forced onto trailers pulled by cars driven by Americans. The men were driven at high speed, which would cause many to fall off and even get run over. When a worker was punished, he was dumped in a barrel and hot tar was poured over him. Others were dragged along the road with their legs tied to a jeep driving at high speed.

125. Ibid., 52.

126. Tehran to FO, June 12, 1946, no. 89, 371/52714, FO.

127. FO to Tehran, June 15, 1946, 371/52714, FO.

128. Tehran to FO, June 21, 1946, no. 65–66, 371/52715, FO.

129. As Ladjevardi explains, the formation of the Ministry of Labor marked "the beginning of organized and systematic suppression of the labor movement by the governments" (Ladjevardi, *Labor Unions and Autocracy in Iran*, 67).

130. The Indian workers, like the Iranians, had grievances concerning the hierarchical distribution of benefits and other biased treatment. The Union of Indian Workers informed AIOC of its intent to strike outside the new labor office. In a letter from the union to the British Consulate at Khorramshahr, the workers (approximately 1,400 to 1,500 Indian artisans) complained of "misbehaviour, ill-treatment, and victimization by their Employers." Grievances concerned the Indian Club, welfare, general treatment, accommodations, transportation, food, medical aid, and a demand for formal recognition of the Indian Workers' Union by including one of its representatives in the newly established labor office. In another petition signed by over 1,300 Indian artisans, the union demanded the replacement of the labor and welfare officers responsible for representing the grievances and claims of Indian workers to AIOC management. The governor of Abadan immediately expelled M. A. Faruki, the leader of the union, and the

demonstration was called off. See Khorramshahr to Tehran, June 28, 1946, no. 71, 371/52716, FO; M. A. Faruki to Willoughby (British Consulate, Khorramshahr), June 26, 1946, no. 50, 371/52719, FO.

131. H. J. U. (General Manager Staff) to General Manager at Abadan, July 19, 1946, 130264, BP. The commission, set up by the Iranian government, involved meetings between representatives of the company, the CCUTU and the Iranian government.

132. Ladjevardi, *Labor Unions and Autocracy in Iran*, 129.

133. Halliday, "Iran," 8.

134. Ibid. There were over 150 casualties, 50 deaths, and many arrests of labor leaders, Tudeh Party organizers, and other alleged strike agitators whose cases were submitted before a military court. Also see Ladjevardi, *Labor Unions and Autocracy in Iran*, 129.

135. Extract of letter from Prince Firouz to AIOC, "B. Diary of Events (14/18 July 1946)," July 1946, 130264, BP. The major stipulations: (1) Friday wages of all company workmen must be paid as of the day the national labor law went into effect (in May 1946). (2) The wages of company workmen must be paid according to the standard set by the new labor law. (3) The company would fully cooperate with the Iranian government's progressive, reform agenda in enforcing the law and providing a comfortable, hygienic standard of living for the workers.

136. Crinson, "Abadan," 348.

137. Ibid., 347.

138. Ibid., 357. This unreformed "point system" "continued to ensure an implicit apartheid with a few token exceptions."

139. Gass to Fraser, November 29, 1947, 80908, BP.

140. Around this time, labor worked 44.5 hours per week in the winter and 45 hours per week in the summer. Overtime was paid in accordance with the new labor law at the rate of 35 percent above straight time. Shift work was eight hours per day, six days per week. In terms of education, eleven additional schools were added until 1945, with 6,000 pupils attending since 1938. World War II brought a setback in general education in the province, but the company agreed in 1943 to provide all teachers at schools in its areas an allowance equal to 75 percent of their basic salaries. See "Social and Municipal Development in Abadan and the South Persian Oilfields," Fraser to Chancellor Dalton, August 28, 1946, 43762, BP.

141. Gass to Fraser, December 2, 1947, 80908, BP.

142. Gobey to Gass, August 23–24, 1948, 80908, BP.

143. Gobey to Gass, August 24–25, 1948, 80908, BP. This interview was circulated to all AIOC directors.

144. Ibid. See the next chapter for a detailed discussion of the role of the Majlis in nationalization.

145. Ibid.

146. Mitchell, *Carbon Democracy*, 77–78. Along with Article 22, which created the system of mandates in the Middle East and Africa, Article 23 represented traces of the demands of the British left and their attempts to propose new instruments for a democratized international order that were ultimately "reduced to appendages," representing instead the "consent of the governed," not any real democratization.

147. "Labour Conditions in the Oil Industry in Iran" by the International Labor Office, 1950, no. 4, 129288, BP.

148. Ibid.

149. Ibid.

150. Elkington to Northcroft (marked personal and confidential), August 9, 1950, BP 71068.

151. According to Abrahamian, Fateh was a "veteran of the early labor movement" who had studied economics at Columbia University. After thirty years at AIOC, he was appointed deputy director of labor relations (Abrahamian, *The Coup: 1953, the CIA, and the Roots of Modern U.S.-Iranian Relations*, 73).

152. "Note on Talk with Mr. Finch, American Labor Attaché" by F. W. Leggett, November 20, 1950, 71068, BP.

153. "Minutes of Meeting Held at Britannic House on Wednesday, 24 September, 1947," 79663, BP.

154. Ibid.

155. Ibid.

156. Ibid.

157. Telegram from Gobey to Jones, April 24, 1948, 72327, BP. For a comparative study of training and education programs for the replacement of foreign labor with domestic labor at Aramco and AIOC/BP, see Dobe, *A Long Slow Tutelage in Western Ways of Work*.

158. Nationalization was formalized in May 1951.

159. Halliday, "Iran," 10.

160. According to Halliday, there were more than 200 strikes over economic issues in the 1951–1953 period (Halliday, "Iran," 10).

161. Ibid. In 1952, unions fell in line with Mosaddiq and while he did not lift the ban on the CCUTU, Halliday says it was able to revive its activities in a "semi-legal manner."

162. Ibid., 9.

163. "Diary of Strike Situation at Abadan/Bandar Mashur 1951," n.d., 9932, BP. All loadings were suspended with only two out of four berths in use the following day. If old-scale allowances were not restored, the shift drivers threatened not to report to work, followed by process men in the tank farm due to work later in the evening. See Drake to Northcroft/Elkington, March 25, 1951, 66248, BP.

164. This included the provision of a cemetery, but no promise was made about the restoration of allowances.

165. Davies to Fowler, May 25, 1951, 68908, BP.

166. Le Page to Drake/Elkington, March 25, 1951, 1–26, 66248, BP.

167. "D-Organization of the Oil Co. Apprentice Trainees," in "List of Proclamations and Pamphlets Published during the Recent Strike and Disturbances," March/April 1951, 68908, BP. As in 1946, the apprentices demanded, in a written declaration, that they be trained under experienced teachers; that students whose average scores on final exams fell below a particular standard be allowed to continue with their studies for one more year; that all trainees who completed the second educational year be promoted to junior status; that staff passes for use of the swimming pool and club be issued; that all trainees be eligible for night classes at the Education Department; that the present means of transportation on industrial buses be replaced by staff buses; and that a Society of the Students of the Technical Institute be formed immediately.

168. Le Page to Drake/Elkington, March 25, 1951, 1–26, 66248, BP. The workers demanded a 1.80 percent increase in wages, cancellation of required rent for housing, cash compensation for unhoused labor, and the cancellation of the company code of disciplinary action so that workers could only be punished with the consent of workers' representatives.

169. Ibid. They also complained about a lack of housing facilities, missing shelters over cooking stoves, and a lack of electric lights and fans.

170. Ibid. Grievances concerned the (1) cut in outstation allowances, (2) cut in annual increases, and (3) inadequate or problematic housing, water, electricity, minimum wages, and cost of living.

171. Ladjevardi says that the labor law, for all intents and purposes, made strikes illegal (Ladjevardi, *Labor Unions and Autocracy in Iran*, 83).

172. Le Page to Drake/Elkington, March 25, 1951, 1–26, 66248, BP. By April 9, only 3 percent of laborers at Agha Jari were collecting their pay.

173. Elkington to Drake, April 18, 1951, 71068, BP.

174. Drake to Elkington/Seddon, April 18, 1951, 71068, BP. By mid-April, all major units of the refinery were closed down due to a lack of manpower. Throughput was reduced from 18.1 to 4.5 gallons per day. The production of one million gallons per day of export crude also ceased.

175. Drake to Elkington, April 14, 1951, 71068, BP.

176. Elkington to Drake, April 18, 1951, 71068, BP.

177. Smith to Snow, April 27, 1951, 66248, BP. Also see "Record Note: General Effect of the Recent Strike on Abadan Refinery," H. L. to Brown, May 9, 1951, 91032, BP.

178. "Notes on Meeting Held at Britannic House on 20th April 1951 to Discuss Refinery Throughput" by Hoffert, April 30, 1951, 91032, BP. The quality would change in terms of viscosity, flash point, and T.E.L. content of motor gasoline.

179. Mitchell, "Carbon Democracy," 409.

180. Podobnik, *Global Energy Shifts*, 49.

181. Shahriari (Managing Body of Abadan Strikers Workers Representative) to Industrial Relations Department, April 19, 1951, Appendix XII, 68908, BP.

182. "List of Proclamations and Pamphlets Published during the Recent Strike and Disturbances," March/April 1951, 68908, BP. April 23 marked a turning point in the strike when 11,000 men returned to work and throughput rose to 9 million from 4.6 million gallons per day. In what it called a "gesture of good will," the company had issued its own proclamation to the workers to make payment equivalent to the loss of their wages to workmen absent from work during the strike. "Proclamation to All the AIOC Workers of Khuzistan," in "General Strike—Abadan," April 1951, 1–14 plus appendixes, 68908, BP. Payment of allowances to workers at all oil production sites would be continued until settlement of the question and other requests submitted to the commission could be examined and implemented "if required by Law."

183. Enclosure of "Noor-e-Islam," 68908, BP. The minimum wage of oil workers should be based on a changeable minimum cost of living, and this should be determined in conjunction with the legitimate representatives of the workers. Suitable houses must be built for all workmen. AIOC's police, which had taken the form of a "Government in this country," must be disbanded immediately.

184. "General Strike—Abadan," April 1951, 68908, BP.

185. Snow to Heath Eves, May 18, 1951, 66248, BP.

186. Mitchell, "Carbon Democracy," 410. Here Mitchell is citing Hanna Batatu, *The Old Social Classes and the Revolutionary Movements of Iraq: A Study of Iraq's Old Landed and Commercial Classes and of Its Communists, Ba'thists, and Free Officers* (London: Saqi Books, 2004).

187. Ibid., 412. On the Mexican oil workers and nationalization, see Santiago, *Ecology of Oil*.

188. Bowker, *Science on the Run*, 82, 91.

189. As mentioned in chapter 2, AIOC sponsored the training of Iranians in petroleum engineering at the University of Birmingham's Department of Petroleum Technology, founded by Cadman in the 1920s. Article 16 (IV) of the 1933 concession stipulated that "the company shall make a yearly grant of £10,000 sterling in order to give in Great Britain, to Persian nationals, the professional education necessary for the oil industry."

190. Works in subaltern studies tend to assert the inner mental, rational, or cultural autonomy of the modern subject in making his or her own history (rather than being subjugated to the history of the bourgeois nation-state). This overlooks the modern techniques of power at work, the collectives of people, knowledge, technologies, and coercion that do not respect the divide between the mind and body of the modern political subject (or the natural versus the social world) but help produce this divide. See Partha Chatterjee, *Nationalist Thought and the Colonial World: A Derivative Discourse* (London: Zed Books,

1986); Ranajit Guha and Gayatri Chakravorty Spivak, eds., *Selected Subaltern Studies* (New York: Oxford University Press, 1988).

Chapter 5: Assembling Intractability: Managing Nationalism, Combating Nationalization

1. For example, see Ervand Abrahamian, "The 1953 Coup in Iran," *Science and Society* 65, no. 2 (2001): 182–215; James E. Bill and William R. Louis, eds., *Musaddiq, Iranian Nationalism, and Oil* (Austin: University of Texas Press, 1988); William Roger Louis, "The Persian Oil Crisis," in *The British Empire in the Middle East, 1945–1951: Arab Nationalism, the United States, and Postwar Imperialism*, 632–690 (Oxford: Clarendon Press, 1984); William Roger Louis, "Britain and the Overthrow of the Mosaddeq Government," in *Mohammad Mosaddeq and the 1953 Coup in Iran*, ed. Mark J. Gasiorowski and Malcolm Byrne, 126–177 (Syracuse, NY: Syracuse University Press, 2004); Homa Katouzian, *Musaddiq and the Struggle for Power in Iran*, 2nd ed. (New York: I. B. Tauris, 1999).

2. Mary Ann Heiss, "The United States, Great Britain, and the Creation of the Iranian Oil Consortium, 1953–1954," *International History Review* 16, no. 3 (1994): 441–600. See chapter 6 on the formation of the Iranian oil consortium.

3. See Ferrier, *History of British Petroleum*; Bamberg, *History of British Petroleum*; Bamberg, *British Petroleum and Global Oil*; Elm, *Oil, Power, and Principle*; Mary Ann Heiss, "The International Boycott of Iranian Oil and the Anti-Mosaddeq Coup of 1953," in *Mohammed Mosaddeq and the 1953 Coup in Iran*, ed. Mark J. Gasiorowski and Malcolm Byrne, 178–200 (Syracuse, NY: Syracuse University Press, 2004).

4. Marsh says that the oil majors were "extremely useful tools of government policies." But they were "far from compliant" and could often reshape the policies of the American and British governments, which used the companies to manage the Anglo-American relationship. Anglo-American policies toward the Iranian oil crisis were a "fine blend of Cold War strategy, commercial calculation, and an unusually pronounced periodic dependence on non-state actors" (Steve Marsh, "Anglo-American Crude Diplomacy: Multinational Oil and the Iranian Oil Crisis, 1951–53," *Contemporary British History* 21, no. 1 (2007): 25–53, esp. 39, 45).

5. For the text of the nationalization laws (nine articles), see *Asnad-e Naft* (*Documents on Oil*) (Tehran: Iranian Ministry of Foreign Affairs Publications, 1951), 68–69.

6. Louis, "Britain and the Overthrow of the Mosaddeq Government," 141.

7. This legislation was backed by Mosaddiq, who was elected to the Majlis in 1944 and, as Elm says, had two aims: to end Iran's subjection to foreign powers and to establish parliamentary rule so that representatives of the people, not a single sovereign ruler, would control affairs of state (Elm, *Oil, Power, and Principle*, 59).

8. Ibid. Some other important points included: (1) AIOC had not abided by the gold guarantee clause in the 1933 concession (Article 5). The royalty figure in 1933 was 4 shillings per ton, representing one-eighth of the price of Iran's crude oil, whereas in 1947, considering the gold guarantee, it represented less than one-sixteenth. Thus, Iran's royalties in relation to the price of oil exported had dropped from 33 percent in 1933 to 9 percent in 1947. (2) Iran should not have been subjected to the limitations set by the British government on the distribution of dividends, which resulted in a major part of Iran's share being held in AIOC's general reserves. (3) AIOC had concluded agreements with the British Admiralty and American oil companies, selling oil products at high discounts and refining a major portion of Iranian oil abroad. (4) AIOC had let gas go to waste in its operations, paying no attention to Iran's demands to either retain the gas in wells or construct gas pipelines to cities.

9. Fraser was himself "born into oil," having "inherited from his father the biggest company in the ... Scottish oil-shale industry, and later merging it with six other companies into BP, to provide them with Scottish outlets" (Sampson, *The Seven Sisters*, 113).

10. Elm, *Oil, Power, and Principle*, 55. The Iranian government had originally asked for £1 per ton and considered the company's offer unacceptably low. Fraser refused to improve the company's offer and called for the matter to be referred to arbitration. In the meantime, Golshayan reported to the government that the AIOC proposals did not respect Iran's rights, but the shah instructed his cabinet to accept AIOC's offer of £12/6 per ton with no rights to revise the concession.

11. Northcroft to Rise, December 3, 1949, 80908, BP.

12. Heiss, "The United States, Great Britain, and the Creation of the Iranian Oil Consortium," 512.

13. Ibid.

14. Bin Cheng, "The Anglo-Iranian Dispute," *World Affairs* 5 (1951): 387–405, esp. 389.

15. Elm, *Oil, Power, and Principle*, 56.

16. Enclosure aide-mémoire from Persian Prime Minister to British Secretary of State, Chadwick (FO) to Young (Treasury), October 19, 1949, no. 43c–f, T 236/2818. Also see "Some Background Notes on Persian Oil," July 23, 1951, 142642, BP.

17. Galpern, *Money, Oil, and Empire in the Middle East*, 87.

18. Elm, *Oil, Power, and Principle*, 62. Also see *Asnad-e Naft* (*Documents on Oil*), 41.

19. Stephen Kinzer, *All the Shah's Men: An American Coup and the Roots of Middle East Terror* (Hoboken, NJ: Wiley, 2008), 69.

20. Elm, *Oil, Power, and Principle*, 63.

21. Ibid., 65.

22. Ibid. Razmara's negotiating points were as follows: (1) A ten-year program devised by AIOC to train Iranians for technical jobs and reduce British and Indian workers. (2) Permission for Iran to examine company books and check its oil exports to settle allegations that the company was exporting more oil than it claimed. (3) Transparent information on the quantity and price of Iranian oil products sold to the British Admiralty and others at high discounts. (4) Advance payments from AIOC to the Iranian government on the basis of the draft Supplemental Agreement to help the country's development plan and to encourage the Majlis to ratify the agreement.

23. Vitalis, *America's Kingdom*, 130.

24. Washington to British Foreign Office (Franks), April 11, 1951, 371/91470, FO.

25. Elm, *Oil, Power, and Principle*, 67.

26. Vitalis, *America's Kingdom*, 130.

27. See "Origins of the 50–50 Oil Profits Split," December 5, 1957, 1–3, Box 56, Reports 1957, Walter J. Levy Papers, Collection Number 082428, American Heritage Center (AHC), University of Wyoming.

28. On the 50–50 principle in Saudi Arabia, see Irvine H. Anderson, *Aramco, the U.S., and Saudi Arabia: A Study of the Dynamics of Foreign Oil Policy 1933–1950* (Princeton, NJ: Princeton University Press, 1981). Also see Irvine H. Anderson, "The American Oil Industry and the Fifty-Fifty Agreement of 1950," in *Musaddiq, Iranian Nationalism, and Oil*, ed. James A. Bill and William R. Louis (Austin: University of Texas Press, 1988), 143–159; Louis, *The British Empire in the Middle East*; and Mitchell, "McJihad."

29. Vitalis, *America's Kingdom*, 130.

30. "Some Background Notes on Persian Oil," July 23, 1951, 142642, BP.

31. Elm, *Oil, Power, and Principle*, 184.

32. Heiss, "The International Boycott of Iranian Oil and the Anti-Mosaddeq Coup of 1953," 183.

33. Louis, "Persian Oil Crisis," 646; Galpern, *Money, Oil, and Empire in the Middle East*, 93.

34. See Walden, "The International Petroleum Cartel in Iran—Private Power and the Public Interest," *Journal of Public Law* 11 (1962): 64–121, esp. 80. Also see Memorandum of Meeting of the Foreign Ministers of the United States and United Kingdom at the Department of State, January 9, 1952, no. 142 in *Foreign Relations of the United States 1952–54, Iran 1951–54: Volume 10* (Washington, DC: US Government Printing Office, 1989), 312–313. The British refused to convert Iran's sterling balances into dollars for repayment of the loan.

35. Elm, *Oil, Power, and Principle*, 72. Also see the Oil Committee's report in *Asnad-e Naft* (Documents on Oil), 51. Five members of the Majlis Oil Committee advocated nationalization, while the remaining thirteen called for new negotiations to take place.

36. Ibid., 73.

37. Cheng, "The Anglo-Iranian Dispute," 394. It should be recalled that, according to the Iranian position, the royalty figure in 1933 was 4 shillings per ton, representing one-eighth of the price of Iran's crude oil, whereas in 1947, considering the gold guarantee, it represented less than one-sixteenth.

38. Elm, *Oil, Power, and Principle*, 73–74.

39. Cheng, "Anglo-Iranian Dispute," 388. In the meantime, the Oil Committee instructed Razmara to study the problems of nationalization with a panel of experts, whom he ordered to conclude that nationalization would be catastrophic. First, Iran lacked the requisite financial or technical facilities to run the industry; second, it would be illegal; third, a compensation of between £300 million and £500 million would have to be paid. Razmara's controversial oil proposals were made public and the National Front accused him of wanting to establish a military dictatorship and not demanding enough from Britain. Kashani also encouraged all Muslims and patriotic citizens to fight against the enemies of Islam and Iran by joining the nationalization struggle (see Abrahamian, *Iran between Two Revolutions*, 266–267). After the contents of Razmara's offensive report were made public, a member of the *Feda'iyan-i Islam* assassinated the prime minister on March 7.

40. Ford, *Anglo-Iranian Oil Dispute*, 51.

41. Cheng, "Anglo-Iranian Dispute," 387.

42. Ibid., 389–390.

43. Ford, *Anglo-Iranian Oil Dispute*, 56.

44. Ibid., 50.

45. Ibid., 58.

46. Ibid.

47. Ibid.

48. Ibid., 59.

49. Ibid.

50. Ibid., 60.

51. Ibid., 61.

52. "Application Instituting Proceedings," by Eric Beckett, May 26, 1951, Anglo-Iranian Oil Company Case (United Kingdom v. Iran), 12–13, International Court of Justice (ICJ). (1) The Iranian government was not entitled to refuse to submit the dispute to arbitration as provided for in Article 22 of the 1933 concession. (2) The Iranian government had effected a unilateral annulment or alteration of the concession terms, contrary to Articles 21 and 26 of the 1933 concession. (3) "In so purporting to effect a unilateral annulment, or alteration of the terms of the 1933 concession," contrary to the said articles, "the Iranian government thereby committed a wrong against the Anglo-Iranian Oil Company, Limited, a British national." (4) The Iranian government was attempting to deny AIOC the legal remedy provided for in the concession (to seek arbitration provided by Article 22).

53. Ibid.

54. Ibid.

55. Ford, *Anglo-Iranian Oil Dispute*, 180.

56. "Application Instituting Proceedings," by Eric Beckett, May 26, 1951, Anglo-Iranian Oil Company Case (United Kingdom v. Iran), 13, International Court of Justice (ICJ).

57. Ibid.

58. Ibid.

59. Ibid.

60. Ford, *Anglo-Iranian Oil Dispute*, 181.

61. Ibid.

62. On the history of the corporation in British colonial India, see Stern, *The Company State*.

63. Ford, *Anglo-Iranian Oil Dispute*, 181.

64. O'Connell concludes that "the juridical situation presented to us constitutes a challenge to traditional modes of reasoning. Because precedent is virtually non-existent" (D. P. O'Connell, "A Critique of the Iranian Oil Litigation," *British Institute of International and Comparative Law* 4, no. 2 (April 1955): 267–293, esp. 293). Also see Homayoun Mafi, "Iran's Oil Concession Agreements and the Role of the National Iranian Oil Company: Economic Development and Sovereign Immunity," *Natural Resources Journal* 48 (2008): 407–430, esp. 412–414.

65. Ibid., 61.

66. Ibid., 62.

67. Ibid.

68. Basil Jackson, deputy chair of AIOC's board of directors, Sir Thomas Gardiner, one of the two government-appointed members of AIOC's board, N. A. Gass, negotiator of the Supplemental Agreement in 1949, and E. H. O. Elkington, well known for his role in past labor and concession disputes, headed the mission.

69. Ford, *Anglo-Iranian Oil Dispute*, 66–77.

70. Ibid. The distribution business within Iran would also be transferred to an entirely Iranian owned and operated company.

71. According to Ford, there were only two cases prior to this in which the ICJ indicated interim measures of protection (Ford, *Anglo-Iranian Oil Dispute*, 85).

72. "Request for Interim Measures of Protection," June 22, 1951, 46, ICJ. This damage included but was not limited to: (1) a loss of skilled personnel; (2) interference with management; (3) consequences of disrupting an integrated enterprise; (4) loss of markets and goodwill. See p. 49.

73. Ibid., 51.

74. Ibid., 52. The British government's request for interim measures of protection included the following points: (1) To produce and sell petroleum without interference calculated to impede operations by the Iranian government, their servants, and/or agents; (2) oblige the Iranian government not to, by any executive or legislative act or judicial process, sequester or seize the property of AIOC including property it had already purported to nationalize or otherwise to expropriate; (3) oblige the Iranian government not to seize or sequester any monies earned by AIOC; (4) oblige the Iranian government not to attempt to dispose of monies of (3) other than in accordance with the 1933 concession or measures of the Court; (5) oblige the Iranian government to ensure that no steps were taken to prejudice the right of the British government to have a decision of the Court in its favor on the merits of the case; and (6) oblige both the Iranian and British governments to ensure that no step was taken to aggravate or extend the dispute submitted to the Court and that the Iranian government should abstain from all "propaganda calculated to inflame opinion in Iran" against AIOC and the United Kingdom.

75. "Oral Proceedings," June 30, 1951, 408, 417, ICJ. Soskice argued that the two issues were separate and should be judged accordingly, even though he was requesting that the Court indicate "precisely measures for the protection of the rights thus asserted in the Application."

76. Ford, *Anglo-Iranian Oil Dispute*, 78. Ford says that Soskice's argument deals with three basic questions that confront every application for the indication of interim measures: (1) whether the Court can properly indicate such measures without having previously determined that it has jurisdiction to try the case on its merits; (2) whether the request can be granted in light of general principles governing interim measures; and (3) whether interim measures are necessary—that is, "whether irreparable damage will result, or the position of either party be prejudiced, or the dispute extended, if the measures are not indicated before the Court makes its final judgment on the case's merits."

77. Also present were Hossein Navab (envoy), Mosaddiq, N. Entezam (ambassador, former minister), and H. Rolin (professor of international law in Brussels).

78. Ford, *Anglo-Iranian Oil Dispute*, 74–75.

79. Anghie, *Imperialism, Sovereignty, and the Making of International Law*, 223. Known as "transnational law," Anghie explains that this novel legal framework was particularly relevant for "Third World states because it was precisely in those states that the activities of these corporations generated new and complex problems that required legal resolution."

80. "ICJ Order Indicating Interim Measures of Protection," July 5, 1951, 2, ICJ.

81. "Dissenting Opinions by Judges Winiarski and Badawi Pasha," 96–97, ICJ.

82. Ibid.

83. Ibid.

84. Ibid.

85. Ibid.

86. Ibid.

87. Ford, *Anglo-Iranian Oil Dispute*, 88–89. As Ford explains, the declaration had by its terms bound Iran to recognize the compulsory jurisdiction of the Court for a period of six years and thereafter, until notice was given of its abrogation, which effectively happened here.

88. For a discussion of legal metrology, see chapter 3.

89. Recall that AIOC's submission to the ICJ is "deferred," and this makes the dispute a state-to-state issue, even though the Iranian government rejects this framing and argues that this is a domestic issue to be resolved between itself and a private company.

90. "Economic Sanctions against Persia" by HM Treasury, May 5, 1951, 142642, BP.

91. "Economic Sanctions against Persia" by HM Treasury, May 5, 1951, 142642, BP. The British government would: (1) block Iran's sterling balances (£25 million as of May 1, 1951), of which £9 million was already committed and the remainder was required as currency backing, which could be accomplished under the Exchange Control Act; (2) denounce the memorandum of understanding between the two countries' central banks and remove the dollar convertibility; (3) cut off Iran's supplies of essential materials from the United Kingdom through an administrative act; (4) deny Iran in the future such credit facilities as British banks, etc.; (5) refuse import licenses for Iranian goods other than oil, cutting off about one-eighth of Iran's remaining trade.

92. Heiss, "International Boycott of Iranian Oil," 180.

93. "B," June 4, 1952, 118974, BP.

94. Ibid.

95. Heiss, "International Boycott of Iranian Oil," 180.

96. Galpern, *Money, Oil, and Empire in the Middle East*, 107–114. The Abadan refinery produced about twenty million tons of chemicals annually, helping to satisfy the bulk of the Eastern Hemisphere's (Indian Ocean area's) postwar demand by producing items such as heating oil, gasoline, airplane fuel, and bitumen, a substance used in road surfacing.

97. Ibid.

98. Heiss, "International Boycott of Iranian Oil," 182.

99. Ibid., 185.

100. Ibid., 187.

101. Walden is citing specific court cases concerning antitrust law (Walden, "The International Petroleum Cartel in Iran," 102n203).

102. "Memorandum: Supplying Petroleum to Free World without Iran," Petroleum Administration for Defense, July 12, 1951, in Replacement Problem, Supply and Distribution Subcommittee, 1951, Box 16, Folder 2, Walter J. Levy Papers, Collection Number 082428, AHC.

103. Ibid. Also see Walden, "The International Petroleum Cartel in Iran," 104.

104. For the specific details and risks of the arrangement, see "Memorandum: Supplying Petroleum to Free World without Iran," Petroleum Administration for Defense, July 12, 1951, in Replacement Problem, Supply and Distribution Subcommittee, 1951, Box 16, Folder 2, Walter J. Levy Papers, Collection Number 082428, AHC.

105. Walden, "The International Petroleum Cartel in Iran," 97–98. Walden is citing "1951 PAD, Report to the Secretary of the Interior and Petroleum Administrator for Defense on the Foreign Voluntary Aid Program from the Assistant Deputy Administrator, Foreign Petroleum Operations," 9–10. Walden remarks that when "analyzed in its true light," the arrangement "presented an anomalous picture of the mobilization by both Britain and the United States of all their economic power to prevent any oil from escaping Iranian shores to any markets in the world, and at the same time the taking of drastic anti-competitive action which they justified on the ground that the defense efforts of the free world had become endangered by the iniquitous acts of the Persians!"

106. Ibid.

107. Walden, "The International Petroleum Cartel in Iran," 103. "From the very beginning," Walden argues, American policy in Iran "paralleled the interests of the international petroleum cartel."

108. Elm, *Oil, Power, and Principle*, 124. Also see *Asnad-e Naft (Documents on Oil)*, 172.

109. Ford, *Anglo-Iranian Oil Dispute*, 99. Mosaddiq also requested that the Majlis authorize a public bond issue of two billion rials and approve the acceptance of a $25 million loan from the US Export-Import Bank. The former but not the latter was approved, authorizing the government to withdraw £14 million from its sterling balances in London toward relieving the shortage of foreign exchange by financing essential imports.

110. As the nationalization bill was passed in the Majlis, Nemazee formally recommended Levy as consultant and assistant to the Iranian government. See Levy to Nemazee, March 8, 1951, Box 15, Folder 3, Walter J. Levy Papers, Collection Number 082428, AHC. Earlier, in August 1948, Nemazee had provided Levy with a memorandum to justify the Iranian case for a revision of the existing concession. He believed that once "the big stumbling block" of the financial and commercial dispute was eliminated, an atmosphere would be created that would enable a settlement of "the somewhat different character" of "the social and political differences" between the government and the company. These he said could only be settled by direct negotiations (Levy to Nemazee, February 21, 1951, Box 15, Folder 3, Walter J. Levy Papers, Collection Number 082428, AHC).

111. Levy to Nemazee, February 21, 1951, Box 15, Folder 3, Walter J. Levy Papers, Collection Number 082428, AHC.

112. "Memorandum," Levy, Butler, Willoughby, July 28, 1951, RG 59, Box 5505A, 888.2553/8–3151, NARA.

113. Harriman to Secretary of State, August 12, 1951, RG 59, Box 5505A, 888.2553/8–1251, NARA.

114. Harriman to President and Secretary of State, July 24, 1951, RG 59, Box 5505A, 888.2553/7–2451, NARA.

115. Nemazee to Levy, April 2, 1951, Box 15, Folder 3, Walter J. Levy Papers, Collection Number 082428, AHC.

116. Ibid. Nemazee also commented that Iran's mining law of February 7, 1939, guaranteed that subsoil resources belonged to the nation and that the transfer of property rights to a national Iranian corporation was possible without involving any payments as compensation. Nemazee calculated, however, that the fair compensation value of the company's installations in Iran (considering the book value of AIOC's assets after deducting them from the value of its interests in Kuwait and Iraq concessions, as well as the value of its interests in the tanker fleet and subsidiary companies) would not exceed about £35 million or $100 million. After providing for compensation installments, the net revenue to the Iranian government over the period of the sale contract would be at least $120 million per annum, an amount, Nemazee calculated, that a 50–50 division would produce. Levy disagreed with Nemazee on whether Iran's mining law of 1939 stipulated that subsoil deposits did or did not belong to the owner of the surface land. The law was clear, however, in giving the right to exploit petroleum exclusively to the government. Levy argued that the Iranian constitution did authorize the transfer of the right to exploit petroleum (with Majlis approval) "as it had done with the Anglo-Iranian concession of 1933."

117. "Summary of Conversation with Engineer Hassibi, Under Secretary of Ministry of Finance and Mr. Salah, Chairman of National Oil Commission," July 16, 1951, 1–9, Box 16, Folder 5, Walter J. Levy Papers, Collection Number 082428, AHC.

118. Ibid.

119. Ibid.

120. Ibid. The complex organization would provide the "technical knowledge and complex equipment required for the production and refining of oil." Such an energy system must also be responsible for the availability and scheduling of tankers to lift the oil from the Port of Abadan to the worldwide markets. Import and export schedules were necessary to facilitate the production of oil and its products in the correct quantities and qualities required by consumers. Further, the complex organization must provide a distribution system in various importing countries to bring the oil "from the port to the smallest village."

121. Ibid. It must make arrangements for the conversion of money it received from consumers into "those currencies which are needed to pay for the production, local expenses, for transportation, etc."

122. "Summary of Conversation with Engineer Hassibi, Under Secretary of Ministry of Finance and Mr. Salah, Chairman of National Oil Commission," July 16, 1951, 1–9, Box 16, Folder 5, Walter J. Levy Papers, Collection Number 082428, AHC.

123. Ibid.

124. Callon, "Some Elements of a Sociology of Translation," 70.

125. "Summary of Conversation with Engineer Hassibi, Under Secretary of Ministry of Finance and Mr. Salah, Chairman of National Oil Commission," July 16, 1951, 1–9, Box 16, Folder 5, Walter J. Levy Papers, Collection Number 082428, AHC.

126. Ibid.

127. Ibid.

128. Ibid.

129. See Heiss, "The United States, Great Britain, and the Creation of the Iranian Oil Consortium," 199.

130. "Second Levy-Hassibi Conversation," July 20, 1951, 1–8, Box 16, Folder 5, Walter J. Levy Papers, Collection Number 082428, AHC.

131. Ibid.

132. Ibid.

133. Ibid.

134. Ibid.

135. The only means of solving these interrelated problems was with the continued availability of British technicians and Iran's use of the substantial tonnage owned or chartered by the major oil companies. Harriman also encouraged Iranian officials to "establish satisfactory relations with the world oil industry or face very serious consequences" (see "Mr. Harriman's Meeting with the Joint Oil Committee," July 20, 1951, Box 16, Folder 5, Walter J. Levy Papers, Collection Number 082428, AHC, UW). Americans present were: Walter Levy, Harriman, William M. Rountree (US Department of State), and C. H. Walter Howe (US Embassy). Iranian members of the Joint Oil Committee present were: Morteza Bayat (senator), Ahmad Matin Daftari (senator), Rezazadeh Shafaq (senator), Mohammad Soruri (senator), Nasr Qoli Ardalan (deputy), Mohammad Moazami (deputy), Seyid Ali Shayegan (deputy), Kazem Hassibi (undersecretary of the ministry of finance). The three major problems facing the Iranian oil industry were: (1) maintaining the technical competence essential for the production of oil and the operation of the Abadan refinery; (2) obtaining the necessary tankers for the transportation of oil from Iran to the customers; (3) finding customers for Iranian crude oil and products in foreign countries.

136. Ibid.

137. Ibid.

138. Ford, *Anglo-Iranian Oil Dispute*, 100–101. The negotiating terms put forward by the Iranian government were as follows: (1) The Iranian government was prepared to enter negotiations with representatives of the British government on behalf of the former AIOC, in the event that the British government, on behalf of AIOC, recognized the principle of nationalization of the oil industry. (2) Before sending any representatives to Tehran, the British government must

make a formal statement of consent to the principles of nationalization of the oil industry on behalf of the former AIOC. (3) The principle of nationalization referred to the proposal approved by the Special Oil Committee and confirmed by law on March 20, 1951. (4) The Iranian government was prepared to negotiate the manner in which the law would be carried out insofar as it affected British interests.

139. "Memorandum of a Telephone Conversation between Mr. Walter Levy and Mr. Howard Page," July 28, 1951, Box 16, Folder 5, Walter J. Levy Papers, Collection Number 082428, AHC; "Meeting with Mr. John Walker, Commercial Counselor for British Embassy and Mr. Walter J. Levy—July 27, 1951," Box 16, Folder 5, Walter J. Levy Papers, Collection Number 082428, AHC.

140. Ibid.

141. Ibid.

142. "Memorandum of Conversation," July 28, 1951, Box 16, Folder 5, Walter J. Levy Papers, Collection Number 082428, AHC. Also see "Summary of a Conversation at the Ministry of Fuel and Power," July 29, 1951, 1–9, Box 16, Folder 5, Walter J. Levy Papers, Collection Number 082428, AHC.

143. The British government could not be a direct participant or majority shareholder in the operating company in Iran, without affecting its interest in AIOC's other activities. The operating company might include AIOC's customers and some Dutch and Belgian companies.

144. Ibid.

145. Ibid.

146. Ibid.

147. Ibid.

148. Nathan J. Citino, "Defending the 'Postwar Petroleum Order': The US, Britain and the 1945 Saudi-Onassis Tanker Deal," *Diplomacy & Statecraft* 11, no. 2 (2000): 137–160, esp. 139.

149. Mitchell, "Carbon Democracy," 408.

150. Ibid. As Mitchell explains, "Prices in Texas ... following the passage of the Texas Market Demand Act of 1932 were protected by production quotas set by a state body, the Texas Railroad Commission, and later by federal import quotas." This US government regulation combined with the 1929 agreement to divide the world's oil resources among the major oil companies and to limit production to maintain prices "prevented the emergence of a competitive market and thus assured extraordinary profits to those who controlled the cheaply produced oil of the Middle East." The pursuit of an antimarket arrangement based on an artificial scarcity and exclusive control of oil production by major oil companies occurred at a particular moment, beginning in the 1930s. This is the fourth feature of the (twentieth-century) political economy of oil, in which "old methods for producing global antimarkets—colonialism—were in the process of collapse" (Mitchell, "McJihad," 6–8).

151. See "Memorandum of Conversation," July 28, 1951, Box 16, Folder 5, Walter J. Levy Papers, Collection Number 082428, AHC, UW. Also see "Summary of a Conversation at the Ministry of Fuel and Power," July 29, 1951, 1–9, AHC. Flett of the British Treasury asked Levy whether the arrangements envisaged American participation in the operating company. Levy responded that although he told the Iranians that no responsible American company would be willing to participate and that he, along with Harriman, had discouraged further American participation, perhaps a 5 or 10 percent American interest "might be helpful" from the British point of view. The Iranians had expressed the desire to conclude an operating contract immediately with a responsible American company. Flett agreed with Butler that such a minority US interest was a possibility, pointing to Standard Oil of New Jersey as a qualified participant, since it was a large customer of AIOC. The point of contention, Flett noted, was between AIOC and British government interests. AIOC would offer "stiff resistance" to bringing in any other companies, to which Levy reiterated that a group of interests forming an operating company would be the best solution and that the British government should not be involved because AIOC was seen as an instrument of the government. Finally, Flett expressed his government's concerns about how the balance-of-payment issue would be affected by any new arrangement in Iran. Levy explained his attempt to persuade Iranians that they would be better off under the present convertibility arrangement with the United Kingdom. Flett agreed that it was very important for oil to be distributed and sold for sterling, and Levy concurred that the operating company should be able to sell oil for sterling worldwide.

152. Louis, "Persian Oil Crisis," 677.

153. Ibid., 676.

154. Ford, *Anglo-Iranian Oil Dispute*, 105, 107. See also *Asnad-e Naft (Documents on Oil)*, 377–379.

155. Ford, *Anglo-Iranian Oil Dispute*, 112–114.

156. In the view of the US government, the Iranian government had confiscated foreign-owned property without paying prompt and adequate compensation, or working out an arrangement mutually satisfactory to the foreign owner and governments. This, Harriman explained, constituted "confiscation" rather than nationalization.

157. "Memorandum of Conversation," August 17, 1951, Box 16, Folder 5, Walter J. Levy Papers, Collection Number 082428, AHC.

158. See for example Louis, "Persian Oil Crisis," 680–681. Abrahamian also discusses this framing (see Abrahamian, *The Coup*, 98–108).

159. Louis, "Persian Oil Crisis," 682. Instead, they would pursue individual contracts with AIOC's former customers and permit them to hire AIOC or one of its subsidiaries as a carrying agent. The British government would pay market price plus carrying charges from the Iranian port.

160. Ford, *Anglo-Iranian Oil Dispute*, 115.

161. "W. A. Harriman (Special Assistant to the President) to Walter J. Levy, September 14, 1951," Box 16, Folder 6, Walter J. Levy Papers, Collection Number 082428, AHC, UW.

162. On British arguments for and against the use of force as well as the consequences of a military intervention in Abadan, see Louis, "Persian Oil Crisis," 686–687. If force was not used, some officials argued, Egypt might take action to end its military threat and nationalize the Suez Canal. If force was used, others argued, the British bargaining position would be weakened by losing the support of the United Nations, and this would further encourage South American and Asian governments to back the US government position, which strongly opposed military intervention.

163. Shams al-Din Amir-Alai, *Khaterat-e Man (My Memoirs)* (Tehran: 1984), 126–128. In a speech to the local business community of Abadan, Amir-Alai expressed his concerns about the harsh living conditions he had witnessed. The workers and poor people had constructed their homes using old tin (*halabi*). In another area, Hasseer Abad, paper was everywhere and had been used to construct homes comparable to chicken coops. The dwellings also lacked basic necessities such as water, even though the "former oil company" was collecting 90 rials per month as rent on each tent and the total cost of the tents was not more than 1,000 rials each.

164. "The Effect of the Withdrawal of British Staff on Producing and Refining Operations in Persia," from Hedley-Miller (Treasury), June 1, 1951, 100738, BP.

165. Ibid.

166. Ibid.

167. Ibid.

168. Ibid.

169. Ibid.

170. See Louis, "Persian Oil Crisis," 687. Louis is quoting Clement Attlee, the British prime minister, in Cabinet Minutes 60 (51), September 27, 1951, 128/20, CAB.

171. Heiss, "International Boycott of Iranian Oil," 180.

172. Ibid., 182.

173. "The Effect of the Withdrawal of British Staff on Producing and Refining Operations in Persia," from Hedley-Miller (Treasury), June 1, 1951, 100738, BP. Pipelines depended on the operation of three pumping stations driven by a steam turbine. The continued safe operations of the steam boilers required water treatment to be kept within "narrow limits," the failure of which would quickly lead to the boilers becoming inoperable. "Water treatment," as the report revealed, "is one of the specialist operations carried out solely by the British staff ... While operating staff in the Pumping Stations are largely Persian, the technical control is British."

174. Heiss, "International Boycott of Iranian Oil," 183.

175. Sampson, *The Seven Sisters*, 118.

Chapter 6

1. Walden's account of the legal aspects of the consortium arrangement is perhaps the only study we have that identifies the overwhelming success of the British and the "catastrophic defeat" suffered by Iranian aspirations for national control. See Walden, "The International Petroleum Cartel in Iran."

2. See Heiss, "The United States, Great Britain, and the Creation of the Iranian Oil Consortium." Also see Elm, *Oil, Power, and Principle*. Elm discusses Iran's nationalization and its aftermath, but summarizes the final form that the international consortium takes in relation to the nationalization crisis, which takes up the bulk of the study. There is no careful discussion of the technical details (e.g., meetings, negotiations, legal battles, organizational forms) of the process through which the consortium's final shape was constituted.

3. Ford, *Anglo-Iranian Oil Dispute*, 122–123.

4. Ibid., 124.

5. On the work of translation in STS, see Callon, "Some Elements of a Sociology of Translation"; Latour, *Reassembling the Social*. More specifically applied to law, see Levi and Valverde, "Studying Law by Association."

6. See Cheng, "Anglo-Iranian Dispute."

7. Ford, *Anglo-Iranian Oil Dispute*, 124–125.

8. Ibid., 126.

9. Ibid., 128.

10. Ibid.

11. Ibid.

12. Mitchell, *Rule of Experts*, 56.

13. Anghie, *Imperialism, Sovereignty, and the Making of International Law*, 225.

14. See Mitchell, *Rule of Experts*, 56, 78–79, on the discussion of the rule of private property and twentieth-century jurisprudence. Mitchell says that "the principle of abstraction on which the order of law depends can be generated only as the difference between order and violence ... the universal and exceptional." But the violent and the exceptional, "all of which the law denounces and excludes, ruptures itself from and supersedes, are never gone. They make possible the rupture, the denunciation, and the order. They are the condition of its possibility."

15. Anghie, *Imperialism, Sovereignty, and the Making of International Law*, 212, 216. This principle embodying the link between sovereignty and natural resources was formally established in the UN General Assembly Resolution 1803 of 1962.

16. Ibid.

17. Anghie, *Imperialism, Sovereignty, and the Making of International Law*, 197.

18. Cheng, "Anglo-Iranian Dispute," 390.

19. Ibid., 391.

20. Ibid., 392. Regarding expropriation proceedings affecting the property of aliens, Cheng explains that: (1) The measure must be for public utility, which means that nationalization is "certainly a legitimate object of expropriation." (2) Compensation must be paid, and the Iranian Law of Implementation of May 1 "clearly provides for compensation." (3) The measure must not be "discriminatory against foreigners," which the British government had argued was precisely the case in its request for interim measures of protection.

21. Cheng, "Anglo-Iranian Dispute," 393. Anghie explains that nationalization was legitimate, according to the West, "provided that a number of conditions were met, the most significant of these being payment of compensation according to internationally determined standards." National standards and laws "asserted by the Third World lacked any such legal foundations." See Anghie, *Imperialism, Sovereignty, and the Making of International Law*, 214.

22. Ibid., 394.

23. Ford, *Anglo-Iranian Oil Dispute*, 129.

24. "British Memorial," October 10, 1951, 64, ICJ.

25. Ibid., 85. The British memorial made the following points: (1) The cancellation of the 1933 concession was a violation of an "express renunciation of the right of unilateral termination." (2) Iran's nationalization laws were directed exclusively against a foreign national, AIOC, and not shown to be necessary for the security of any "vital public interest." (3) The expropriation of the property of foreigners was unlawful unless in the public interest and of vital importance.

26. Ibid.

27. Ibid., 65–66.

28. Ibid., 74.

29. Anghie, *Imperialism, Sovereignty, and the Making of International Law*, 232–235.

30. Ibid.

31. "British Memorial," October 10, 1951, 75, ICJ.

32. Ibid., 76–78.

33. Ibid.

34. I am following Latour's account of "law by association" and considering its implications for politics. See the discussion in chapter 3.

35. Ford, *Anglo-Iranian Oil Dispute*, 134.

36. The provision that former customers of AIOC should have the right to purchase at current prices the same quantities of oil that they previously imported from Iran.

37. Ford, *Anglo-Iranian Oil Dispute*, 134.

38. Ibid.

39. Ibid., 135.

40. Anghie, *Imperialism, Sovereignty, and the Making of International Law*, 222.

41. Ford, *Anglo-Iranian Oil Dispute*, 137–138, 149. He noted that the power given to the Security Council by Article 94 of the Charter to recommend or decide on measures to be taken to give effect to a judgment of the ICJ was limited to cases in which the judgment was both final and binding. The provision in paragraph 2 of Article 41 that the ICJ shall notify the Council of measures indicated by the Court had been designed, Saleh argued, to further cooperation between the two organs of the UN. Article 2, paragraph 7 limited the jurisdiction of the Council and the ICJ when a dispute was found to arise out of a matter by which *international law was solely within the domestic jurisdiction of that party*, and the statute of the Court was not exclusively concerned with the rights and duties of the ICJ and could not confer powers on the Security Council by implication.

42. Ibid., 306. Hossein Navab, representative of the Iranian government, made the case for the incompetence of the Court and his government's refusal of its jurisdiction. Navab argued that the United Kingdom had forced the dispute into "international organs," even though his government rejected their competence.

43. "Preliminary Objection; Ruling That the Court Has No Jurisdiction," July 22, 1952, 93, ICJ. The claims of the British government were based on treaties concluded between Iran and other powers the benefit of which could by invoked only by the British application of the "most favored nation" clause. This clause appeared only in the treaties concluded between Iran and the United Kingdom in 1857 and 1903—that is, prior to the ratification of the Iranian Declaration, or upon the exchange of notes, which did not possess the character of a treaty or convention (dated May 10, 1928). This occurred prior to the ratification of the Iranian Declaration.

44. Ibid. With regard to the alleged tacit agreement between the two governments in connection with the renewal of the 1933 concession, Navab denied such an agreement existed. He further claimed that the British government's "prima facie examination" sufficed to show that the British claims had no relation to the treaties or alleged treaties invoked, as they did not possess the scope that the British state attributed to them. AIOC had not yet exhausted all local remedies provided by Iranian law, particularly concerning the amount of compensation due to AIOC. Finally, that Iran and the United Kingdom had reserved questions in their declarations that "according to international law are within the exclusive jurisdiction of States," must be understood as extending to questions that were "essentially within the domestic jurisdiction of States." Alternatively, Navab asked to place on record for the Iranian government its declaration that it availed itself of the right reserved in its declaration to require the suspension of the proceedings, since the dispute before the Court was submitted to the Security Council and was under examination by that body.

45. "Summary of the Judgment of the Court," July 22, 1952, 1–2, ICJ. The Court's judgment referred to the principle according to which "the will of the Parties is the basis of the Court's jurisdiction." The jurisdiction for this case also depended on "the Declarations accepting the compulsory jurisdiction of the Court," made by Iran and the United Kingdom under Article 36, paragraph 2, of the Statute.

46. Anghie, *Imperialism, Sovereignty, and the Making of International Law*, 230.

47. The declarations contained the "condition of reciprocity, and as that of Iran is more limited," the judgment ruled, "it is upon that Declaration that the Court must base itself." According to Iran's declaration, the Court had jurisdiction when "a dispute relates to the application of a treaty or convention accepted by Iran." Iran argued that jurisdiction was limited to treaties signed "subsequent to the Declaration." Contrary to the United Kingdom's view that earlier treaties could also be considered, the Court judged that both contentions could "strictly speaking be regarded as compatible with the text." However, the Court could not base its judgment on a "purely grammatical interpretation," and must seek a ruling with regard to the intention of Iran at the time it formulated the declaration. The "natural and reasonable way of reading the text," ruled the Court, was that only treaties subsequent to the ratification of the declaration came into consideration.

48. On the United Kingdom's claim to treatment under the "most favored nation" clause signed between Iran and other countries in subsequent treaties to the declaration, the Court ruled that the United Kingdom's only legal connection to these relied on its participation in treaties signed prior to the ratification of the declaration (e.g., 1857 and 1903). Furthermore, Iranian law by which the Majlis approved the declaration confirms Iran's intention that "the treaties and conventions which come into consideration are those which the Government will have accepted after the ratification." Therefore, it could not rely on the treaties anterior to the declaration.

49. "Summary of the Judgment of the Court," July 22, 1952, 1–2, ICJ.

50. Ibid.

51. Heiss, "The United States, Great Britain, and the Creation of the Iranian Oil Consortium," 195.

52. Ibid.

53. Louis, "Britain and the Overthrow of the Mosaddeq Government," 153.

54. Under the Truman administration, the Americans refused to provide budgetary assistance to Iran in the absence of an oil settlement. This policy, which was continued under the Eisenhower administration, went against Mosaddiq's hopes for American financial assistance to support Iran's national budget and help offset the loss of foreign exchange that accompanied AIOC's boycott. As Heiss points out, however, "US policymakers attributed far more importance to oil revenues as a percentage of the Iranian national budget than was really the case." In 1950, for example, oil royalties totaled only 12 percent of government

revenue and only 4 percent of national income. Thus, "the bulk of Iran's income was coming from non-oil sources." Iran did make some sales to Italian and Japanese buyers, but total oil exports during 1952 and the first six months of 1953 amounted to only 118,000 tons compared to 31 million tons in 1950, the last complete year of AIOC's Iranian operations. See Heiss, "The United States, Great Britain, and the Creation of the Iranian Oil Consortium," 189, 195, 198.

55. Amy Staples, "Seeing Diplomacy through Bankers' Eyes: The World Bank, the Anglo-Iranian Oil Crisis, and the Aswan High Dam," *Diplomatic History* 26, no. 3 (2002): 397–418. Also see Anghie, *Imperialism, Sovereignty, and the Making of International Law*, 192.

56. Ibid., 397.

57. Anghie, *Imperialism, Sovereignty, and the Making of International Law*, 204.

58. Staples, "Seeing Diplomacy through Banker's Eyes," 398, 400–401.

59. Ibid., 402.

60. Ibid. Also see the text of the letter from Gardner to Mosaddiq, December 28, 1951, and Mosaddiq's reply to Garner, January 3, 1952, 129–133, cited in Hossein Maki, *Ketab-e Seeyah (Black Book: Years of Nationalization from Azar 1330 to Shahrivar 1330)*, vol. 6 (Tehran: Amir Kabir, 1363 (1984–1985)).

61. Staples, "Seeing Diplomacy through Bankers' Eyes," 403. Staples says that the British government and AIOC were anxious that any interim pricing schedule established by the bank not exceed a 50–50 profit split as this had been the basis of AIOC's abortive final offer to the Mosaddiq government.

62. Mosaddeq to Garner, January 3, 1952, 118975, BP.

63. Louis, "Britain and the Overthrow of the Mossadeq Government," 153.

64. Oliver Franks (FO) to ?, January 1952, 118975, BP. This consisted of a small mixed board of "neutral" Iranian directors; a supply of free oil to AIOC for five years; a discounted sale contract to AIOC for fifteen years; and an agreement negotiated through the World Bank as intermediary.

65. Staples, "Seeing Diplomacy through Bankers' Eyes," 403.

66. Gardener to Mosaddiq, December 28, 1951, 130, cited in Maki, *Black Book*.

67. Staples, "Seeing Diplomacy through Bankers' Eyes," 403. As Gardner had suggested in his initial letter to Mosaddiq, the bank fully intended to rehire much of the former AIOC staff, including British nationals, because they were the most familiar with Iranian operations and could enable an immediate resumption. To avoid the possibility of political disputes, the bank also agreed to provide complete information on Iranian operations to both parties, to make concessions AIOC had refused to make, and above all, to give the Iranian government responsibility for providing social and municipal services to the employees of the fields and refinery.

68. Ibid.

69. Ibid.

70. Staples, "Seeing Diplomacy through Banker's Eyes," 404.

71. Abrahamian, *The Coup*, 161–163.

72. Staples, "Seeing Diplomacy through Bankers' Eyes," 405.

73. Levy to Garner, January 29, 1952, Box 15, Folder 6, Walter J. Levy Papers, Collection Number 082428, AHC.

74. Ibid.

75. Staples, "Seeing Diplomacy through Bankers' Eyes," 407.

76. In his personal account of the oil nationalization crisis, Maki concluded that the American government was not interested in providing technical aid or pursuing friendly relations with his country. During his visit to the United States, Maki also met with the Mexican ambassador, who advised him to follow Mexico's successful attempt at nationalization of its oil industry. He urged Maki to reduce Iran's reliance on oil revenues by making cuts in other parts of the national budget. The ambassador warned that as long as Iran showed itself dependent on oil revenues, it would remain caught in Britain's economic trap. Maki recalled his meeting with the head of the American government's Point IV program as well. He concluded that political motives shaped this program in Tehran after nothing came of his requests for economic and technical assistance for the construction of a dam and the development of farming tools and factories in Khuzistan (Maki, *Black Book*, 11–12, 20).

77. For a discussion of the role of Bretton Woods Institutions such as the UN, the IMF, and the World Bank in undermining "Third World" sovereignty, see Anghie, *Imperialism, Sovereignty, and the Making of International Law*, chap. 5.

78. On the Suez Canal crisis, see Staples, "Seeing Diplomacy through Bankers' Eyes."

79. "Persia," April 10, 1952, 118974, BP. Alternative political outcomes included the following: (1) A successor government would either reopen negotiations directly with the British, or (2) a successor government would opt for additional time to convert public opinion into pro-British sentiment that would enable the World Bank to resume negotiations.

80. "Note of Interview with Sir William Fraser Who Was Accompanied by Mr. Gass on 15 July 1952," July 15, 1951, 118974, BP. Fraser feared that if new proposals were agreed on with Americans at the current stage, Tehran would find out, and this would provide the Iranian government with a basis for demanding further concessions and preventing a quick settlement. He indicated a willingness to accept the idea of a "purchasing contract" to purchase Iranian oil at a price that would include compensation, although the amount to be purchased and marketed in the current situation would require "careful consideration." He expressed no objection to a "managing agency" composed of leading British and American oil companies operating in the Middle East. There was a danger, however, that the leading American companies would not want to participate and AIOC would strongly oppose the participation of "second or third rate American companies." The complete exclusion of AIOC from participation in a managing agency would have serious effects on AIOC throughout the Middle East and serious ramifications for its commercial prestige.

81. Louis, "Britain and the Overthrow of the Mosaddeq Government," 153.

82. Ibid., 154.

83. Ibid., 157.

84. Ibid., 151.

85. Ibid., 152.

86. Ibid., 134–135.

87. Ibid., 139–140.

88. Ibid., 145.

89. Ibid., 156.

90. ? to the Chairman (AIOC), "Joint HMG/USA Proposals," August, 11, 1952, 91032, BP. Eden agreed with Acheson's assessment of the threat of communism in Iran, the extent to which Mosaddiq could be regarded as a barrier against it, and the effects on British interests elsewhere in the world. But Eden urged that neither his government nor the US government should rush into "hasty action in going to Mossadeq's aid." Eden expressed his government's willingness to join in an offer based on Acheson's proposals but subject to modifications: (1) US financial aid must be conditional on the Iranian government's agreement to arbitration on acceptable terms. (2) The conclusion with AIOC of suitable arrangements for taking stocks of oil. (3) AIOC would not be asked to abandon its right to take legal action against illicit purchasers of Iranian oil before the conclusion of a final settlement.

91. Butler (Ministry of Fuel and Power), "Persia," August 16, 1952, 91032, BP. The British government would offer to progressively relax some restrictions placed on exports to Iran and on the use of Iranian sterling. Within British government circles, the view was that a "disguised 50/50" was the only way for AIOC to achieve adequate compensation while avoiding "disastrous repercussions elsewhere." The only solution, Butler argued, was for AIOC to negotiate the purchase of oil to which it claimed to have rights, or alternatively, to enroll an international bank as an arbitrator and buy oil from Iran on the basis of a price at which Iran could sell it to AIOC, "who are really the only practicable distributors of the oil."

92. Louis, "Britain and the Overthrow of the Mosaddeq Government," 158.

93. Ibid., 129.

94. Abrahamian, The Coup; Abrahamian, "The 1953 Coup in Iran"; Louis, "Britain and the Overthrow of the Mosaddeq Government."

95. After the overthrow of the Mosaddiq government, "the last covert interventions by US agencies in oil-producing states during the era when a handful of companies still managed worldwide price and supply stability, an era that was coming to an end, were under the Kennedy administration in Iraq in 1963 and the Johnson administration in Indonesia in 1965" (Vitalis, America's Kingdom, 14).

96. Mitchell, Carbon Democracy, 121–122.

97. See note 37.

98. Memorandum of Conversation, May 14, 1951, RG 59 Box 5514, 888.2553-AIOC/5–1451, US (N)ational (A)rchives and (R)ecords (A)administration. Representatives from Standard of Oil of New Jersey, Aramco, and the Gulf Oil Company participated in this meeting.

99. See the testimony of David I. Haberman, attorney-at-law in the Antitrust Division of the Department of Justice, in *Transnational Oil Corporations and U.S. Foreign Policy: Report Together with Individual Views to the Committee on Foreign Relations, United States Senate, by the Subcommittee on Transnational Corporations* (Washington, DC: US Government Printing Office, 1975), part 7, 16–17.

100. Ibid., 25–27.

101. Memorandum by the Assistant Secretary of State for Economic Affairs (Thorp) to the Secretary of State, March 31, 1952, in *Foreign Relations of the United States, 1952–54, General: Economic and Political Matters*, vol. I, part 2, 1259–1261.

102. Sampson, *The Seven Sisters*, 123. Put together by a "radical economist," John M. Blair, the 400-page report published the details of the cartel arrangements of the "Seven Sisters" starting with the Achnacarry and Red Line agreements of 1928 and analyzed the complex sharing arrangements of the largest transnational oil corporations around the world. The findings revealed that the seven companies controlled all the "principal oil-producing areas outside the United States, all foreign refineries, patents, and refining technology: they divided the world markets between them, and shared pipelines and tankers throughout the world; and ... they maintained artificially high prices for oil."

103. Ibid.

104. Memorandum, April 11, 1952, RG 59, Box 5508A, 888.2553/4–1152 NARA.

105. Secretary of State (Dean Acheson) to Chairman of the FTC (Meade), April 25, 1952, in *Foreign Relations of the United States, 1952–54, General: Economic and Political Matters*, vol. I, part 2, 1261.

106. Ibid.

107. Ibid.

108. Consequences of the Future Revelation of the Contents of Certain Government Documents, May 6, 1952 (attachment), *FRUS*, vol. I (1952–54), 1272.

109. President Truman to FTC (Mead), August 15, 1952, *FRUS*, vol. I (1952–54), 1281.

110. Byroade (NEA) to Under Secretary of State, October 2, 1952, RG 59, Box 5509 888.2553/10–252 NARA.

111. Memorandum by Deputy Assistant Secretary of State for Economic Affairs (Linder) to Acting Secretary of State, December 16, 1952, *FRUS*, vol. I (1952–54), 1289.

112. Memorandum by Assistant Legal Adviser for Economic Affairs (Metzger), *FRUS*, vol. I (1952–54), 1304–1305. Report to the NSC by the Departments of State, Defense, Interior, and Justice, January 6, 1953, "Report by Department of State, Defense and the Interior on Security and International Issues Arising from Current Situation in Petroleum," *FRUS*, vol. I (1952–54), 1318–1328.

113. Draft Aide-Mémoire to British, October 7, 1952, RG 59, Box 5509, 888.2553/10–752 NARA.

114. See Emmerglick's testimony, which contains Truman's note to Attorney General McGranery on January 12, 1953, in *Transnational Oil Corporations and US Foreign Policy*, 102. In his testimony, Emmerglick explains that the criminal proceeding would have shortened the process to a matter of months rather than the four to eight years required by a civil action against international cartel arrangements. Only Congress, not the president, could have made the kind of determination about whether national security overrides civil/criminal action because Congress had established antitrust laws as the primary instrument of public policy involving international trade.

115. Ibid., 107.

116. Memorandum of Discussion at the 140th Meeting of the NSC on April 22, 1953, *FRUS*, vol. I (1951–54), 1351. The success of US oil companies in framing their strategy in terms of "national security" has been reproduced in the scholarship. For example, see Stephen D. Krasner, *Defending the National Interest: Raw Materials Investments and U.S. Foreign Policy* (Princeton, NJ: Princeton University Press, 1978).

117. See Haberman's testimony in *Transnational Oil Corporations and US Foreign Policy*, 30–32.

118. On this point, also see Sampson, *The Seven Sisters*, 134. He explains that the case against the cartel continued to be built, but it was not until 1961, under the Kennedy administration, that a case was put forward against the five American "sisters." As in the complaints made eight years prior, the US antitrust chief argued the existence of monopoly power. However, the Iranian consortium ensured that the prosecution's case was less concerned with the joint production agreements, "which were the heart of the collusion." Exxon, Texaco, and Gulf agreed to a settlement without admitting guilt, and the two other cases against Mobil and Socal were "later dismissed without prejudice."

119. On the history of intergovernment-business cooperation in the context of the oil cartel case, see Burton I. Kaufman, *The Oil Cartel Case* (Westport, CT: Greenwood Press, 1978).

120. David Bruce (Acting Secretary of State) to John Foster Dulles, December 2, 1952, RG 59, Box 5510, 888.2553/12–252 NARA.

121. Memorandum of Discussion at the 181st Meeting of the National Security Council, January 21, 1954, *FRUS*, vol. X (1952–54), 907–911. Representatives of the Texas Oil Company expressed concern over the impact of such a solution on Saudi Arabia, the currency problems involved should Iranian oil be marketed as dollar oil instead of sterling oil, and the threat of antitrust action. The

representative of the Socony-Vacuum Company, Charles Harding, argued that it might be easier for the American oil companies to take the lead in negotiations, "provided it were pursuant to a request from the State Department in which section 708 of the Defense Production Act was referred to." See Memorandum, November 21, 1952, RG 59, Box 5510, 888.2553/11–2152 NARA.

122. Ibid.

123. Memorandum, December 11, 1952, RG 59, Box 5510, 888.2553/12–1652 NARA.

124. Linder to Chapman (Petroleum Administrator for Defense), December 17, 1952, RG 59, Box 5510, 888.2553/12–1752 NARA.

125. Ibid.

126. Minutes of Meeting in Secretary's Office, December 9, 1952, RG 59, Box 5510, 888.2553/12–952 NARA.

127. John Jernegan (NEA) to Secretary of State, February 28, 1953, RG 59, Box 5511, 888.2553/2–2853 NARA.

128. AIOC played no role in the actual execution of the coup events. See Louis, "Britain and the Overthrow of the Mosaddeq Government."

129. "The Iranian Situation," September 5, 1953, RG 59, Box 511A, 888.2553/9–253 NARA.

130. September 11, 1953, RG 59, Box 5511A, 888.2553/9–1153 NARA.

131. Memorandum of Discussion at 162nd Meeting of the National Security Council, Washington, September 17, 1953, no. 367, *FRUS*, vol. X (1952–54), 794–796.

132. Secretary of State to Embassy in Iran (signed by Dulles; same message drafted by Hoover and signed by Byroade), September 23, 1953, RG 59 Box 5511A 888.2553/9–2153 NARA.

133. "Proposed Iranian Consortium Plan," January 1954, RG 59, Box 5512, 888.2553/1–?53 NARA.

134. Memorandum of Conversation, September 26, 1953, Box 5511A, 888.2553/9–2653 NARA.

135. Ibid.

136. Henderson to Secretary of State, October 22, 1953, RG 59, Box 5511A, 888.2553/10–2253 NARA.

137. This was not entirely the case as the giant Abqaiq field in Saudi Arabia was suffering from the problem of overproduction.

138. See Howard Page's testimony in *Transnational Oil Corporations and US Foreign Policy*, 289–290.

139. Memorandum of Conversation, November 24, 1953, RG 59, Box 5512, 888.2553/11–2453 NARA.

140. Memorandum by the Legal Adviser (Phleger), December 8, 1953, RG 59, Box 5512, 888.2553/12–853 NARA. Also see Memorandum of Conversation, December 7, 1953, ibid. (Koegler of Standard Oil of New Jersey, Foster of

Socony-Vacuum, Phleger, and Jernegan present). For Hoover's discussion with representatives of AIOC and the British Foreign Office, see Hoover to Department of State, December 4, 1953, 848 (601.4188/12–453).

141. Attorney General (Brownell) to the NSC, January 20, 1954, RG 59, Box 5509, 888.2553/1–2054 NARA.

142. Memorandum of Discussion at the 178th Meeting of the National Security Council, December 30, 1953, *FRUS* vol. X, 858–861.

143. The act authorized government officials to consult with business and other groups to encourage voluntary action to carry out national defense objectives: "The President may request members of an industry to enter into a voluntary agreement or program, upon a finding that it is vital to the national defense. No act or omission to act pursuant to any such voluntary agreement, the Act provides, shall be construed to be within the prohibitions of the antitrust laws while the Act is in effect." See Memorandum of Discussion at the 223d Meeting of the National Security Council, November 9, 1954, *FRUS*, vol. I (1952–54), 1375–1378.

144. Citino, Defending the "Postwar Petroleum Order," 138.

145. With regard to the question of oil and the coup, Mitchell, *Carbon Democracy*, and Abrahamian, *The Coup*, both make this point.

146. Heiss, "The United States, Great Britain, and the Creation of the Iranian Oil Consortium," 515.

147. Ibid., 513.

148. Ibid., 514.

149. "Draft for Discussion: Outline of Corporate Set-Up for Oil Purchases from Iran by a Consortium," January 14, 1952, 91032, BP.

150. FO to Washington, August 23, 1953, no. 395, 11/726, PREM.

151. FO to AIOC (?), "'Skeleton' Memorandum on Middle East Oil," July 31, 1953, 1–6, 91032, BP.

152. See Irene Gendzier, *Notes from the Minefield: United States Intervention in Lebanon, 1945–1958*, 2nd ed. (New York: Columbia University Press, 2006).

153. FO to AIOC (?), "'Skeleton' Memorandum on Middle East Oil," July 31, 1953, 1–6, 91032, BP.

154. Enclosure "Document Handed to Hoover and Ambassador Henderson by Persian Foreign Minister, Entezam, on 1 November 1953 (Translation from the Persian)," in Belgrave to Beeley, November 4, 1953, 371/104616, FO.

155. Entezam also expressed his government's wishes that the World Bank act as an intermediary in the conclusion of the consortium arrangement to examine accounts and supervise the agreement's enforcement. It is important to note here that in January 1953, the Iranian government approved a "Persian Mining Law," to be managed by the Ministry of National Economy, and replacing the annulled mining law of 1938–1939. AIOC studied the law carefully. The articles stated that petroleum was the absolute property of the government and that exploration and exploitation of these mines could be effected only by

the government. According to AIOC, "There appears nothing in this 1953 Act, which will prohibit the employment by the Government or by NIOC of foreign concern to operate the industry." See "Persian Legislation Affecting Negotiations with the Persian Government or NIOC," December 15, 1953, appendixes A–F, 58252, BP.

156. "Telegram from Washington to London," March 5, 1954, no. 341, 79661, BP.

157. Heiss, "The United States, Great Britain, and the Creation of the Iranian Oil Consortium," 519.

158. Washington to FO, February 19, 1954, no. 232–235, 11/726, PREM.

159. "Conclusions Minute 3," February 22, 1954, no. 225, 11/726, PREM.

160. FO to Washington, March 2, 1954, FO to Washington, no. 210–211, 11/726, PREM.

161. Heiss, "The United States, Great Britain, and the Creation of the Iranian Oil Consortium," 522.

162. Ibid., 523.

163. Mitchell, "Carbon Democracy," 408.

164. Ibid., 531.

165. Memorandum of Conversation, January 12, 1954, RG 59 Box 5512, 888.2553/1–1254 NARA. Also on the formulation of APQ to avoid an oil glut, see Sampson, *The Seven Sisters*, 131.

166. See Fereidun Fesharaki, *Development of the Iranian Oil Industry: International and Domestic Aspects* (New York: Praeger, 1976), 54–55.

167. Mitchell, "McJihad."

168. Iricon Agency Ltd. grouped six American oil companies that together held the remaining 5 percent of the consortium arrangement: 1/6 each to American Independent Oil, Getty Oil, and Charter Oil; 1/3 to Atlantic Richfield (Arco— later owned by BP); and 1/12 each to Continental (Conoco) and Standard Oil of Ohio. The signing of the agreement on October 29, 1954, marked this increase in (American) participants from eight companies to fourteen. Walden has a slightly different breakdown: 3/12 to Richfield Oil Corp., 2/12 to American Independent Oil Co., 2/12 to Signal Oil Co., 1/12 to Standard Oil Co. (Ohio), 1/12 to San Jacinto Petroleum Corp., 1/12 to Getty Oil, 1/12 to Tidewater Oil Co., 1/12 to Atlantic Refining Co. See Walden, "The International Petroleum Cartel in Iran," 111n239.

169. Fesharaki, *Development of the Iranian Oil Industry*, 54–55.

170. See Levinson's reply to Shafer in *Transnational Oil Corporations and US Foreign Policy*, 254–256.

171. Fesharaki, *Development of the Iranian Oil Industry*, 54–55.

172. *Transnational Oil Corporations and US Foreign Policy*, 264. To elaborate, Shafer explains that, "… unfortunately, each participant will not have the same requirements normally year to year to off-lift crude oil. So they must have some

sort of an arrangement between themselves for fixing the basic levels and which lifting will be determined. So that each participant has equal use of the facilities and the capital input that he has made into developing the property."

173. "Notes on Interview," May 15, 1954, 371/110061, FO. Also see FO to Baghdad, May 28, 1954, 371/110061, FO.

174. *Transnational Oil Corporations and US Foreign Policy*, part 7, 269.

175. Memorandum of Conversation, May 17, 1954, RG 59, Box 5512, 888.2553/5–1754 NARA.

176. Ambassador in Iran (Henderson) to Department of State, June 30, 1954, *FRUS*, vol. X (1952–54), 1033.

177. Ambassador in Iran (Henderson) to Department of State, July 6, 1954, *FRUS*, vol. X, 1038.

178. Ibid.

179. Henderson to Secretary of State, September 29, 1954, RG 59, Box 5513, 888.2553/9–2954 NARA.

180. Henderson to Secretary of State, October 1, 1954, RG 59, Box 5513, 888.2553/10–154 NARA.

181. The final agreement saw CFP, the French company, settling for a 6 percent share, 14 percent to Shell, and 40 percent each to AIOC and the group of five US oil majors. The British and American governments agreed that AIOC should receive a sum on the order of £100 million as compensation.

182. See note 153.

183. On the battle between the American oil independents and oil majors for a percentage share in the consortium, see Ralph Davies of American Independent Oil Co. to J. F. Dulles (Secretary of State), October 15, 1954, and Page's testimony in *Transnational Oil Corporations and U.S. Foreign Policy*, 249–251, 295.

184. Heiss, "The United States, Great Britain, and the Creation of the Iranian Oil Consortium," 532.

185. See Mitchell, "Carbon Democracy," 419, 421, for a discussion of the sociotechnical mechanisms that tied together democracy in the West, oil, and the US dollar. As he explains, the commercialization of weapons exports from the United States was "made possible by establishing a series of linkages between the Western import of oil from the Middle East, the flow of dollars to producer countries" such as Iran and Saudi Arabia, "the production of political vulnerabilities and military threats to the further flow of oil," and finally, "the use of petrodollars to purchase arms from the West as a protection against those threats."

186. Walden says that "viewed from any standpoint," the Iranian consortium "represented a resounding triumph for the British and a catastrophic defeat for the national aspirations of Iran." The arrangement was particularly remunerative for American interests, seeing as how most of the net income from operations in Iran would be tax free (Walden, "The International Petroleum Cartel in Iran," 112–113).

187. Sampson, *The Seven Sisters*, 135. Unlike the consortium arrangements in Iraq and Saudi Arabia, all seven sisters were involved in limiting oil production from Iran through the Iranian consortium. Two or more of the seven were in control of each of the major oil-producing countries of the Middle East. Four of the partners of Aramco were also in Iran.

Conclusion

1. MacKenzie, "An Equation and Its Worlds," 252.

2. For a technical study of how STS can be used to explore the devices through which politics and policy are done, see Jan-Peter Voss and Basil Bornemann, "The Politics of Reflexive Governance: Challenges for Designing Adaptive Management and Transition Management," *Ecology and Society* 16, no. 2 (2011): 9. Lucy Suchman proposes the device of "configuration" as a "method assemblage" for recovering "heterogeneous relations that technologies fold together," which captures what I am doing by following the construction of sociotechnical devices in the reconfiguration of politics (Lucy Suchman, "Configuration," in *Inventive Methods: The Happening of the Social*, ed. Celia Lury and Nina Wakeford, 48–60 (London: Routledge, 2014)).

3. Barry, "Technological Zones," 250.

4. Sampson, *The Seven Sisters*, 163. OPEC succeeded in preventing further reductions in the posted price of oil and loss of control to the oil corporations over the unilateral fixing of prices. However, OPEC members could not agree on how to fix prices or restrict production on their own. According to Robert Mabro, the posted price of oil refers to a statement of the oil price announced by a firm indicating what it will pay for the oil or the price at which it will sell it. It is circulated publicly. In the 1950s and 1960s, they were "used to indicate a company's selling price." For a discussion of the formulas, see Robert Mabro, *On Oil Price Concepts* (Oxford: Oxford Institute for Energy Studies, 1984), 6.

5. See "How We Organized Strike That Paralyzed Shah's Regime: First-Hand Account by Iranian Oil Workers," in *Oil and Class Struggle*, ed. Petter Nore and Terisa Turner, 293–301 (London: Zed Press, 1980).

Bibliography

Archives

Asnad-i Melli-i Iran [Iranian National Archives], Tehran
British Petroleum Archive, University of Warwick, Coventry
India Office Records, British Library, London
National Archives, Kew Gardens, London
National Archives and Records Administration, College Park, MD

Private Manuscript Collections

Walter J. Levy Papers, American Heritage Center, Laramie, WY

Court Proceedings and Law Journals

I.C.J. Pleadings, Anglo-Iranian Co. Case (United Kingdom v. Iran)
League of Nations Official Journal

Published Government Documents in English and Farsi

Asnad-e Naft, Tehran: Ministry of Foreign Affairs publications.
Foreign Relations of the United States, 1952–1954, Vol. I, Part 2: General: Economic and Political Matters. Washington, DC: US Government Printing Office, 1983.
Foreign Relations of the United States, 1952–1954, Vol. X: Iran, 1951–1954. Washington, DC: US Government Printing Office, 1989.
Multinational Oil Corporations and U.S. Foreign Policy: Report Together with Individual Views to the Committee on Foreign Relations, United States Senate / by the Subcommittee on Multinational Corporations. Washington, DC: US Government Printing Office, 1975.
Naft dar Durah-i Reza Shah. Tehran: Sazman-i Chap va Intisharat-i Vizarat-i Farhang va Irshad-i Islami, 1378.

Newspapers and Periodicals in Farsi and English

Bulletin of the American Association of Petroleum Geologists
Habl al-Matin
Journal of the Institute of Petroleum
Peyk-e Naft
Shafagh-i Surkh
World Petroleum Congress Proceedings

Books and Articles in Farsi and English

Abbasi-Shahni, Danesh. *Tarikh-i Masjid Suleiman (History of Masjid Suleiman)*. Tehran: Hirmand Press, 1995.

Abdelrehim, Neveen, Josephine Maltby, and Steven Toms. "Imperialism, Employment, and Racial Discrimination: The Anglo-Iranian Oil Company, 1933–51." Paper presented at the Business History Conference, University of Georgia, Athens, March 2010. http://www.thebhc.org/annmeet/program10.html#E5.

Abrahamian, Ervand. *The Coup: 1953, the CIA, and the Roots of Modern U.S.-Iranian Relations*. New York: The New Press, 2013.

Abrahamian, Ervand. "The 1953 Coup in Iran." *Science and Society* 65, no. 2 (2001): 182–215.

Abrahamian, Ervand. *Iran between Two Revolutions*. Princeton, NJ: Princeton University Press, 1982.

Abtahi, Ali Reza. *Naft va Bakhtiyari-ha (Oil and the Bakhtiari)*. Tehran: Studies in the Contemporary History of Iran, 2005.

Allen, Michael T., and Gabrielle Hecht, eds. *Technologies of Power: Essays in Honor of Thomas Parke Hughes and Agatha Chipley Hughes*. Cambridge, MA: MIT Press, 2001.

Alsharhan, A. S., and A. E. M. Nairn. *Sedimentary Basins and Petroleum Geology of the Middle East*. New York: Elsevier, 2003.

Amir-Alai. *Shams al-Din: Khaterat-e Man*. Tehran: My Memoirs, 1984.

Anderson, Irvine H. "The American Oil Industry and the Fifty-Fifty Agreement of 1950." In *Musaddiq, Iranian Nationalism, and Oil*, ed. James A. Bill and William R. Louis, 143-159. Austin: University of Texas Press, 1988.

Anderson, Irvine H. *Aramco, the U.S., and Saudi Arabia: A Study of the Dynamics of Foreign Oil Policy 1933–1950*. Princeton, NJ: Princeton University Press, 1981.

Anghie, Antony. *Imperialism, Sovereignty, and the Making of International Law*. Cambridge: Cambridge University Press, 2007.

Atabaki, Touraj. "Far from Home, but at Home: Indian Migrant Workers in the Iranian Oil Industry." *Studies in History* 31, no. 1 (2015): 85–114.

Atabaki, Touraj. "From '*Amaleh* (Labor) to *Kargar* (Worker): Recruitment, Work Discipline and Making of the Working-Class in the Persian/Iranian Oil Industry." *International Labor and Working Class History* 84 (Fall 2013): 159–175.

Avery, Peter. G. R. G. Hambly, and C. P. Melville, eds. "Iran Under the Later Qajars, 1848-1922." In *The Cambridge History of Iran*, Volume 7: *From Nadir Shah to the Islamic Republic*, 187–191. Cambridge: Cambridge University Press, 1991.

Avery, Peter. G. R. G. Hambly, and C. P. Melville, eds. "The Iranian Oil Industry." In *The Cambridge History of Iran*, Volume 7: *From Nadir Shah to the Islamic Republic*, 639–705. Cambridge: Cambridge University Press, 1991.

Bakhtiari, Ali Morteza Samsam. *Last of the Khans*. New York: iUniverse, 2006.

Bamberg, James H. *British Petroleum and Global Oil, 1950–1975: The Challenge of Nationalism*. Vol. 3 of *The History of the British Petroleum Company*. Cambridge: Cambridge University Press, 2000.

Bamberg, James H. *The History of the British Petroleum Company, Volume 2: The Anglo-Iranian Years, 1928–1954*. Cambridge: Cambridge University Press, 1994.

Banani, Amin. *The Modernization of Iran, 1921–1941*. Stanford, CA: Stanford University Press, 1961.

Barry, Andrew. *Material Politics: Disputes along the Pipeline*. Malden, MA: Wiley, 2013.

Barry, Andrew. "Technological Zones." *European Journal of Social Theory* 9, no. 2 (2006): 239–253.

Batatu, Hanna. *The Old Social Classes and the Revolutionary Movements of Iraq: A Study of Iraq's Old Landed and Commercial Classes and of Its Communists, Ba'thists, and Free Officers*. London: Saqi Books, 2004.

Bayat, Kaveh. "With or without Workers in Reza Shah's Iran: Abadan, May 1929." In *The State and the Subaltern: Modernization, Society and the State in Turkey and Iran*, ed. Touraj Atabaki, 111–122. New York: I. B. Tauris, 2007.

Beblawi, Hazem, and Giacomo Luciani, eds. *The Rentier State, Nation, State, and Integration in the Arab World*. New York: Croom Helm, 1987.

Beck, Peter J. "The Anglo-Persian Oil Dispute 1932–33." *Journal of Contemporary History* 9, no. 4 (1974): 123–151.

Bennett, Tony, and Patrick Joyce, eds. *Material Powers: Cultural Studies, History and the Material Turn*. New York: Routledge, 2010.

Best, Jacqueline. "Bureaucratic Ambiguity." *Economy and Society* 41, no. 1 (2012): 86–106.

Biddleston, A. J., and J. Bridgewater. "From Mining to Chemical Engineering at the University of Birmingham." In *One Hundred Years of Chemical*

Engineering, ed. N. A. Peppas, 237–244. Dordrecht, The Netherlands: Kluwer Academic Publishers, 1989.

Bill, James E., and William R. Louis, eds. *Musaddiq, Iranian Nationalism, and Oil*. Austin: University of Texas Press, 1988.

Bowden, Gary. "The Social Construction of Validity in Estimates of US Crude Oil Reserves." *Social Studies of Science* 15, no. 2 (1985): 207–240.

Bowker, Geoffrey. "How to Be Universal: Some Cybernetic Strategies, 1943–70." *Social Studies of Science* 23 (February 1993): 107–127.

Bowker, Geoffrey. *Science on the Run: Information Management and Industrial Geophysics at Schlumberger, 1920–1940*. Cambridge, MA: MIT Press, 1994.

Bowker, Geoffrey, and Susan Leigh Star. *Sorting Things Out: Classification and Its Consequences*. Cambridge, MA: MIT Press, 2001.

Brockway, Thomas. "Britain and the Persian Bubble, 1888–1892." *Journal of Modern History* 13, no. 1 (1941): 36–47.

Busk, H. G. and H. T. Mayo. "Some Notes on the Geology of the Persian Oil-fields." *Journal of the Institute of Petroleum* 5 (1918–1919): 3–26.

Cadman, John. "Science in the Petroleum Industry." *World Petroleum Congress Proceedings* 1 (1933): 56 3-570.

Caliskan, Koray, and Michel Callon. "Economization, Part 1: Shifting Attention from the Economy towards Processes of Economization." *Economy and Society* 38, no. 3 (2009): 368–398.

Callon, Michel, ed. *The Laws of the Markets*. Malden, MA: Blackwell, 1998.

Callon, Michel. "Society in the Making: The Study of Technology as a Tool for Sociological Analysis." In *The Social Construction of Technological Systems: New Directions in the Sociology and History of Technology*, ed. Wiebe E. Bijker, Thomas P. Hughes, and Trevor J. Pinch, 83–103. Cambridge, MA: MIT Press, 1987.

Callon, Michel. "Some Elements of a Sociology of Translation: Domestication of Scallops and the Fishermen of St. Brieuc Bay." In *The Science Studies Reader*, ed. Bruno Biagioli, 67–94. New York: Routledge, 1999.

Callon, Michel. What Does It Mean to Say That Economics Is Performative? In *Do Economists Make Markets? On the Performativity of Economics*, ed. Donald A. MacKenzie, Fabian Muniesa and Lucia Siu. 311–357. Princeton, NJ: Princeton University Press, 2007.

Callon, Michel, Pierre Lascoumes, and Yannick Barthe. *Acting in an Uncertain World: An Essay on Technical Democracy*. Trans. Graham Burchell. Cambridge, MA: MIT Press, 2009.

Callon, Michel, Yuval Millo, and Fabian Muniesa, eds. *Market Devices*. Malden, MA: Blackwell, 2007.

Carse, Ashley. *Beyond the Big Ditch: Politics, Ecology, and Infrastructure at the Panama Canal*. Cambridge, MA: MIT Press, 2014.

Cattan, Henry. *The Evolution of Oil Concessions in the Middle East and North Africa*. Dobbs Ferry, NY: Oceana Publications, 1967.

Chatterjee, Partha. *The Nation and Its Fragments: Colonial and Postcolonial Histories*. Princeton, NJ: Princeton University Press.

Chatterjee, Partha. *Nationalist Thought and the Colonial World: A Derivative Discourse*. London: Zed Books, 1986.

Cheng, Bin. "The Anglo-Iranian Dispute." *World Affairs* 5 (1951): 387–405.

Citino, Nathan J. "Defending the 'Postwar Petroleum Order': The US, Britain and the 1945 Saudi-Onassis Tanker Deal." *Diplomacy & Statecraft* 11, no. 2 (2000): 137–160.

Clawson, Patrick. "Capital Accumulation in Iran." In *Oil and Class Struggle*, ed. Terisa Turner and Petter Nore, 143–171. London: Zed Press, 1980.

Corley, T. A. B. *A History of the Burmah Oil Company 1886–1924*. London: Heinemann, 1983.

Coronil, Fernando. *The Magical State: Nature, Money, and Modernity in Venezuela*. Chicago: University of Chicago Press, 1997.

Crinson, Mark. "Abadan: Planning and Architecture under the Anglo-Iranian Oil Company." *Planning Perspectives* 12, no. 3 (1997): 341–359.

Crinson, Mark. *Modern Architecture and the End of Empire*. Burlington, VT: Ashgate, 2003.

Cronin, Stephanie. "The Politics of Debt: The Anglo-Persian Oil Company and the Bakhtiyari Khans." *Middle Eastern Studies* 40, no. 4 (2004): 1–34.

Cronon, William. *Nature's Metropolis: Chicago and the Great West*. New York: Norton, 1991.

Curzon, George N. *Persia and the Persian Question*. 2 vols. London: Longmans, Green, 1892.

Damluji, Mona. "The Oil City in Focus: The Cinematic Spaces of Abadan in the Anglo-Iranian Oil Company's Persian Story." *Comparative Studies of South Asia, Africa and the Middle East* 33, no. 1 (2013): 75–88.

De Laet, Marianne, and Annemarie Mol. "The Zimbabwe Bush Pump: Mechanics of a Fluid Technology." *Social Studies of Science* 30, no. 2 (2000): 225–263.

Dobe, Michael E. "A Long Slow Tutelage in Western Ways of Work: Industrial Education and the Containment of Nationalism in Anglo-Iranian and Aramco, 1923–1963." Doctoral dissertation, Rutgers University, 2008.

Dunstan, A. E., and George Sell, eds. *World Petroleum Congress Proceedings* 1 (1933).

Edwards, Paul N. *A Vast Machine: Computer Models, Climate Change, and the Politics of Global Warming*. Cambridge, MA: MIT Press, 2010.

Ehsani, Kaveh. "Social Engineering and the Contradictions of Modernization in Khuzestan: A Look at Abadan and Masjed-Soleyman." *International Review of Social History* 48, no. 3 (2003): 361–399.

Elm, Mostafa. *Oil, Power, and Principle: Iran's Oil Nationalization and Its Aftermath*. Syracuse, NY: Syracuse University Press, 1992.

Elwell-Sutton, Laurence Paul. *Persian Oil: A Study in Power Politics*. History and Politics of Oil Series. Westport, CT: Hyperion Press, 1976.

Enos, John. *Petroleum Progress and Profits: A History of Process Innovation*. Cambridge, MA: MIT Press, 1962.

Fateh, Mostafa. *Panjah Saleh-i Naft-i Iran (Fifty Years of Iranian Oil)*. Tehran: Payam Press, 1970.

Ferrier, R. W. *The History of the British Petroleum Company, Volume 1: The Developing Years, 1901–1932*. Cambridge: Cambridge University Press, 1982.

Ferrier, R. W. "Petrol Advertising in the Twenties and Thirties." *Journal of Advertising History* 9, no. 1 (1986): 29–51.

Fesharaki, Fereidun. *Development of the Iranian Oil Industry: International and Domestic Aspects*. New York: Praeger, 1976.

Fitzgerald, Edward Peter. "Business Diplomacy: Walter Teagle, Jersey Standard, and the Anglo-French Pipeline Conflict in the Middle East, 1930–31." *Business History Review* 67, no. 2 (1993): 207–245.

Fleig, Allison. *Oil Empire*. Cambridge, MA: Harvard University Press, 2005.

Forbes, R. J., and D. R. O'Beirne. *The Technical Development of the Royal Dutch / Shell 1890–1940*. Leiden: Brill, 1957.

Ford, Alan W. *The Anglo-Iranian Oil Dispute of 1951–1952: A Study of the Role of Law in the Relations of States*. Berkeley: University of California Press, 1954.

Frank, Alison. *Oil Empire: Visions of Prosperity in Austrian Galicia*. Cambridge, MA: Harvard University Press, 2007.

Fursenko, A. A. *The Battle for Oil: The Politics and Economics of International Corporate Conflict over Petroleum, 1860–1930*. Greenwich, CT: JAI Press, 1990.

Galpern, Steven G. *Money, Oil, and Empire in the Middle East: Sterling and Postwar Imperialism 1944–1971*. Cambridge: Cambridge University Press, 2009.

Garthwaite, Gene R. *Khans and Shahs: A Documentary Analysis of the Bakhtiyari in Iran*. Cambridge: Cambridge University Press, 1983.

Gelb, Alan H. *Oil Windfalls: Blessing or Curse?* New York: Oxford University Press, 1988.

Gendzier, Irene. *Notes from the Minefield: United States Intervention in Lebanon, 1945–1958*. 2nd ed. New York: Columbia University Press, 2006.

Gibson, H. S. "Scientific Unit Control." Paper presented at the World Petroleum Congress, Imperial College of Science and Technology, South Kensington, London, July 19–25, 1933.

Goldberg, Ellis, Erik Wibbels, and Eric Mvukiyehe. "Lessons from Strange Cases: Democracy, Development, and the Resource Curse in the U.S. States." *Comparative Political Studies* 41, nos. 4–5 (2008): 477–514.

Gross, Matthias. "Give Me an Experiment and I Will Raise a Laboratory." *Science, Technology & Human Values* 41, no. 3 (2015): 1–22.

Guha, Ranajit, and Gayatri Chakravorty Spivak, eds. *Selected Subaltern Studies*. New York: Oxford University Press, 1988.

Hager, Dorsey. *Fundamentals of the Petroleum Industry*. New York: McGraw-Hill, 1939.

Halliday, Fred. "Iran: Trade Unions and Working Class Opposition." *MERIP Reports*, no. 71 (1978): 7–13.

Hausmann, Ricardo. "State Landed Property, Oil Rent and Accumulation in Venezuela: An Analysis in Terms of Social Relations." Doctoral dissertation, Cornell University, 1981.

Hecht, Gabrielle. *The Radiance of France: Nuclear Power and National Identity after World War I*. Cambridge, MA: MIT Press, 1998.

Hecht, Gabrielle. "Rupture-Talk in the Nuclear Age: Conjugating Colonial Power in Africa." *Social Studies of Science* 32, nos. 5–6 (2002): 691–727.

Heiss, Mary Ann. "The International Boycott of Iranian Oil and the Anti-Mosaddeq Coup of 1953." In *Mohammed Mosaddeq and the 1953 Coup in Iran*, ed. Mark J. Gasiorowski and Malcolm Byrne, 178–200. Syracuse, NY: Syracuse University Press, 2004.

Heiss, Mary Ann. "The United States, Great Britain, and the Creation of the Iranian Oil Consortium, 1953–1954." *International History Review* 16, no. 3 (1994): 441–600.

Holm, Petter, and Kare Nolde Nielsen. "Framing Fish, Making Markets: The Construction of Individual Transferable Quotas (ITQs)." In *Market Devices*, ed. Michel Callon, Yuval Millo, and Fabian Muniesa, 173–195. Malden, MA: Blackwell, 2007.

Home, R. K. "Town Planning and Garden Cities in the British Colonial Empire, 1910–1940." *Planning Perspectives* 5, no. 1 (1990): 23–37.

Hooglund, Eric J. *Land and Revolution in Iran, 1960–1980*. Austin: University of Texas Press, 1982.

"How We Organized Strike That Paralyzed Shah's Regime: First-Hand Account by Iranian Oil Workers." In *Oil and Class Struggle*, Petter Nore and Terisa Turner, eds., 292–301. London: Zed Press, 1980.

Hughes, Thomas. *Networks of Power: Electrification in Western Society, 1880–1930*. Baltimore: Johns Hopkins University Press, 1983.

Islamoglu, Huri. *State and Peasant in the Ottoman Empire*. Leiden: Brill, 1994.

Jack (Kent), Marian. "The Purchase of British Government's Shares in the British Petroleum Company, 1912–1914." *Past & Present* 39, no. 1 (April 1968): 139–168.

Jones, Geoffrey. *The State and the Emergence of the British Oil Industry.* London: Macmillan, 1981.

Jones, Geoffrey, Judith Nichol, Grigori Gerenstein, and Frances Bostock. *Banking and Empire in Iran: The History of the British Bank of the Middle East.* 2 vols. New York: Cambridge University Press, 1986.

Jones, Toby. *Desert Kingdom: How Oil and Water Forged Modern Saudi Arabia.* Cambridge, MA: Harvard University Press, 2010.

Karl, Terry Lynn. *The Paradox of Plenty: Oil Booms and Petro-States.* Berkeley: University of California Press, 1997.

Karshenas, Massoud. *Oil, State, and Industrialization in Iran.* Cambridge: Cambridge Univeristy Press, 1990.

Kasravi, Ahmad. *Tarikh-i Pansad Saleh-i Khuzistan (Five-Hundred-Year History of Khuzistan).* Tehran: Payam Press, 1950.

Katouzian, Homa. *Musaddiq and the Struggle for Power in Iran.* 2nd ed. New York: I. B. Tauris, 1999.

Katouzian, Homa. *The Political Economy of Modern Iran: Despotism and Pseudo-Modernism, 1926–1979.* New York: New York University Press, 1981.

Kaufman, Burton I. *The Oil Cartel Case.* Westport, CT: Greenwood Press, 1978.

Kazemi, Ranin. "The Tobacco Protest in Nineteenth-Century Iran: The View from a Provincial Town." *Journal of Persianate Studies* 7 (2014): 251–295.

Keddie, Nikkie R. *Modern Iran: Roots and Results of Revolution.* New Haven, CT: Yale University Press, 2003.

Kent, Marian. *Oil and Empire: British Policy and Mesopotamian Oil, 1900–1921.* London: Harper & Row, 1976.

Khazeni, Arash. "Opening the Land: Tribes, State, and Ethnicity in Qajar Iran, 1800–1911." Doctoral dissertation, Yale University, 2005.

Khazeni, Arash. *Tribes and Empire on the Margins of Nineteenth-Century Iran.* Seattle: University of Washington Press, 2009.

Kinzer, Stephen. *All the Shah's Men: An American Coup and the Roots of Middle East Terror.* Hoboken, NJ: Wiley, 2008.

Krasner, Stephen D. *Defending the National Interest: Raw Materials Investments and U.S. Foreign Policy.* Princeton, NJ: Princeton University Press, 1978.

Ladjevardi, Habib. *Labor Unions and Autocracy in Iran.* Syracuse, NY: Syracuse University Press, 1985.

Lambton, Ann K. S. *Landlord and Peasant in Persia: A Study of Land Tenure and Land Revenue Administration.* London: Oxford University Press, 1953. Reprint New York: I. B. Tauris, 1991.

Latour, Bruno. *Aramis or the Love of Technology.* Trans. Catherine Porter. Cambridge, MA: Harvard University Press, 1996.

Latour, Bruno. *Pandora's Hope: Essays on the Reality of Science Studies.* Cambridge, MA: Harvard University Press, 1999.

Latour, Bruno. *Politics of Nature: How to Bring the Sciences into Democracy.* Trans. C. Porter. Cambridge, MA: Harvard University Press, 2004.

Latour, Bruno. *Reassembling the Social: An Introduction to Actor-Network-Theory.* New York: Oxford University Press, 2005.

Latour, Bruno. *Science in Action: How to Follow Scientists and Engineers through Society.* Cambridge, MA: Harvard University Press, 1987.

Layard, Austen Henry. *Early Adventures in Persia, Susiana, and Babylonia.* London: John Murray, 1894.

Lees, G. M. "Reservoir Rocks of Persian Oilfields." *Bulletin of the American Association of Petroleum Geologists* 17, no. 3 (1933): 229–240.

Lees, G. M. "The Source Rocks of Persian Oil." Paper presented at the World Petroleum Congress, Imperial College of Science and Technology, South Kensington, London, July 19–25, 1933.

Lepinay, Vincent. "Parasitic Formulae: The Case of Capital Guarantee Products." In *Market Devices*, ed. Michel Callon, Yuval Millo, and Fabian Muniesa, 261–283. Malden, MA: Blackwell, 2007.

Levi, Ron, and Mariana Valverde. "Studying Law by Association: Bruno Latour Goes to the Conseil d'État." *Law & Social Inquiry* 33, no. 3 (2008): 805–825.

Lilley, Ernest Raymond. *The Geology of Petroleum and Natural Gas.* New York: Van Nostrand, 1928.

Longhurst, Henry. *Adventure in Oil: The Story of British Petroleum.* London: William Clowes and Sons, 1959.

Louis, William Roger. "Britain and the Overthrow of the Mosaddeq Government." In *Mohammad Mosaddeq and the 1953 Coup in Iran*, ed. Mark J. Gasiorowski and Malcolm Byrne, 126–177. Syracuse, NY: Syracuse University Press, 2004.

Louis, William Roger. *The British Empire in the Middle East, 1945–1951.* Oxford: Clarendon Press, 1984.

Louis, William Roger. "The Persian Oil Crisis." In *The British Empire in the Middle East, 1945–1951*, 632–690. Oxford: Clarendon Press, 1984.

Lucier, Paul. *Scientists and Swindlers: Consulting on Coal and Oil in America, 1820–1890.* Baltimore: Johns Hopkins University Press, 2008.

Lussac, Samuel. "The State as a (Oil) Company? The Political Economy of Azerbaijan," *GARNET Working Paper* 74/10. Warwick, UK: Center for the Study of Globalisation and Regionalisation, University of Warwick, 2010.

Mabro, Robert. *On Oil Price Concepts.* Oxford: Oxford Institute for Energy Studies, 1984.

MacKenzie, Donald. "An Equation and Its Worlds: Bricolage, Exemplars, Disunity and Peformativity in Financial Economics." *Social Studies of Science* 33, no. 6 (2003): 831–868.

MacKenzie, Donald. "Missile Accuracy: A Case Study in the Social Processes of Technological Change." In *The Social Construction of Technological Systems: New Directions in the Sociology and History of Technology*, ed. Wiebe E. Bijker, Thomas P. Hughes, and Trevor J. Pinch, 195–222. Cambridge, MA: MIT Press, 1987.

MacKenzie, Donald. "Nuclear Missile Testing and the Social Construction of Accuracy." In *The Science Studies Reader*, ed. Mario Biagioli, 343–357. New York: Routledge, 1999.

Mafi, Homayoun. "Iran's Oil Concession Agreements and the Role of the National Iranian Oil Company: Economic Development and Sovereign Immunity." *Natural Resources Journal* 48 (2008): 407–430.

Mahdavy, Hossein. "The Patterns and Problems of Economic Development in Rentier States: The Case of Iran." In *Studies in the Economic History of the Middle East: From the Rise of Islam to the Present Day*, ed. M. A. Cook, 428–467. London: Oxford University Press, 1970.

Maki, Hossein. *Ketab-E Seeyah (Black Book: Years of Nationalization from Azar 1330 to Shahrivar 1330)*. Vol. 6. Tehran: Amir Kabir, 1363 (1984–1985).

Marsh, Steve. "Anglo-American Crude Diplomacy: Multinational Oil and the Iranian Oil Crisis, 1951–53." *Contemporary British History* 21, no. 1 (2007): 25–53.

Marsh, Steve. *Anglo-American Relations and Cold War Oil: Crisis in Iran*. New York: Palgrave Macmillan, 2003.

McGoey, Linsey, ed. "Strategic Unknowns: Towards a Sociology of Ignorance." Special issue, *Economy and Society* 41, no. 1 (2012).

Mellinger, Philip J. *Race and Labor in Western Copper: The Fight for Equality, 1896–1918*. Tucson: University of Arizona Press, 1995.

Mitchell, Timothy. "Carbon Democracy." *Economy and Society* 38, no. 3 (2009): 399–432.

Mitchell, Timothy. *Carbon Democracy: Political Power in the Age of Oil*. London: Verso, 2011.

Mitchell, Timothy. *Colonising Egypt*. Berkeley: University of California Press, 1991.

Mitchell, Timothy. "Culture and Economy." In *The Sage Handbook of Cultural Analysis*, ed. Tony Bennett and John Frow, 447–466. London: Sage, 2008.

Mitchell, Timothy. "Limits of the State: Beyond Statist Approaches and Their Critics." *American Political Science Review* 85 (1991): 77–96.

Mitchell, Timothy. "McJihad." *Social Text* 73 (2002): 1–18.

Mitchell, Timothy. "The Properties of Markets." In *Do Economists Make Markets? On the Performativity of Economics*, ed. Donald MacKenzie, Fabian Muniesa, and Lucia Siu, 244–275. Princeton, NJ: Princeton University Press, 2007.

Mitchell, Timothy. *Rule of Experts: Egypt, Techno-Politics, Modernity*. Berkeley: University of California Press, 2002.

Mundy, Martha, and Richard Saumarez Smith. *Governing Property, Making the Modern State: Law, Administration and Production in Ottoman Syria.* New York: I. B. Tauris, 2007.

Najmabadi, Afsaneh. *Land Reform and Social Change in Iran.* Salt Lake City: University of Utah Press, 1987.

Nore, Petter, and Terisa Turner, eds. *Oil and Class Struggle.* London: Zed Press, 1980.

Nowell, Gregory P. *Mercantile States and the World Oil Cartel, 1900–1939.* Ithaca, NY: Cornell University Press, 1994.

O'Connell, D. P. "A Critique of the Iranian Oil Litigation." *British Institute of International and Comparative Law* 4, no. 2 (April 1955): 267–293.

Owen, Edgar Wesley. *Trek of the Oil Finders: A History of the Exploration for Petroleum.* Tulsa, OK: American Association of Petroleum Geologists, 1975.

Painter, David. *Oil and the American Century: The Political Economy of U.S. Foreign Oil Policy 1941–1954.* Princeton, NJ: Princeton University Press, 1986.

"'Pay-e Kalam Delneshine Sagha.' *Peyk-e Naft.*" *Bahman* 1367 (1989): 52–53.

Podobnik, Bruce. *Global Energy Shifts: Fostering Sustainability in a Turbulent Age.* Philadelphia: Temple University Press, 2006.

Pomeranz, Kenneth. *The Great Divergence: China, Europe, and the Making of the Modern World.* Princeton, NJ: Princeton University Press, 2000.

Pritchard, Sara B. *Confluence: The Nature of Technology and the Remaking of the Rhone.* Cambridge, MA: Harvard University Press, 2011.

Richardson, R. K. "The Geology and Oil Measures of South-West Persia." *Journal of the Institute of Petroleum Technology* 10 (1924): 1–3.

Ross, Michael. "Does Oil Hinder Democracy?" *World Politics* 53, no. 3 (2001): 325–361.

Rosser, Andrew. "Escaping the Resource Curse: The Case of Indonesia." *Journal of Contemporary Asia* 37, no. 1 (2007): 38–58.

Rouhani, Fouad. *Tarikh-E Melli Shodan-E Sanat-E Naft-E Iran (History of the Nationalization of Iran's Oil).* Tehran: Sherkat-e Sahamy-e Ketabhay-e Jeeby, 1353.

Sabin, Paul. *Crude Politics: The California Oil Market, 1900–1940.* Berkeley: University of California Press, 2005.

Sachs, Jeffrey D., and Andrew M. Warner. "Natural Resource Abundance and Economic Growth." *National Bureau of Economic Research Working Paper Series*, no. 5398 (1995): 1–47.

Sampson, Anthony. *The Seven Sisters: The Great Oil Companies and the World They Shaped.* New York: Viking Press, 1975.

Santiago, Myrna I. *The Ecology of Oil: Environment, Labor, and the Mexican Revolution, 1900–1938.* Cambridge: Cambridge University Press, 2009.

Schwantes, Carlos A. *Vision & Enterprise: Exploring the History of Phelps Dodge Corporation*. Tucson: University of Arizona Press; Phelps Dodge Corporation, 2000.

Scott, James C. *Seeing Like a State: How Certain Schemes to Improve the Human Condition Have Failed*. New Haven, CT: Yale University Press, 1998.

Shulman, Peter A. "'Science Can Never Demobilize': The United States Navy and Petroleum Geology, 1898–1924." *History and Technology* 19 (4) (2003): 365–395.

Smith, F. E. "Obituary Notices of Fellows of the Royal Society: John Cadman, Baron Cadman, 1877–1941." *Royal Society* 3, no. 10 (1941): 915–928.

Staples, Amy. "Seeing Diplomacy through Bankers' Eyes: The World Bank, the Anglo-Iranian Oil Crisis, and the Aswan High Dam." *Diplomatic History* 26, no. 3 (2002): 397–418.

Stern, Philip. *The Company State: Corporate Sovereignty and the Early Modern Foundations of the British Empire in India*. Oxford: Oxford University Press, 2011.

Steyn, Phia. "Oil Exploration in Colonial Nigeria, c. 1903–58." *Journal of Imperial and Commonwealth History* 37, no. 2 (2009): 249–274.

Stork, Joe. "Oil and the Penetration of Capitalism in Iraq." In *Oil and Class Struggle*, ed. Petter Nore and Terisa Turner, 172–198. London: Zed Press, 1980.

Strunck, William Theodore. "The Reign of Shaykh Khaz'al Ibn Jabir and the Suppression of the Principality of 'Arabistan: A Study in British Imperialism in Southwestern Iran, 1897–1925." Doctoral dissertation, Indiana University, 1977.

Suchman, Lucy. "Configuration." In *Inventive Methods: The Happening of the Social*, ed. Celia Lury and Nina Wakeford, 48–60. London: Routledge, 2014.

Tilley, Helen. *Africa as a Living Laboratory: Empire, Development, and the Problem of Scientific Knowledge, 1870–1950*. Chicago: University of Chicago Press, 2011.

Tolf, Robert W. *The Russian Rockefellers: The Saga of the Nobel Family and the Russian Oil Industry*. Stanford, CA: Hoover Institution Press, 1976.

Tripp, Charles. *A History of Iraq*. 2nd ed. Cambridge: Cambridge University Press, 2002.

Verran, Helen. *Science and an African Logic*. Chicago: University of Chicago Press, 2001.

Vitalis, Robert. *America's Kingdom: Myth-Making on the Saudi Oil Frontier*. Stanford, CA: Stanford University Press, 2006.

Voss, Jan-Peter, and Basil Bornemann. "The Politics of Reflexive Governance: Challenges for Designing Adaptive Management and Transition Management." *Ecology and Society* 16, no. 2 (2011): 9.

Walden, Jerrold L. "The International Petroleum Cartel in Iran—Private Power and the Public Interest." *Journal of Public Law* 11 (1962): 64–121.

Watts, Michael. "Resource Curse? Governmentality, Oil, and Power in the Niger Delta, Nigeria." *Geopolitics* 9 (2004): 50–80.

White, Richard. *Railroaded: The Transcontinentals and the Making of Modern America.* New York: Norton, 2011.

Williamson, Harold. *The American Petroleum Industry: The Age of Energy, 1899–1959.* Vol. 2. Evanston, IL: Northwestern University Press, 1963.

Williamson, J. W. *In a Persian Oil Field.* London: Ernest Benn, 1927.

Yergin, Daniel. *The Prize.* New York: Free Press, 1992.

Zirinsky, Michael P. "Imperial Power and Dictatorship: Britain and the Rise of Reza Shah, 1921–1926." *International Journal of Middle East Studies* 24, no. 4 (1992): 639–663.

Index

Najmabadi, Afsaneh, 33
National Front Party, 115, 165–166
National Iranian Oil Company
(NIOC), 52, 198, 199–200, 237
Nationalization
arrangement, 188
controversy, 212
of Iran's oil industry, 167
law, 109
of oil industry, 148
right wording for, 233–235
Nationalizing labor(atory), 189–193
National Ministry of Labor, 140
National Oil Commission, 181
National Oil Company, 172
National Security Council (NSC),
217–219
New Oil Refining Process Ltd., 66
Nichols, H. E., 93
1978–1979 Islamic Revolution,
242
1933 concession agreement
Article 10, 109
Article 13, 57
Article 16, 111, 121
Article 26, 202
Article 14B, 57
NIOC. See National Iranian Oil
Company (NIOC)
Non-Persian employees, 112
reduction of, 116

Ohm's law, 110
Oil
development, 6
economic facts, 179–189
industry in Middle East, 4
Masjid Suleiman reserves of, 57
in Middle East, technical
development of, 6
Middle East infrastructure and its
limits in terms of, 12–15
nationalization crisis and, 236
pricing the properties of, 41–46
properties of, 46–52
Oil industry nationalization, 157–160
collectives of, 193–195

crisis, 158
financial restrictions and technical
blockages, 175–179
international law and national
sovereignty, 166–175
militant nationalism, threat of,
160–166
nationalizing labor(atory), 189–193
oil, economic facts, 179–189
technologies of profit sharing,
160–166
Oil workers, 121–123
beneficence and violence, 127–130
"conception of the rights of the
Persian," 135–145
coordinating labor nationally and
internationally, 135–148
diffusing worker's power, 148–153
history of (oil) nationalism,
153–155
(in)calculability, 111–117
points of vulnerability within energy
system, 124–127
1929 Strike, 130–135
Operation Ajax, 215
Organization of the Petroleum
Exporting Countries (OPEC),
242
Ottoman Empire, 15, 30

Pakistan, 185
Paternalism, 102, 122
Permanent Court of International
Justice (PCIJ), 17, 96–97, 99
Persia and the Persian Question, 31
Persia Memorandum, 100
Persian Bank Mining Rights
Corporation, 31, 32
Persian Geology, 78
Persian Gulf, 4, 25, 26, 28, 127
Persianization, 111–112, 121, 134
British-controlled oil industry,
121
Persian province of Khuzistan, 101
"Persian Scarf Dancer" (stone
sculpture by Francis Derwent
Woods), 1, 3

Petroleum, properties of, 21–26
Article 6 and Southwest Iran, 26–30
concessionary property, 31–35
mineral rights, 35–38
porosity of concessionary property,
38–41
pricing, 41–46
reassembling, 46–52
Petroleum Administration for Defense
(PAD), 178
Petroleum geology, 69
Petroleum knowledge, 57–61
AIOC in public arena, 79–84
oilfield and politics of unknowability,
71–79
origins of, 84–85
petroleum laboratory, 67–71
properties and practices of control,
61–67
Petroleum technologists, 83
Pilgrim, G. E., 69
Pipelines, 124
Pirnia, Husayn, 145
Podobnik, Bruce, 151
Political construction, 11
Pomeranz, Kenneth, 13
"Pool Association," 92
Porosity of concessionary property,
38–41
Post–World War II petroleum order,
157
Pre–World War I oil industry, 26
Pricing equation, Black-Scholes, 10
Private ownership of subsoil, 23
The Prize (Daniel Yergin), 6
Process of invention, 61
and innovation, 85
Profit royalty, 105, 107
Profit sharing
arrangement, 226–227
technologies of, 160–166
Prud'homme, Hector, 212
Public and private interests, 179
Pucka bungalow, 125

Qajar government, 27, 28
Qavam-al-Saltaneh, 160

Qishlaq (garmsir), 42
Quantitative formulas, 239
Quasi-sovereign status, 203

Racial mixing, 136
Racial-technical organization, 193
Ra'iyat, 32
Ramsbotham, Peter, 213
Rangoon, Burma, 62, 66, 127
Razmarra, Ali, 116, 162
"Red Line Agreement," 93
Redwood, Boverton, 29, 65
Rentier state, 7
*Report on the International Petroleum
Cartel*, 217
Reuter Concession, 15
Reynolds, George, 29, 36, 37, 39
Reza Pahlavi, 109
Rieber, Torkild, 212
Rights of the Persian, conception of,
135–145
Rockefeller, David, Sr., 23, 148
Rothschild firm, 27
Royal Dutch Shell groups, 4, 7, 79,
83, 193, 226
Royalty formula, 105, 110
Royalty schemes, 105
"Rule of capture," 23
Russian oil industry, 27

Sa'ed, Mohammad, 161
Safavid Empire (1501–1722), 33
Saleh, Allahyar, 181, 205–206
Salt movements, 4
Saudi Arabia, 23, 157
Scott, John, 220
Seddon, Richard, 172
"Seven Sisters," 19
Shafagh-i Sorkh, 132
Shah, Mohammad Reza, 231
Shah, Nadir, 33
Shah, Reza, 33, 50, 87
Shand, S. J., 61
Shaykh Khaz'al, 35–36, 40, 128
Shell Petroleum Company Limited, 4,
28, 147
Shepherd, Francis, 162

Printed in the United States
by Baker & Taylor Publisher Services